CARBOHYDRATES IN THE BODY AND FUNCTIONS

Carbohydrates in the Body and Functions

Edited by:

Shiv Sanjeevi Sripathi

Delve PUBLISHING

www.delvepublishing.com

Carbohydrates in the Body and Functions

Shiv Sanjeevi Sripathi

Delve Publishing
224 Shoreacres Road
Burlington, ON L7L 2H2
Canada
www.delvepublishing.com
Email: orders@arclereducation.com

ISBN: 978-1-77407-694-1 (Hardcover)

DECLARATION

ABOUT THE EDITORS

ShivSanjeevi Sripathi completed his Masters in Biotechnology from Mumbai University in 2008. He was awarded for academic excellence in both his Bachelors and Masters for securing second rank in Mumbai University in 2006 and first rank in his college: Kishinchand chellaram College. For his Masters he secured first rank in his college KET's V.G.Vaze College. He qualified CSIR and NET and and TOEFL in September 2008. He then worked on a stem cell project at the Specialized Centre for Cell Based Therapy (SCCT),KEM Hospital at Mumbai on a project entitled, : Isolation & detection of stem cells from Human Umbilical cord/ amniotic membrane" following which he worked at Junior Research Fellow at Microbiology & Cell Biology Department, Indian Institute of Science, Bangalore from August 2009 to November 2011. The field of work involved cloning of cell wall genes and transcription factors in *E.coli & M.smegmatis*. As a writer, he has authored and co-authored 25 books with Delve and Arcler International Press. He loves to read and share on interesting aspects of life sciences in books. In his free time he loves to travel and explore ancient customs and traditions.

TABLE OF CONTENTS

LIST OF CONTRIBUTORS

ShivSanjeevi Sripathi

Eric R. Vimr
Laboratory of Sialobiology, Department of Pathobiology, University of Illinois at Urbana-Champaign, Urbana, IL 61802, USA

Alan Kirwan
Glycoscience Group, National Centre for Biomedical Engineering Science, National University of Ireland Galway, Galway, Ireland

Marta Utratna
Glycoscience Group, National Centre for Biomedical Engineering Science, National University of Ireland Galway, Galway, Ireland

Michael E. O'Dwyer
Department of Hematology, National University of Ireland Galway, Galway, Ireland

Lokesh Joshi
Glycoscience Group, National Centre for Biomedical Engineering Science, National University of Ireland Galway, Galway, Ireland

Michelle Kilcoyne
Glycoscience Group, National Centre for Biomedical Engineering Science, National University of Ireland Galway, Galway, Ireland
Carbohydrate Signalling Group, Microbiology, School of Natural Sciences, National University of Ireland Galway, Galway, Ireland

Elena Franco-Robles
Centro de Investigación y de Estudios Avanzados del IPN, Unidad Irapuato, Km 9.6 Libramiento Norte Carretera Irapuato-León, 36821 Irapuato, GTO, Mexico

Mercedes G. López
Centro de Investigación y de Estudios Avanzados del IPN, Unidad Irapuato, Km 9.6 Libramiento Norte Carretera Irapuato-León, 36821 Irapuato, GTO, Mexico

Darrell Cockburn
Enzyme and Protein Chemistry, Department of Systems Biology, Technical University of Denmark, Lyngby, Denmark

Casper Wilkens
Enzyme and Protein Chemistry, Department of Systems Biology, Technical University of Denmark, Lyngby, Denmark

Adiphol Dilokpimol
Enzyme and Protein Chemistry, Department of Systems Biology, Technical University of Denmark, Lyngby, Denmark

Hiroyuki Nakai
Enzyme and Protein Chemistry, Department of Systems Biology, Technical University of Denmark, Lyngby, Denmark

Anna Lewińska
Enzyme and Protein Chemistry, Department of Systems Biology, Technical University of Denmark, Lyngby, Denmark

Maher Abou Hachem
Enzyme and Protein Chemistry, Department of Systems Biology, Technical University of Denmark, Lyngby, Denmark

Birte Svensson
Enzyme and Protein Chemistry, Department of Systems Biology, Technical University of Denmark, Lyngby, Denmark

Kauko K. Makinen
Institute of Dentistry, University of Turku, Lemminkaisenkatu 2, 20520 Turku, Finland

Astrid Kolderup
Faculty of Public Health, Hedmark University College, P.O. Box 400, 2418 Elverum, Norway

Birger Svihus
Norwegian University of Life Sciences, P.O. Box 5003, 1432 Aas, Norway

Ivone Lima Santana
Department of Dentistry I, School of Dentistry, Federal University of Maranhão, São Luis, Maranhão, Brazil,

Letícia Machado Gonçalves
Department of Prosthodontics and Periodontology, Piracicaba Dental School, University of Campinas, Piracicaba, São Paulo, Brazil,

Andréa Araújo de Vasconcellos
Department of Prosthodontics and Periodontology, Piracicaba Dental School, University of Campinas, Piracicaba, São Paulo, Brazil,

Wander José da Silva
Department of Prosthodontics and Periodontology, Piracicaba Dental School, University of Campinas, Piracicaba, São Paulo, Brazil,

Jaime Aparecido Cury
Department of Physiological Sciences, Piracicaba Dental School, University of Campinas, Piracicaba, São Paulo, Brazil,

Altair Antoninha Del Bel Cury
Department of Prosthodontics and Periodontology, Piracicaba Dental School, University of Campinas, Piracicaba, São Paulo, Brazil,

M. Laura Ferrando
Host-Microbe Interactomics, Animal Sciences, Wageningen University, Wageningen, The Netherlands,
Department of Medical Microbiology, Academic Medical Center, Amsterdam, The Netherlands,

Peter van Baarlen
Host-Microbe Interactomics, Animal Sciences, Wageningen University, Wageningen, The Netherlands,

Germano Orrù
Oral Biotechnology Laboratory, University of Cagliari, Cagliari, Italy,

Rosaria Piga
Host-Microbe Interactomics, Animal Sciences, Wageningen University, Wageningen, The Netherlands,

Roger S. Bongers
NIZO Food Research B.V., Ede, The Netherlands,

Michiel Wels
NIZO Food Research B.V., Ede, The Netherlands,

Astrid De Greeff
Central Veterinary Institute, Animal Sciences Group, Wageningen University, Lelystad, The Netherlands,
University of Oklahoma Health Sciences Center, United States of America,

Hilde E. Smith
Central Veterinary Institute, Animal Sciences Group, Wageningen University, Lelystad, The Netherlands,
University of Oklahoma Health Sciences Center, United States of America,

Jerry M. Wells
Host-Microbe Interactomics, Animal Sciences, Wageningen University, Wageningen, The Netherlands,

LIST OF ABBREVIATIONS

AAL	Aleuria aurantia lectin
AGP	Alpha-1-acid glycoprotein
ARIC	Atherosclerosis Risk in Communities
BOD	Benign ovarian diseases
BPH	Benign prostatic hyperplasia
BPDs	Benign pulmonary diseases
CDCD	Caesarean-derived, colostrum-deprived
CBMs	Carbohydrate binding modules
CCC	Carbon catabolite control
CMC	Carboxymethylcellulose
CEA	Carcinoembryonic antigen
CNS	Central nervous system
CLSM	Confocal laser scanning microscopy
DCs	Dendritic cells
DNL	De novo lipogenesis
DRI	Dietary Reference Intakes
DRE	Digital rectal exam
DDS	Drug delivery systems
EMSA	Electrophoretic mobility shift assays
eNOS	endothelial nitric oxide synthase
EIA	Enzyme immunoassay
ECD	Extracellular domain
ECM	Extracellular matrix
FAO	Food and Agriculture Organization
FDA	Food and Drug Administration
FOS	Fructooligosaccharides
GOS	Galactooligosaccharides
GIT	Gastrointestinal tract

GO	Gene ontology
GCTs	Germ cell tumours
GSTs	Glutathione S-transferases
GI	Glycaemic index
GL	Glycaemic load
GPR	G protein-coupled receptors
GALT	Gut-associated lymphoid tissue
HDLs	High-density lipoproteins
HFCS	High-fructose corn syrup
HPLC	High performance liquid chromatography
hCG	Human chorionic gonadotropin
HIV	Human immunodeficiency virus
HEC	Hydroxyethyl cellulose
IM	Inner membrane
IEPs	Insoluble extracellular polysaccharides
IPP	Insoluble polysaccharide pulldown
iIELs	intraepithelial lymphocytes
ITFs	Inulin-type fructans
ITC	Isothermal titration calorimetry
LP	Lamina propria
LBM	Lean body mass
LCA	Lens culinaris agglutinin
LDLs	Low-density lipoproteins
MOS	Mannanoligosaccharides
MLNs	Mesenteric lymph nodes
MM	Multiple myeloma
NDC	Non-digestible carbohydrates
ONS	Oral nutritional supplements
PTS	Phosphotransferase systems
PAGE	Polyacrylamide gel electrophoresis
PURE	Prospective Urban Rural Epidemiology
PSA	Prostate-specific antigen
SOS	Sialyloligosaccharides
SEPs	Soluble extracellular polysaccharides

ROMA	Risk of Ovarian Malignancy Algorithm
SCFAs	Short chain fatty acids
SEPs	Soluble extracellular polysaccharides
SBSs	Spots termed surface binding sites
SGM	Submaxillary gland mucin
SPR	Surface plasmon resonance
TLRs	Toll-like receptor
VEGF	Vascular endothelial growth factor
VLDLs	Very-low-density lipoproteins
WAP	Whey acidic protein
WHO	World Health Organization

PREFACE

A life without grains: bread or rice would sound challenging?

Potatoes?

What about energy?

These questions have one answer: carbohydrates that are present in whole grains and potatoes apart from other sources. Apart from providing a quick source of energy (remember the rejuvenating effect of glucose –based drinks or eating a fruit?), these molecules assume several roles in organisms. The roles include structural strength (cellulose and pectin), signaling, host-pathogen interactions (glycoproteins), and even as treatment approaches.

As far as their intake is concerned, researchers suggest that more than the quantity or the percentage of carbohydrates consumed, the quality is important. Amongst the various carbohydrate sources, the consumption of whole grains is recommended along with legumes that contain other nutrients like protein. For instance, a good diet suggested according to Walter Willett, Fredrick John Stare professor of epidemiology and nutrition and chair, Department of Nutrition at HSPH, and Frank Hu, professor of nutrition and epidemiology is lowered refined carbs and intake of lean protein, whole grains, fruits, vegetables, legumes and fats.

Chapter 1 describes the various terminologies and definitions required to get a clearer picture of research and news all around. Chapter 2 describes various functions of carbohydrates along with case studies to further highlight the importance of carbohydrates. The next chapter is carbohydrate vaccines where carbohydrate molecules are being used to design vaccines against diseases such as malaria.

An important point here is readers are requested not to plan reading the examples and case studies given. This book is to be treated as a reference and not as a diagnostic guide. To summarize, the ESP of the book is to give research-backed articles to debunk various theories and eat a healthy and normal diet and not blindly follow "advice" available out there in the media.

To summarize the importance of carbohydrates in life, let us recapitulate the fundamental processes in biology that supplies energy for life: Glycolysis that breaks down glucose (a carbohydrate) to release energy. "Biological energy comes from the sun. Light energy harvested by photosynthesis in chloroplasts and phototropic bacteria becomes stored in carbohydrates and fats. This stored energy can be released by oxidative metabolism in the form of adenosine triphosphate (ATP) and used as fuel for other biological processes" -John E. Walker (Nobel prize-winning British biochemist).

We believe that the readers find this book informative as information on all aspects of carbohydrates has been compiled in this single book.

Editors

1

Carbohydrates: Looking at the Basics

Shivsanjeevi Sripathi

ShivSanjeevi Sripathi completed his Masters in Biotechnology from Mumbai University in 2008. He was awarded for academic excellence in both his Bachelors and Masters for securing second rank in Mumbai University in 2006 and first rank in his college: Kishinchand chellaram College. For his Masters he secured first rank in his college KET's V.G.Vaze College. He qualified CSIR and NET and and TOEFL in September 2008. He then worked on a stem cell project at the Specialized Centre for Cell Based Therapy (SCCT),KEM Hospital at Mumbai on a project entitled, : Isolation & detection of stem cells from Human Umbilical cord/ amniotic membrane" following which he worked at Junior Research Fellow at Microbiology & Cell Biology Department, Indian Institute of Science, Bangalore from August 2009 to November 2011. The field of work involved cloning of cell wall genes and transcription factors in *E.coli* & *M.smegmatis*. As a writer, he has authored and co-authored 25 books with Delve and Arcler International Press. He loves to read and share on interesting aspects of life sciences in books. In his free time he loves to travel and explore ancient customs and traditions.

1.1 OVERVIEW

Carbohydrate is a macronutrient that provides energy to between 40 and 80 percent of total energy intake. The food carbohydrates are categorized by degree of polymerization according to the 1999 FAO/WHO report. This system seeks to reach a balance between a chemical classification and a

physiological classification. This circumvents the usage of confusing words such as "complex carbohydrate" that comes under the bracket of all polysaccharides in Europe and starch in North America.

The following is a definition based on the recommendations about the FAO/WHO, 1998 report:

Standardizing the terminology used to describe dietary carbohydrates with carbohydrates classified primarily by molecular size (degree of polymerization or DP) into sugars (DP 1-2), oligosaccharides (DP 3-9) and polysaccharides (DP 10+). Monosaccharide composition is the next step for added subdivision. Physiological properties can then be utilized for nutritional groupings (Nantel, 1999).

As far as their classification is concerned according to the Joint Food and Agriculture Organization (FAO)/World Health Organization (WHO) Expert Consultation on Carbohydrates in human nutrition convened in Rome in 1997. The classification finds its basis on:

- degree of polymerization (DP),
- the type of linkage (α or non-α)
- the character of individual monomers

On the basis of a chemical approach, there are 3 major groups of carbohydrates:

- sugars (DP1–2),
- oligosaccharides (short-chain carbohydrates) (DP3–9)
- polysaccharides (DP more than 10).

Sugars are of the following types:

- monosaccharides,
- disaccharides and
- polyols (sugar alcohols).
- Oligosaccharides are of two types:

(a) malto-oligosaccharides (α-glucans) obtained from the hydrolysis of starch

(b) non-α-glucans are inclusive of other oligosaccharides such as raffinose and stachyose

Polysaccharides include α-1:4 and 1:6 glucans(starch) and nonstarch polysaccharides (NSPs).

Certain carbohydrates don not find a place in this scheme (Example: inulin) as they are found in many molecular forms in nature (Cummings and Stephen, 2007; Roberfroid, 2005).

The following figure and table summarize the classification of dietary carbohydrates:

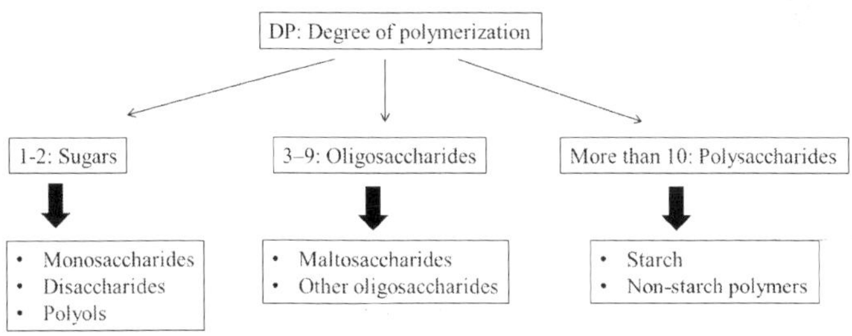

Figure 1.1: Classification of carbohydrates

Table 1.1: Classifying carbohydrates

Degree of polymerization	**Group**	**Examples**
Sugars (1–2)	Monosaccharides	Glucose, galactose , fructose
	Disaccharides	Maltose , Sucrose, lactose, trehalose
	Polyols (sugar alcohols)	Sorbitol, mannitol, maltitol, lactitol, isomalt xylitol, erythritol,
Oligosaccharides (3–9)	Maltooligosaccharides (α-glucans)	Maltodextrins
	Non-α-glucan oligosaccharides	Raffinose, stachyose, fructo and galacto oligosaccharides, inulin polydextrose

Polysaccharides (More than 10)	Starch	Amylose, amylopectin, modified starches
	Non-starch polysaccharides (NSPs)	Cellulose, hemicellulose, pectin, β-glucan, arabinoxylans, glucomannans, plant gums , mucilages, hydrocolloids

There are several aspects to carbohydrates in terms of functions:

- They are involved in energy metabolism
- They influence the levels of blood glucose and insulin
- They affect satiety
- They play a role in various functions of the colon such as bowel habits, the nature of the microbiota, the health of the epithelial cells in the bowels.

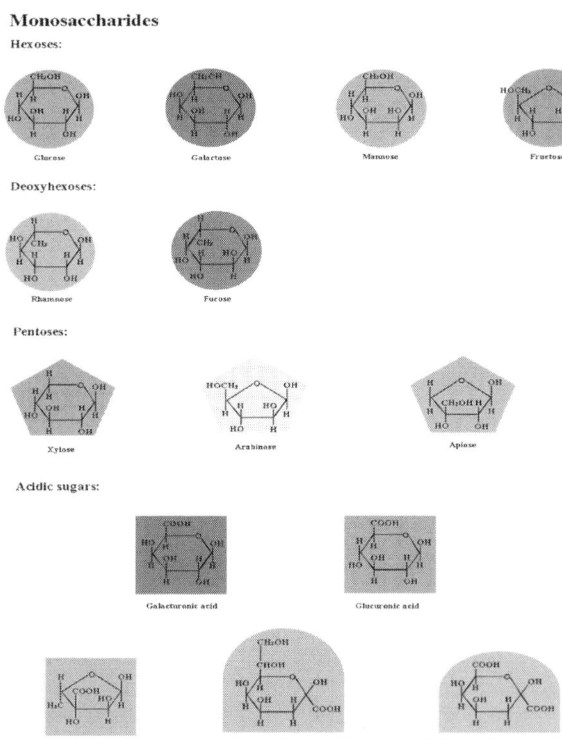

Figure 1.2 below shows monosaccharides in digestible carbohydrates and fiber (Navarro et al, 2019).

1.2 THE TERMINOLOGY USED IN CARBOHYDRATES:

1.2.1 Total carbohydrate:

This term displays a certain level of ambiguity in terms of definition; as reported in food tables. There are two approaches used to explain the concepts of total carbohydrate:

The calculation of carbohydrate 'by difference' has been an approach since the early 20th century that is also extended to current times (United States Department of Agriculture, 2007). This involves the determination of moisture, protein, fat, ash and alcohol that is then subtracted from the total weight of the food under study. The carbohydrate content is the difference. However, this 'by difference' is also inclusive of other elements such as wax, organic acids, lignin, tannins and certain Maillard products. Another challenge in this approach includes any errors in analyses used for tests. A third challenge is a lack of understanding of the nature and amount of each carbohydrate element present in a food article. Such amounts can help in the study of each element and its potential role.

The second approach is a direct analysis of carbohydrate components where the first instance was that of McCance and Lawrence (1929). This involves the estimation of 'available carbohydrate' that in turn refers to the absence of plant cell wall polysaccharides such as fiber. This offers an edge over the above-mentioned approach. Such an approach allows for an easier analysis that facilitates studies on individual carbohydrate in terms of their levels and roles.

The values calculated by the above-mentioned techniques do not show the same values when certain foods are taken into account such as starchy food. Both approaches were used to calculate carbohydrate intakes from 52 dietary records from a Canadian study (Health Canada, 2005). First, the carbohydrate by difference was estimated followed by the calculation of McCance and Widdowson's The Composition of Foods.

The results are shown in Table 1.2 below:

Approach	Energy (kcal)	Protein (g) (%)	Fat (g) (%)	CHO (g) (%)
Canadian nutrient file	2265	95.9 (16.9)	81.4 (31.5)	294.2 (52.9)

Analysis using 'The composition of foods *P>0.001.	1992	89.8 (17.9)	74.5 (32.7)	252.4* (48.5)*

(Stephen, 2006)

This study showed that the "by difference" approach yielded 14% higher carbohydrate intake and 12% increased energy intake (Stephen, 2006).

1.2.2 Sugars

The mono- and disaccharides in the food account for the term 'sugars'; these include glucose, fructose and galactose. They are the basic building blocks of naturally occurring di-, oligo- and polysaccharides. The sweetest is fructose while glucose and fructose are present in fruits or honey to serve as the main energy source.

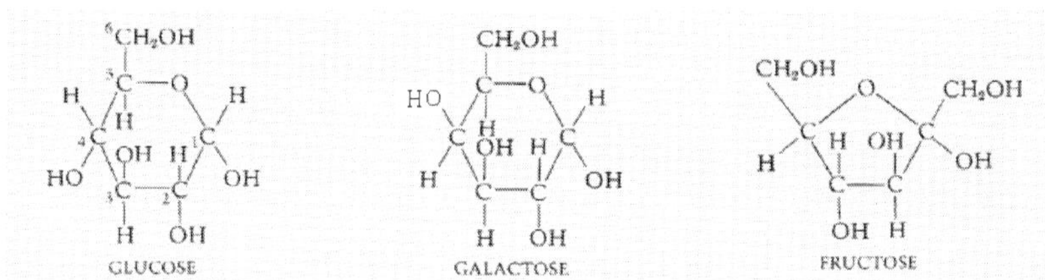

Figure 1.3 shows the structures of these 3 sugars:

Source: http://www.biology-pages.info/C/Carbohydrates.html

The food industry utilizes corn syrup made of glucose and fructose: obtained by corn starch hydrolysis.

- The palatability of many food articles is augmented by the presence of sugars.
- The sugars also function as preservatives as in jellies and jams.
- The qualities of texture, viscosity, body and browning capacity.
- Moisture control by sugars allows for protection from drying out.
- The breakdown of starch and proteins is influenced by sugars.
- The dough yield is increased.

Polyols are alcohols of glucose and other sugars present either naturally or made industrially where aldehyde group of the glucose molecule is converted to the alcohol with aldose reductase. An example is sorbitol that finds application as a sucrose replacement in diabetic patients.

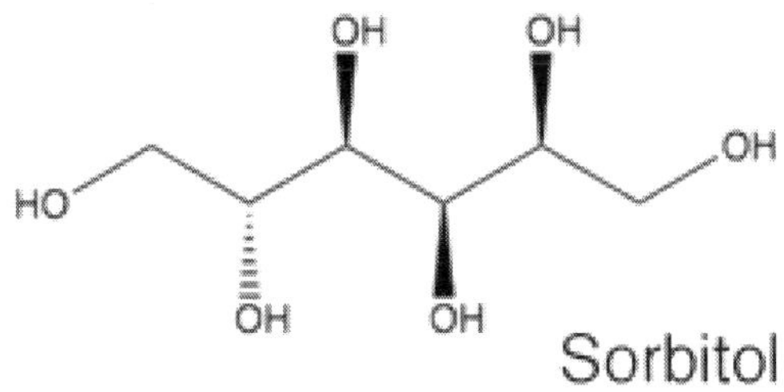

Figure 1.4: Sorbitol

Source: *https://www3.hhu.de/biodidaktik/zucker/sugar/sorbit.html*

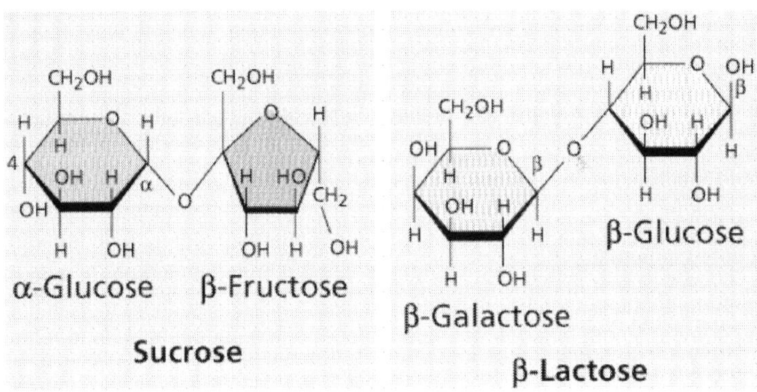

Figure 1.5: The chief disaccharides are sucrose (α-Glc(1-2)b-Fru) and lactose (β-Gal(1-4)Glc). The former is present in berries, fruits and vegetables while extraction is done from sugar cane or beet.

Source: https://www2.estrellamountain.edu/faculty/farabee/biobk/disacc.gif

Sprouted wheat and barley show the presence of maltose while milk contains lactose.

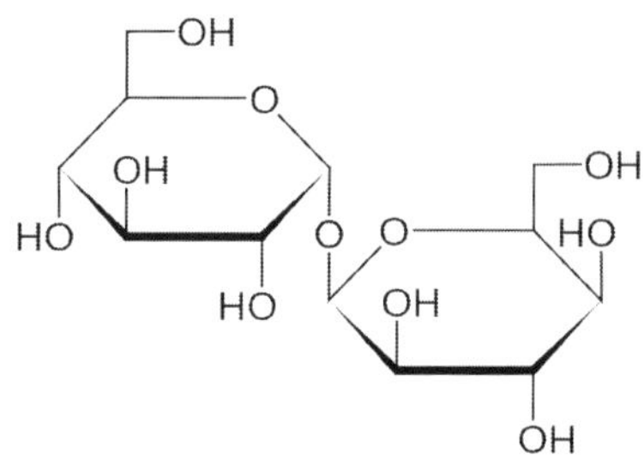

Figure 1.6: Trehalose. Another example is Trehalose (α-Glc(1-4) α-Glc) found in mushrooms, yeast, honey and bread. As its properties are at par with that of sucrose albeit at lower sweetness, it finds use as its replacement.

Source*: https://commons.wikimedia.org/wiki/File:Trehalose_Haworth.svg*

In order to classify the sugars to allow highlighting the sugars and label them, the following classification is in terms of sugar is done:

1.2.2.1 Total sugars

The total sugars include 'all monosaccharides and disaccharides other than polyols' (European Communities, 1990).

1.2.2.2 Free sugars

This involves several concepts: 'free sugars' are any sugars in food that are free and not bound. This term also refers to hydrolyzed carbohydrates in a food that can be determined colorimetry or chromatographically. The definition is 'monosaccharides and disaccharides added to foods by the manufacturer, cook and consumer, plus sugars naturally present in honey, syrups and fruit juices'.

The term was utilized at the WHO/FAO Expert Consultation on 'Diet, Nutrition and the Prevention of Chronic Diseases' (WHO, 2003).

1.2.2.3 Added sugars

This term in the United States refers to sugars and syrups that are added to foods during their preparation (during processing or home preparation). Examples: molasses, honey, brown sugar, fruit juice concentrate, corn sweetener, glucose, sucrose, lactose, high-fructose corn syrup and malt syrup.

Figure 1.7: Various sugars

Source: Assistant Professor, Dr. Satvinder Kaur, UCSI University, Kuala Lumpur

1.2.2.4 Extrinsic and intrinsic sugars

A United Kingdom (UK) Department of Health Committee report in 1989 (Department of Health, 1989) 'to distinguish sugar as *Intrinsic* or naturally integrated into the cellular structure of the food while *Extrinsic* ones that are free in the food or added to it.

Intrinsic sugars refer to sugars located in the cell as elements of few unprocessed foodstuffs such as fructose, glucose and sucrose in whole fruits and vegetables.

Extrinsic sugars refer to those added to processed food such as those found in fruit juices. For example, the cellular structure of milk lacks lactose and is added. Sweeteners in fruit juices and honey or additives to cereals or drinks are extrinsic sugars not sourced from milk excluding lactose.

1.2.3 Oligosaccharides, short-chain carbohydrates

'Oligosaccharides are compounds in which monosaccharide units are joined by glycosidic linkages' with 2-19 DP with the germ disaccharides utilized for DP2 that again are placed as sugars by some scientists.

There is randomness in the division between oligo- and polysaccharides on account of molecular sizes. While the IUB–IUPAC and Joint Commission on Biochemical Nomenclature, 1982 lists the distinction at DP 10, the definition of a polysaccharide according to the IUPAC–IUB Joint Commission on Biochemical Nomenclature, 1996 is 'a macromolecule consisting of a large number of monosaccharide (glucose) residues joined to each other by glycosidic linkages'.

An approach to distinguish oligo- and polysaccharides is treatment with 80% v/v ethanol to cause precipitation from aqueous solutions. Challenges here are between 10 and 100 DP branched-chain carbohydrates that do not precipitate and the presence of chain lengths that are present at the border of oligo and polysaccharides in inulin and polydextrose. This calls for the use of the average DP that allows for classification.

Food oligosaccharides belong to two groups:

(i) Maltodextrins (average DP=8) is obtained from starch. They have α-1:4 and 1:6 linkages. Example: maltotriose and limit dextrins. They can modify food texture apart from serving as fat substitutes and sweeteners.

Figure 1.8 (a): Maltotriose

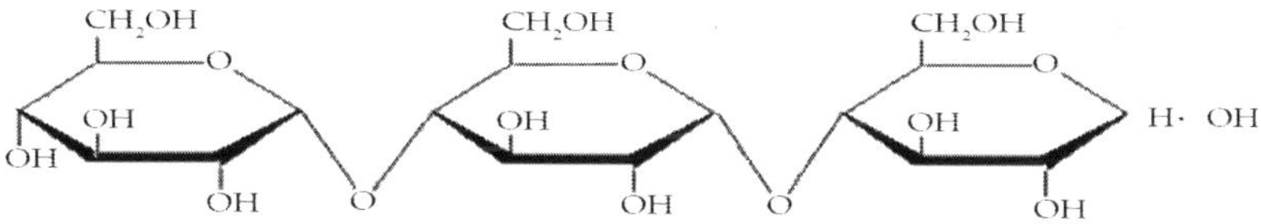

(b) Limit dextrins

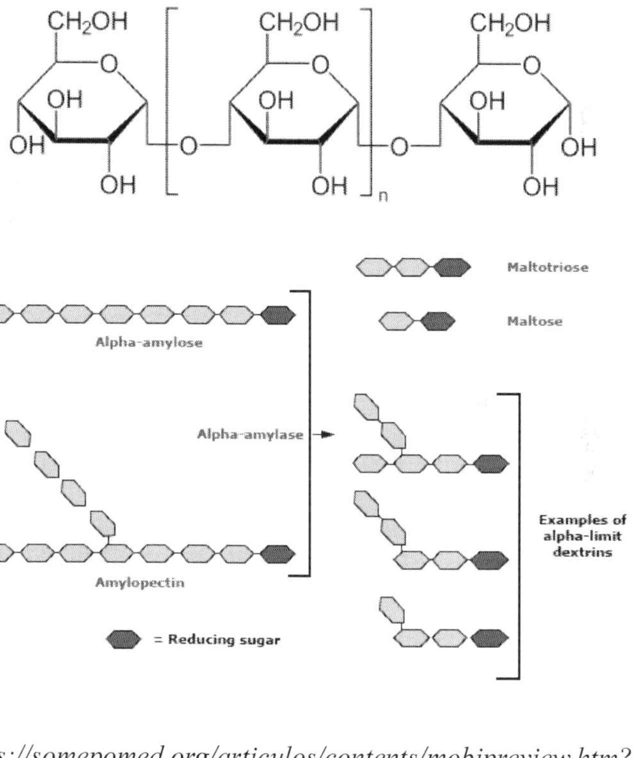

Source: https://somepomed.org/articulos/contents/mobipreview.htm?6/40/6794

(ii) Oligosaccharides that are not α-glucans are present in many seeds (lentils, peas, beans) and are mainly sucrose attached to different numbers of galactose molecules. Examples are stachyose ((Gal)2 1:6 Glu 1:2 Fru), raffinose (α-Gal(1-6)α-Glc(1-2)β-Fru), and verbascose ((Gal)3 1:6 Glu 1:2 Fru).

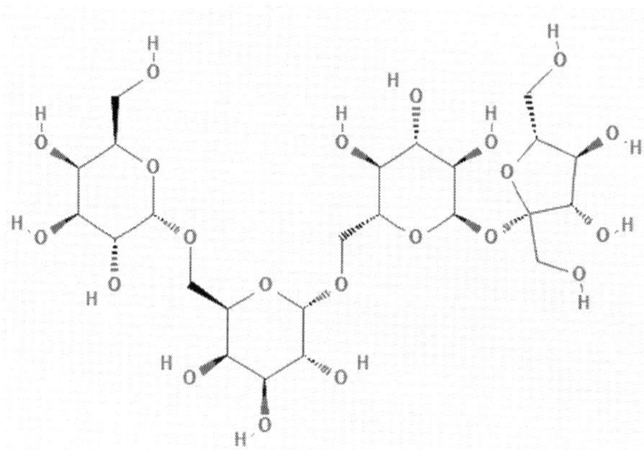

Figure 1.9: Stachyose

Source: https://pubchem.ncbi.nlm.nih.gov/compound/Stachyose#section=Structures

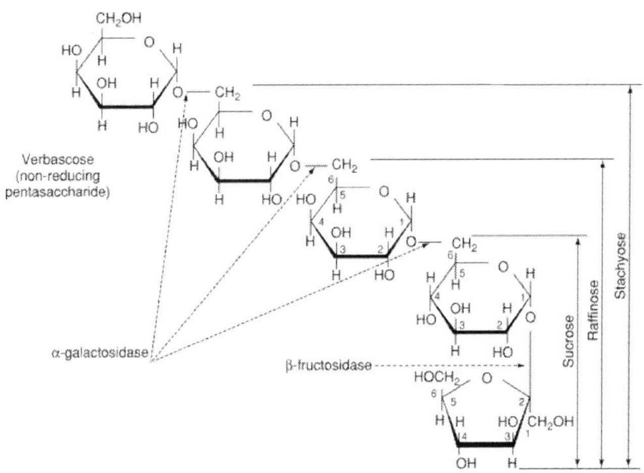

Figure 1.10: Verbacose, Raffinose and stachyose

Source: https://www.sciencedirect.com/topics/nursing-and-health-professions/raffinose

Fructans include inulin α-Glc(1-2)β-Fru(2-1) β-Fru(N) or β-Fru(2-1) b-Fru(N)) that are present in artichokes, chicory, onion, asparagus, rye and wheat (There is an option of industrial production). As these lack α-1,4 or 1,6 glucan bonds, the enzymes of the pancreas or intestine do not digest them hence earning them the term 'non-digestible oligosaccharides'.

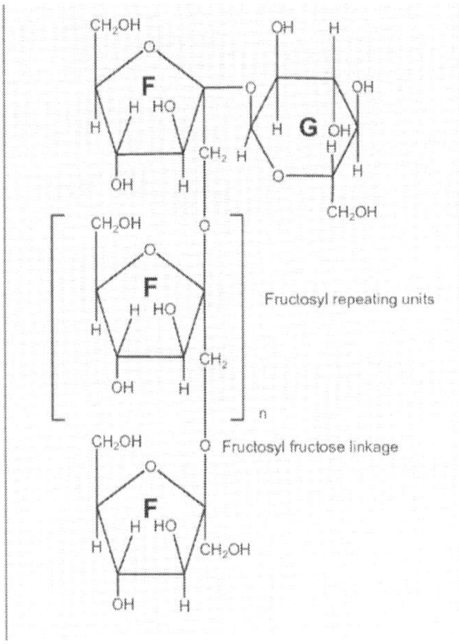

Figure 1.11: Inulin

Source: (F-fructose, G-glucose): *https://www.researchgate.net/figure/Structure-of-inulin-F-fructose-G-glucose_fig2_322135188*

Milk (human) is primarily galactose through the diversity displayed is extensive in terms of the oligosaccharides. The level of oligosaccharides in human milk is 5.0–8.0 g/l against trace levels in cow's milk.

These oligosaccharides have been reported to support the growth of Bifidobacteria in the gut of infants as seen by:

- The growth of the bacterium on milk oligosaccharide-media as the sole carbon source
- The major biota in infants exposed to breastfeeding

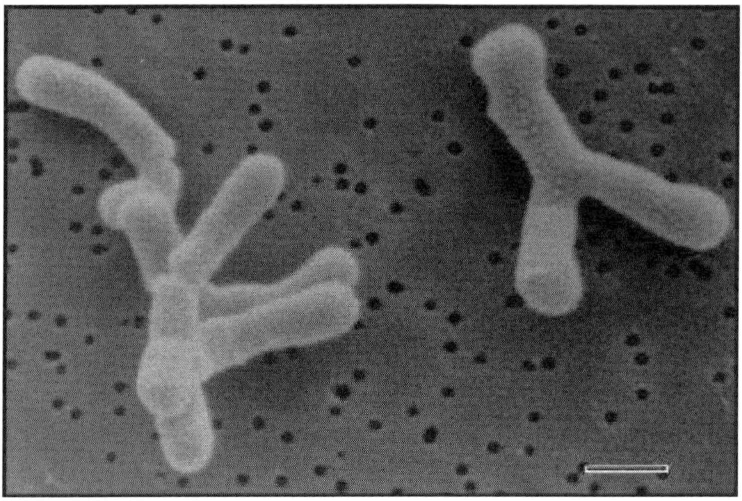

Figure 1.12: Bifidobacteria

Source: https://microbewiki.kenyon.edu/index.php/Bifidobacterium

Apart from their role as prebiotics as mentioned above, milk oligosaccharides are similar to carbohydrates on the cell surface of gut epithelial cells that is suggestive of their role to target pathogens and function in colonization resistance. Their immune-modulatory roles have also been suggested.

1.2.4 Starch and modified starch

The major carbohydrate in most diets is starch that is composed of glucose and is present in legumes, root vegetables and cereals. It is of two polymers: amylose (DP-10^3) and amylopectin (DP-10^4–10^5). The former is a non-branching helical chain of glucose residues connected by α-1,4 glucosidic bonds and accounts for 15–30% of major cereal starches. Both α-1,4 and α-1,6 glucosidic bonds are present in amylopectin that is present in large amounts in rice, maize, barley and sorghum.

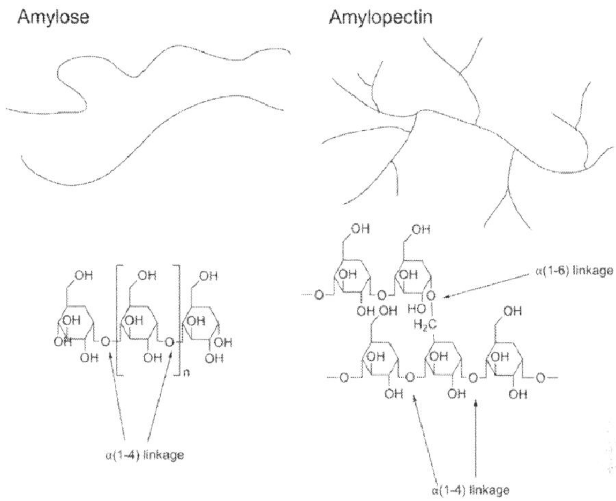

Figure 1.13: Amylose and amylopectin

Source: https://www.researchgate.net/figure/Isomers-of-starch-amylose-and-amylopectin_fig2_326954245

Both polymers display unique X-ray diffraction patterns as:

- The A type is seen in cereals such as maize, wheat and rice
- The B type is seen in banana and potato that display resistance to pancreatic amylase
- The C type is seen in legumes and is between A and B types

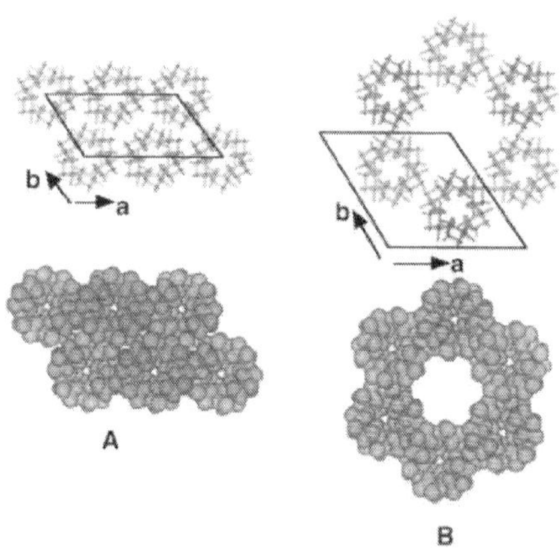

Figure 1.14: Types of starch

Source: https://www.researchgate.net/figure/Crystalline-packing-of-double-helices-in-a-A-type-amylose-and-b-B-type-amylose_fig2_308388990

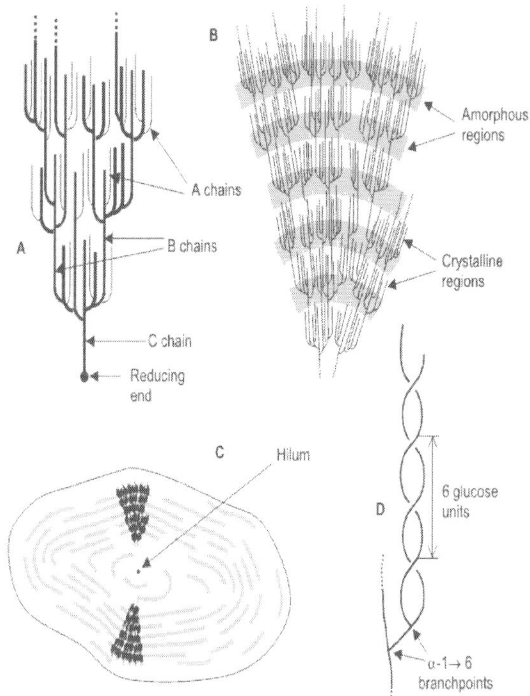

Source: *http://www1.lsbu.ac.uk/water/starch.html*

The process of gelatinization or heating in water causes loss of the crystalline structure due to digestion. The B form is formed when retrogradation or recrystallization is allowed for.

Modified starch

Plant breeding allows for varying the proportions of amylose and amylopectin in food. The genetic modification of different crops allows for generating specific starch molecules to allow for desired nutritional and functional properties. For example, high-resistant starch (RS) food can be formed by the use of starch with higher amylose contents as the temperature for gelatinization is higher causing retrogradation.

Another approach to alter starch is the use of chemicals to alter the properties of:

- Texture

- Appearance
- Mouthfeel
- Stability
- Resistance to heat

The 2 approaches used here are substitution and crosslinking:

The former refers to causing retrogradation by etherification or esterification of few hydroxyl groups on the glucose units of amylose and amylopectin. This allows for:

- staling bread,
- boosts viscosity and
- imparts stability to freeze–thaw

Crosslinking is associated with the formation of linkages at few spots between the chains of amylose and amylopectin. This allows for:

- Imparts stability to acid and heat
- Boosts the gelatinization temperature
- Viscosity control
- Prevents the formation of gels

1.2.5 NSPs

NSPs are 'macromolecules consisting of a large number of monosaccharides (glycose) residues joined to each other by glycosidic linkages' (IUB-IUPAC and Joint Commission on Biochemical Nomenclature, 1982). They non-α-glucan polysaccharides present in plant cell walls in our diet.

The diversity of NSPs is the foremost among carbohydrate groups with a vast array of molecular structures.

- Cellulose is the most copious molecule made of straight-chain β1–4-linked glucan (DP 10^3–10^6) accounting for 10 - 30% of the NSP in foods. The linear nature of this molecule minus branches allows the formation of microfibrils or cellulose fibers as a three-dimensional latticework to form the plant cell wall.

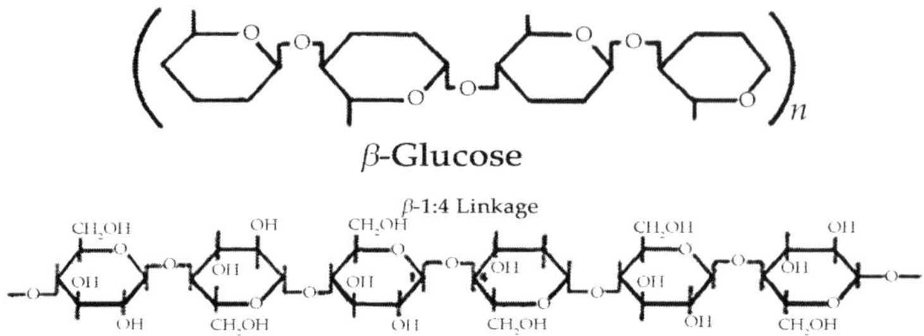

Figure 1.15: Cellulose

Source: http://www.biologydiscussion.com/carbohydrates/polysaccharides/types-of-polysaccharides-3-types/44929

Hemicelluloses are heteropolymers with branched hexose (6C) and pentose (5C) sugar monomers. The DP is 150-200 with arabinose, mannose, galactose and glucose branches among xylose molecule frameworks.

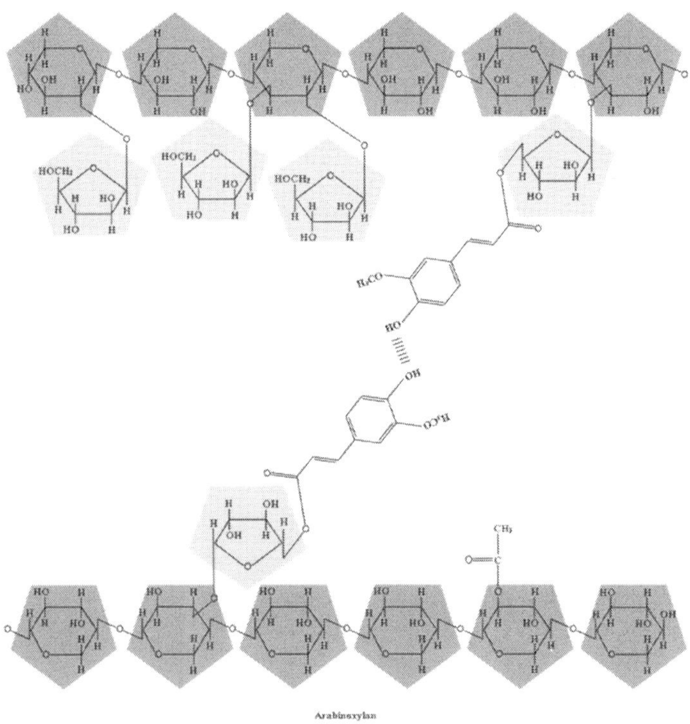

Figure 1.16: Chemical structure of arabinoxylans (Navarro et al, 2019)

Examples include arabinoxylans of cereals and uronic acids made of glucose and galactose subjected to carboxylation that function as carboxylic acids. Pectin is a component of the cell wall in plants and is made of 1–4β-D galacturonic acid residues along with side chains made of rhamnose, galactose and arabinose. The presence of methyl substitutions on the uronic acids allows for gel formation hence finding use in jam making.

Hemicellulase

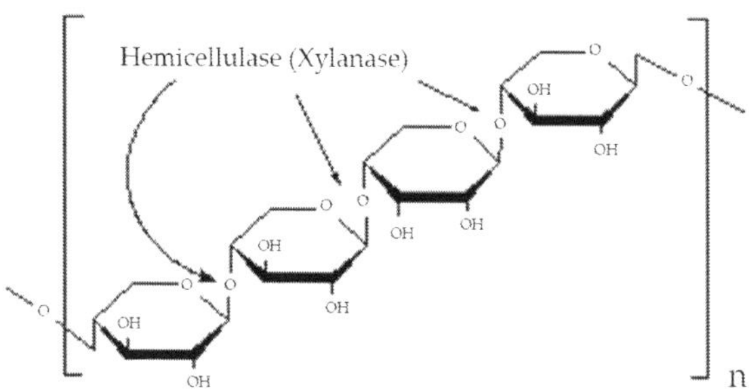

Figure 1.17: Hemicellulose

Pectin

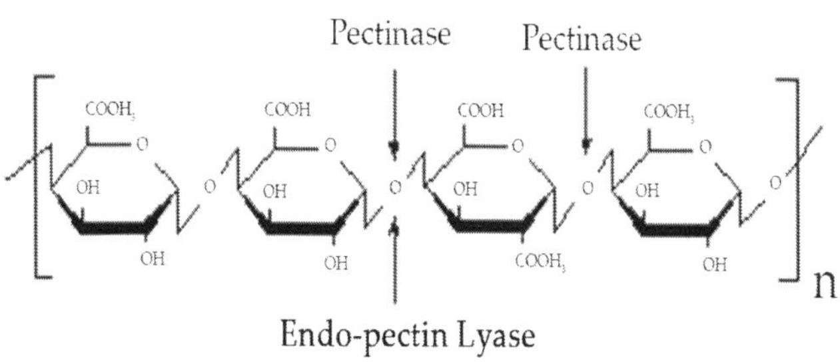

Figure 1.18: Pectin

On the same line are the plant gums and mucilages. The former are branched complex uronic acid complexes that sticky exudates at injury sites. Example: Gum Arabic from Acacia trees used to slow down sugar crystallization and as a thickener. Other gums used as food additives are xanthan, locust bean gum, guar, karaya (sterculia) and tragacanth.

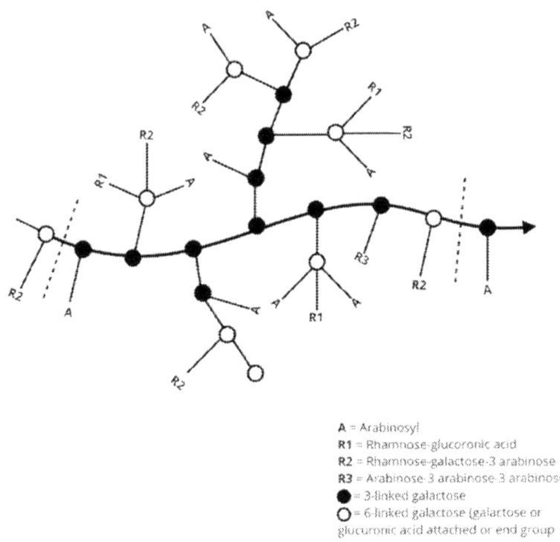

Figure 1.19: Gum Arabic

Plant mucilages are associated with seed endosperm storage carbohydrates. They can store water to circumvent drying out or desiccation. They find use as thickeners and stabilizers in soup, mayonnaise and toothpaste.

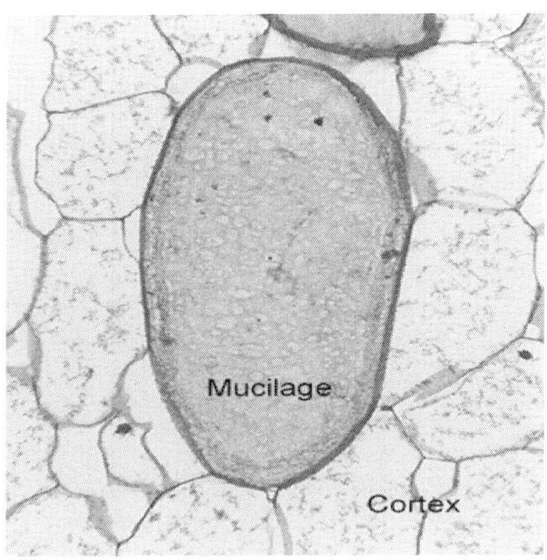

Figure 1.20: Mucilage

NSP from algae includes agar, alginate and carrageenan that serve as plant cell wall polysaccharides. Apart from forming gels, these molecules especially carrageenan can react with milk protein to hence find application in chocolate and in dairy products.

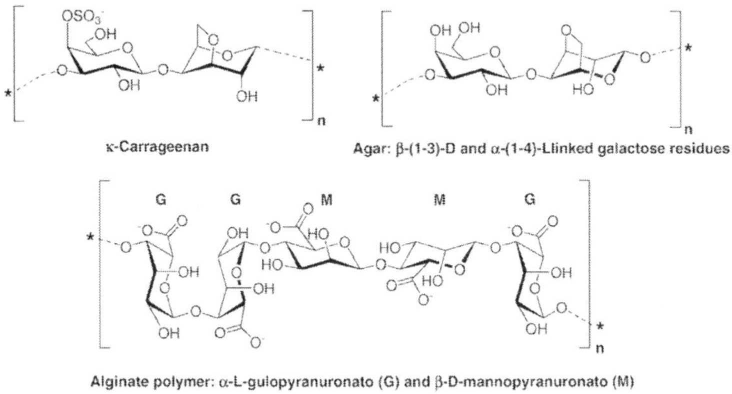

Figure 1.21: Agar, alginate and carrageenan

1.3 TERMINOLOGY USING PHYSIOLOGY

A major hurdle in the classification of carbohydrates using chemistry is the joining together of the classes with physiology and health. Hence, the use of only chemistry in classification cannot permit the studying of nutritional benefits from the chemical divisions as the physiological effects of each class show overlap. The use of physiological properties to coin terms can allow for studying the health benefits of carbohydrates to be used in diets.

Table 1.3 below shows the various physiological properties of dietary carbohydrates:

Carbohydrate	Energy	Satiety	Lower Cholesterol	SCFA source	Boost the absorption of Calcium	Boost stool Output	Prebiotic function	Modulate the immunity	Glycemic nature
Monosaccharides	Yes	-	-	-		-	-	-	Yes
Disaccharides	Yes	-	-	-	Yes	-	-	-	Yes
Polyols	Yes	-	-	Yes Except erythritol.	-	Yes	-	-	

Maltodextrins	Yes	-	-	-	-	-	-	-	Yes
Oligosaccharides (non-α-glucan)	Yes	-	-	Yes	Yes	-	Yes	Yes	-
Starch	Yes	-	-	Yes (Resistant starch)	-	Yes (Resistant starch)	-	-	Yes
NSP	Yes	Yes	Yes (Few)	Yes	-	Yes	-	-	-

Key: SFCA: Short-chain fatty acids (Cummings and Stephen, 2007)

However, many subgroups from the main classification (Table 1.4) display physiological or health benefits that opens up the avenue to repeatedly revise these systems.

Table 1.4 below shows physiological/health groupings of dietary carbohydrates:

Physiological/health groupings	Dietary carbohydrates
Glycaemic	• glucose • fructose • galactose • maltose • lactose • sucrose • trehalose • maltodextrins • starch
Non-glycaemic	• polyols, • non-α-glucan oligosaccharides • resistant and modified starch • NSP

Absence of any effect on stools	• glucose • galactose, • maltose • sucrose, • trehalose, • maltodextrins, • oligosaccharides • majority starches
Boost the output of stool	• Large amounts of fructose • Lactose in Caucasians • Polyols (except erythritol) • some starches • NSP

(Cummings and Stephen, 2007)

There are certain challenges here such as erythritol that is completely absorbed among polyols and the exact nature of starch absorption depends on the processing method of starch or the ability of only select populations to digest lactose.

Table 1.5: Such challenges have allowed for the definition of fractions and sub-fractions of carbohydrates as shown in:

Term	Chemical terminology	Physiological/botanical terminology
Useful ones	• Total carbohydrate • Monosaccharides • Disaccharides • Polyols • Total sugars • Short-chain carbohydrates • Oligosaccharides • Polysaccharides • Starch • Non-starch polysaccharides	• Glycaemic • Dietary fiber • Prebiotic • Resistant starch
Lesser use	• Sugars/ Sugar • Free sugars • Refined sugars • Added sugars • Extrinsic and intrinsic sugars	• Non-digestible oligosaccharides • Available and unavailable • carbohydrate • Complex carbohydrate • Soluble and insoluble fiber

The identification of carbohydrates using chemical means followed by grouping them on the basis of health and physiological effects can allow for the realization of such an integrated classification (Cummings and Stephen, 2007).

1.4 FEW ADDED TERMS

1.4.1 Resistant starch

Resistant starch refers to starch and starch digestion products (such as maltose, maltotriose, α-limit dextrins) not subjected to absorption at the small intestine. While pancreatic a-amylase can digest unmodified starch, the extent of this digestion is dependent on the physical and chemical properties. This has resulted in the following RS classification:

RS I: starch in whole grains that is physically inaccessible

RS II: RS granules (Type B)

RS III: post-processing retrograded starch

RS IV: modified starches

Food	Serving/Amount	Resistant Starch (g/serving)[5]
Banana, slightly green	1 medium, peeled	4.7
Hi-maize* resistant starch	1 Tbsp (approx 9.5 g in weight)* (13 calories)	4.5
Oats, rolled	¼ cup, uncooked	4.4
White Beans	½ cup, prepared	3.7
Lentils	½ cup, cooked	3.4
Pizza	1 slice, 1/8th of a 14" pizza (~100 g)	2.8
Aunt Millie's Whole Grain White Bread	Two slices (45 g)	2.5
Yams	½ cup, cooked	2.5
Chickpeas	½ cup, prepared	2.0
Peas, green	½ cup, prepared	2.0
Barley, pearl	½ cup, cooked	1.9
Rice, brown	½ cup, cooked	1.6
Kidney Beans	½ cup, prepared	1.4
Pasta (cooked or cooled, white or whole wheat)	1 cup, cooked	1.2 (average)
Pumpernickel bread	1 slice	1.1
Rice, white	½ cup, cooked	1.1
Quinoa**	½ cup, cooked	1.0
Corn Flakes	1 cup	0.9
Potato	½ cup, cooked/mashed	0.6-0.8
Oats, cooked	1 cup, cooked	0.5

*To obtain the weight of Hi-maize in a tablespoon, 20 samples of Hi-maize were collected by scooping a tablespoon out of a container of Hi-maize resistant starch, leveling it with a spatula and weighing it. The average weight of the samples was 9.52 g.

**Unpublished data, measured by the modified Englyst method performed by Ingredion and Englyst Carbohydrates, South Hampton, England, 2010.

Figure 1.22: Content of Resistant Starch in foods

Source: *https://resistantstarchresearch.com/posts/rs-in-foods/*

1.4.2 Glycemic carbohydrate

In the context of human health comes the term 'glycemic carbohydrate' that refers to a carbohydrate that provides glucose for metabolism such as mono- and disaccharides along with certain oligosaccharides such as maltodextrins. Those carbohydrates that pass the large intestine before undergoing metabolism are referred to as non-glycemic carbohydrates such as NSPs and resistant starches.

The glycemic index (GI) refers to the extent to which carbohydrates present in food increase the levels of blood glucose using a reference carbohydrate. This is used to classify the dietary carbohydrates and is influenced by many factors such as the diet, meal composition and host biology.

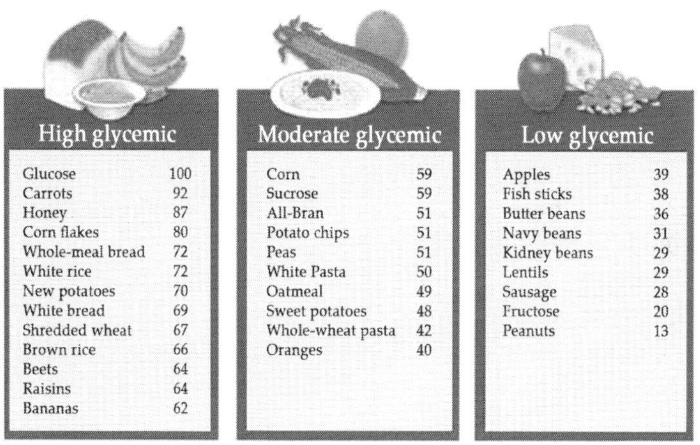

Figure 1.23: GI of several foods

Source: https://infograph.venngage.com/p/217295/glycemic-index

1.4.3 Dietary fiber

The field of nutrition ascribes different connotations to the term fiber or dietary fiber. An initial definition is 'the proportion of food which is derived from the cellular walls of plants which is digested very poorly in human beings' (Trowell, 1972). There has been extensive research in this aspect in terms of analysis, epidemiology and physiology.

At a meeting convened by WHO/FAO and held in Geneva on 17–18 July 2006, the following definition of dietary fiber emerged: 'Dietary fiber consists of intrinsic plant cell wall polysaccharides'.

Functional fiber refers to isolated, non-digestible carbohydrates with beneficial physiological effects in humans.

A total of Dietary fiber and Added fiber is Total fiber.

'Dietary fiber means carbohydrate polymers with a degree of polymerization (DP) not lower than 3 which are neither digested nor absorbed in the small intestine. A degree of polymerization not lower than 3 is intended to exclude mono- and disaccharides. It is not intended to reflect the average DP of a mixture.

Dietary fiber consists of one or more of:

Edible carbohydrate polymers naturally occurring in the food as consumed; carbohydrate polymers, which have been obtained from food raw material by physical, enzymatic or chemical means, synthetic carbohydrate polymers' (http://www.ccfnsdu.de/fileadmin/user_upload/PDF/ReportCCFNSDU2005.pdf).

'Dietary fiber should be defined to include all non-digestible carbohydrates (NDC). Lignin and other non-digestible but quantitatively minor components that are associated with the dietary fiber polysaccharides and may influence their physiological properties should be included as well' European Food Standards Agency Draft Paper on Carbohydrates and Dietary Fiber (Becker and Asp, 2006).

'Dietary fibers are the endogenous components of plant material in the diet which are resistant to digestion by enzymes produced by humans. They are predominantly non-starch polysaccharides and lignin and may include, in addition, associated substances' (Health and Welfare Canada, 1985).

'Dietary fiber is the edible parts of plants or analogous carbohydrates that are resistant to digestion and absorption in the human small intestine with complete or partial fermentation in the large intestine. Dietary fiber includes polysaccharides, oligosaccharides, lignin, and associated plant substances. Dietary fibers promote beneficial physiological effects including laxation, and/or blood cholesterol attenuation, and/or blood glucose attenuation'

(American Association of Cereal Chemists, 2001).

1.5 SUMMARY

This chapter is a concise introduction to allow readers to refresh or update themselves with the various terminologies utilized in the context of carbohydrates. Apart from playing roles as energy sources for organisms, carbohydrates have been shown to give several roles in satiety, health and gut bacterial profile. These shall be examined in the upcoming chapters along with the recommended levels in diets.

1.6 REFERENCES

1. G. Nantel. Carbohydrates in human nutrition. FNA/ANA 24, 1999.
2. JH Cummings and AM Stephen. Carbohydrate terminology and classification. European Journal of Clinical Nutrition (2007) 61 (Suppl 1), S5–S18.
3. Diego M. D. L. Navarro, Jerubella J. Abelilla1and Hans H. Stein. Structures and characteristics of carbohydrates in diets fed to pigs: a review. Journal of Animal Science and Biotechnology (2019) 10:39.
4. Roberfroid M (2005). Inulin-type Fructans. Functional Food Ingredients. CRC Press: Boca Raton.
5. United States Department of Agriculture (2007): http://www.ars.usda.gov/nutrientdata
6. Health Canada (2005). Canadian nutrient file. http://www.hc-sc. gc.ca/fn-an/nutrition/fiche-nutri-data/index_e.html.
7. Stephen AM (2006). Impact of Carbohydrate Methodology on Dietary Intake. Copenhagen, April 2006.
8. IUB–IUPAC Joint Commission on Biochemical Nomenclature (1982). Abbreviated terminology of oligosaccharide chains. J Biol Chem 257, 3347–3351.
9. IUPAC–IUB Joint Commission on Biochemical Nomenclature (1996). Nomenclature of carbohydrates. www.chem.qmul.ac.uk/iupac/2carb/.
10. European Communities (1990). Council directive 90/496/EEC on nutrition labelling for foodstuffs.
11. Trowell H (1972). Dietary fibre and coronary heart disease. Eur J Clin Biol Res 17, 345–349.
12. Becker W, Asp N-G (2006). Carbohydrates and dietary fibre.
13. Health and Welfare Canada (1985). Report of the Expert Advisory Committee a Dietary Fibre.
14. American Association of Cereal Chemists (2001). The definition of dietary fibre. Cereal Foods World 46, 112–126.
15. Institute of Medicine (2001). Dietary Reference Intakes for Energy, Carbohydrate, Fiber, Fat, Fatty Acids, Cholesterol, Protein and Amino Acids. National Academy Press: Washington, DC.
16. Venn BJ, Green TJ (2007). Glycemic index and glycemic load: measurement issues and their effect on diet–disease relationships. Eur J Clin Nutr 61 (Suppl 1), S122–S131.

2

Aspects of Carbohydrates

Shivsanjeevi Sripathi

ShivSanjeevi Sripathi completed his Masters in Biotechnology from Mumbai University in 2008. He was awarded for academic excellence in both his Bachelors and Masters for securing second rank in Mumbai University in 2006 and first rank in his college: Kishinchand chellaram College. For his Masters he secured first rank in his college KET's V.G.Vaze College. He qualified CSIR and NET and and TOEFL in September 2008. He then worked on a stem cell project at the Specialized Centre for Cell Based Therapy (SCCT),KEM Hospital at Mumbai on a project entitled, : Isolation & detection of stem cells from Human Umbilical cord/ amniotic membrane" following which he worked at Junior Research Fellow at Microbiology & Cell Biology Department, Indian Institute of Science, Bangalore from August 2009 to November 2011. The field of work involved cloning of cell wall genes and transcription factors in *E.coli & M.smegmatis*. As a writer, he has authored and co-authored 25 books with Delve and Arcler International Press. He loves to read and share on interesting aspects of life sciences in books. In his free time he loves to travel and explore ancient customs and traditions.

2.1 LOOKING AT SUGARS

There have been several reports about carbohydrates with reports from either end of the spectrum. The 1970s were flooded with reports that weight gain was due to carbohydrates. Later, the "villain" tag in the diet was applied to fat with dietary carbohydrates assuming a positive role mainly in the context of weight control. As far as carbohydrate-specific models are concerned in feeding the satiating efficiency of the macronutrients is given in the order of:

- protein,
- carbohydrate and
- fat

The approximate caloric compensation is generated by carbohydrates while protein induces super caloric compensation and sub caloric compensation in the case of fat. The satiety in the case of proteins is more than that of carbohydrates or fats. The intake of fats is higher as the energy density and palatability of fat foods are high. As the post-absorptive satiety of fat is lower, it causes subsequent compensatory decreases in energy intake. All this finally contributes to high-fat hyperphagia. While the satiety factor of carbohydrate is more than that of fat; the energy density of the former is lesser than that of the latter.

The energetic efficiency of lower in the case of carbohydrates as:

- The metabolizable energy coefficient is different
- Physical activity or non-exercise thermogenesis is different
- The efficiency of storage

The association of energy density and dietary macronutrients were studied in 102 free-living subjects who consumed "ready-to-eat" foods from the British food tables (Blundell and Stubbs, 1999). These foods were foods from all of the categories without any cooking or preparation. It was found that energy density was contributed by all nutrients while there was a negative contribution by water on a plot of water and nutrients as g/100 g of food.

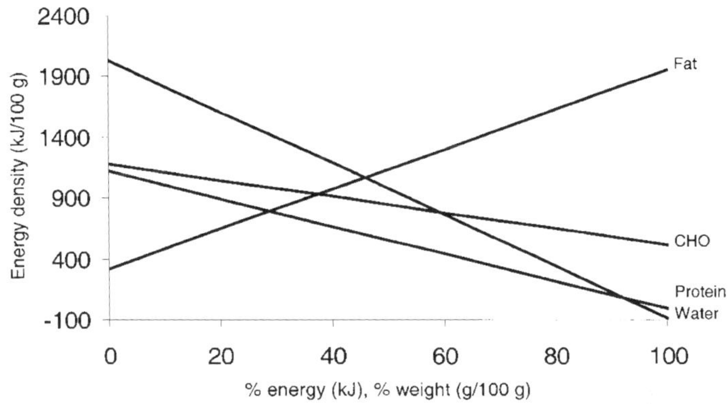

Figure 2.1 shows the plot: percentage weight from dietary macronutrients and water (predictor variables) and energy density (Stubbs et al, 2000).

There are large *R*2 values for fat (energy density [ED (KJ/100 g)] = 462.6 1 35.4 x fat, *R*2 = 75.0) and water (ED(KJ/100 g) = 2034 - 21.2 x water, *R*2 = 66.7). Protein and carbohydrate showed a low relationship with energy density. The energy density can be considerable in high carbohydrate foods with energy density overlap between high fat and high carbohydrate foods.

The overconsumption of glucose and sucrose in solution (overeating) led to obesity in rats (Hirsch and Walsh, 1982). In the case of humans, a parallel experiment was designed where dietary supplements containing sucrose or artificial sweeteners were consumed by overweight subjects (36 women and 6 men) as two groups. At the end of 10 weeks, those who consumed sucrose increased the percent energy from carbohydrates than fat and protein. The intake of sweeteners caused a lowering of the intake of sucrose by 4% of the energy intake. Overall, this is suggestive of the ready occurrence of excess energy intake with the consumption of sweet short-chain carbohydrates in foods (Stubbs et al, 2001).

As far as GI is concerned; carbohydrates are not the only aspect of high- and low-GI foods; satiation in a meal was promoted by low-GI treatments by 66.7%. The subsequent energy intake in short-term interventions was lowered by low glycemic index foods.

Harber et al (1977) compared the same weight of apple puree (minus structural fibers) and juiced apples with 60g of available carbohydrate. In comparison with intact apples, the consumption of juice was 11 times faster while that of puree was four times faster. In terms of satiating, that of juice was lower than that of puree that was lower than intact apples.

The levels of plasma glucose concentration increased with all the three: apples, juice and puree with a conspicuous rebound post-juice that was seen lower puree that was not seen with intact apples. The levels of serum insulin were higher than intact apples than juice and puree.

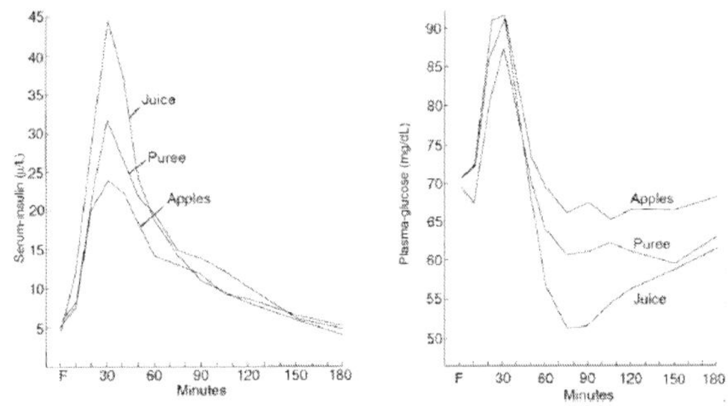

Figure 2.2 shows the trend discussed above: (Harber et al, 1977)

The results were summarized as " "the removal of fiber from food, and also its physical disruption can result in faster and easier ingestion, decreased satiety, and disturbed glucose homeostasis which is probably due to inappropriate insulin release. These effects favor over-nutrition and, if often repeated, might lead to diabetes mellitus" (Harber et al, 1977).

Table 2.1: Appetite and energy balance are mostly influenced by many carbohydrate subtypes with few points summarized below.

Feature	**Monosaccharides**	**Sucrose**	**Oligosaccharides**	**Starch**	**Fiber**
Energy density	++++	+++	+	+++++	
Palatability	+++	++++	++	++	
Glycemic index	++++	+++	+	+++++	+
Digestibility	++++	++++	++	++++++	+
Preabsorptive satiety	++	++	-	++++++	++++
Postabsorptive satiety	+++	+++	-	++++++	+
Osmotic load	++++	++	+++	++++++++	++

(Stubbs et al, 2001)

2.2 ATHLETES AND CARBOHYDRATES

The endogenous carbohydrate stores during an athlete's training or competition cannot suffice for the fuel for the muscle and central nervous system. This calls for the supply of additional carbohydrate that permits sustained high-intensity work or sub-maximal or intermittent high-intensity exercise. In the case of an athlete, restoring the glycogen in the muscle and liver is a major aim. In this context, several guidelines have been published to serve as *guidelines* for refueling carbs as follows:

- The intake of daily carbohydrates should be on par with the fuel needs of training and glycogen restoration when undergoing intense training.
- While the intake for daily carbohydrates depends mainly on the exercise load, the volume of active muscle and body mass- tailor-made suggestions for the diet are warranted. These guidelines would depend on the training feedback and the athlete's overall dietary goals.
- The consumption of carbohydrates should be as soon as possible in the event the period between exercise sessions is less than 8 h to allow for refueling.
- Such refueling can be augmented earlier by frequent small feedings of a higher rate of carbohydrate intake.
- Storage of glycogen can be boosted by the addition of protein to a meal/snack if the carbohydrate intake is sub-optimal for refueling.
- In the event the recovery period is more than 24 h, the carbohydrate-rich meals and snacks can be in lieu of practicality and taste.
- The effectiveness of glycogen storage is the same as that of male athletes when the carbohydrate and energy consumption is adequate (there are only minor changes in storing glycogen across the menstrual cycle).
- The consumption of alcohol should be sensible in the case of athletes especially in the recovery period after exercise.

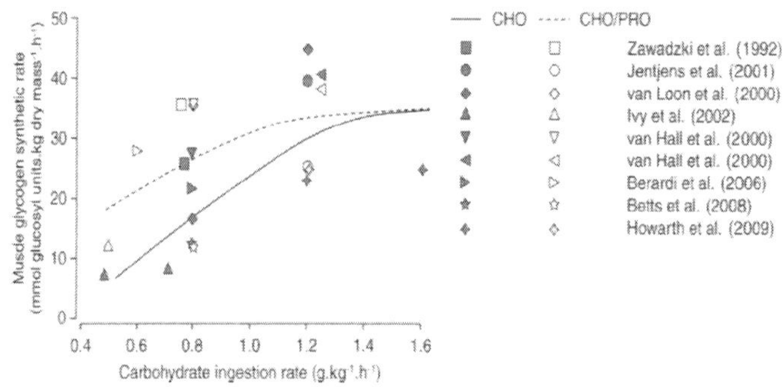

Figure 2.3: Carbohydrate (CHO) intake vs. muscle glycogen storage 2–6 h post-exercise in the presence or absence of protein (PRO); the synthesis of glycogen is increased by the addition of protein (around 20g): (Burke et al, 2011)

Increasing carbohydrate availability for exercise:

The use of approaches to manipulate nutrition and exercise before participating in an event allows for addressing glycogen stores. The stores can be recovered with 24 hours of optimal food and lowered exercise when there is no severe muscle damage.

To summarize, research points out the intestinal absorption by the sodium-dependent glucose transporter SGLT1 as the rate-limiting step in the oxidation of ingested carbohydrates. While this consumption is 1 g per min, the values can go beyond 1.5 g / min when another carbohydrate is added such as fructose that is absorbed by a different transport mechanism. A ratio of 2:1 of glucose: fructose serves this purpose.

When an event or exercise is of 3 hours then the consumption of multiple transportable carbohydrates can offer benefits. Such 'multiple transportable carbohydrates' are present in many fluids/gels/bars that are commercially available.

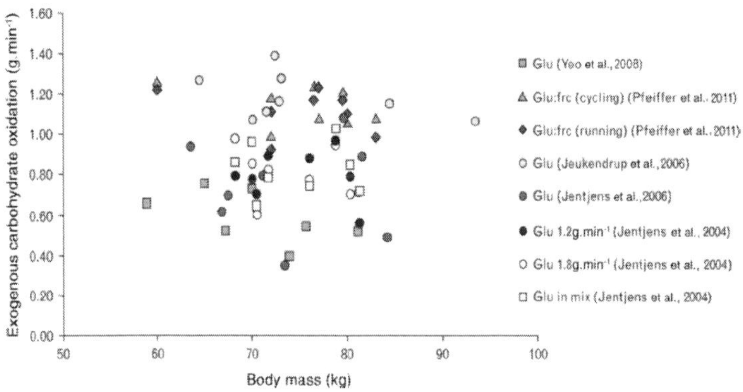

Figure 2.4 below shows body mass vs. peak oxidation rates of consumed exogenous carbohydrate: Glc=glucose, frc=fructose (Burke et al, 2011)

As the body size/body mass has been found to not exert a major effect on the oxidation of exogenous carbohydrate, the guidelines below are in absolute amounts.

Table 2.2: Summary of the carbohydrate intake recommendations for athletes (Burke et al, 2011)

Case in hand	**The target of carbohydrate intake (g) athlete's body mass per day (kg)**	**Points**
Daily needs: While the intake for daily carbohydrates depend mainly on the exercise load, the volume of active muscle and body mass- tailor-made suggestions for the diet are warranted		
Low-intensity or skill-based activities	3–5 g per kg	• When the total needs are being met, the carbohydrates can be consumed as per convenience and individual choice. • The incorporation of protein can aid in achieving other nutrient goals. • The timing of the intake can revolve around the training sessions or refueling needs.
Moderate exercise: 1 h per day	5–7 g	
Endurance: moderate-to-high intensity exercise of 1–3 h each day	6–10 g	
High exercise: an exercise of 4–5 h per day	8–12 g	
ACUTE FUELING: To allow participation in events		

<90 min exercise preparation	7–12 g per kg per 24 h	• The consumption of fiber/ residue-low carb-rich food can allow for gut comfort and meeting the fuel targets. • Small regular snacks can be beneficial. • The fuel targets can be met by consuming carbohydrate-rich foods and drinks: the timing, amount, and type can be in accordance with individual needs, preferences, experience and the event. • The risk of gastrointestinal issues at an event can be lowered by avoiding fat/protein/ fiber-rich foods. • In the event carbohydrate cannot be consumed during exercise, sustained fuel can be met by low GI foods.
Preparation for events: loading carbohydrates > 90 min of sustained or intermittent exercise	36–48 h of 10–12 g per kg body mass per 24 h	
>90 min of sustained/ intermittent exercise	36–48 h of 10–12 g per kg body mass per 24 h	
Less than 8 h recovery between two sessions that require fuel	1.0–1.2 g per kg per h for the first 4 h after which daily needs can be followed.	
> 60 min Pre-event fuelling	1–4 g per kg consumed 1–4 h before the event	
> 45 min brief exercise	Not required	• Easily consumed carbohydrate is available in many foods and drinks. • The consumption of foods and drinks would depend on the event or sport rules. • Such food can be specialized sports products or daily diet habits. • The refueling should match the gut comfort and hydration needs of an athlete.
45-75 min of sustained high-intensity exercise	In little quantities(mouth rinse included)	
1.0–2.5 h of endurance exercises (inclusive of start-stop sports)	30–60 g per hour	
>2.5–3.0 h ultra-endurance exercise	90 g per hour	• Better performance is met by higher intakes of carbohydrate • The intense carbohydrate usage in this activity can be met by glucose: fructose mixtures as these sugars are transported quickly.

2.3 NEW INSIGHTS IN GLYCOBIOLOGY

Living beings display a complicated array of heterogeneity of carbohydrates. This can be attributed to the properties of these molecules that include (Ghazarian et al, 2011):

- Forming glycosidic bonds among various sugars
- The presence or absence of branching

- The anomeric link
- The various structural characteristics

If we consider a single hexose monosaccharide combining with another single hexose monosaccharide to form a single disaccharide molecule: eleven different disaccharides are possible as an outcome. While 4 different amino acids can form 24 different tetrapeptides; 4 four different hexose monosaccharides have the potential to form 35,560 unique tetrasaccharides! ((Sharon and Lis, 1989,1993).

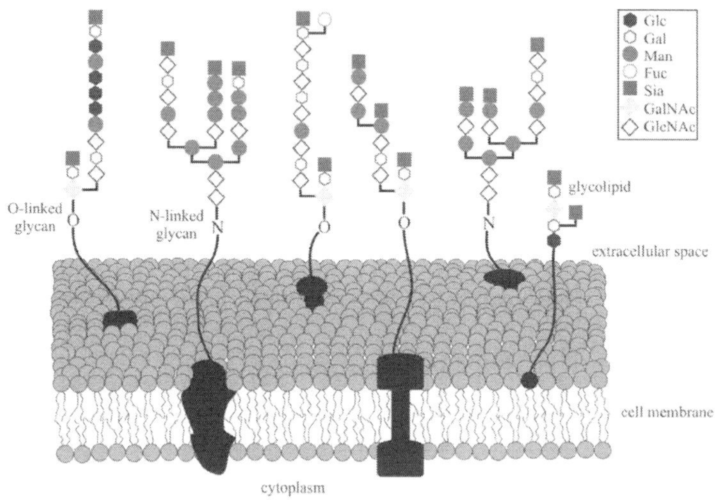

Figure 2.5 below summarizes heterogeneity of carbohydrates: as cell membrane glycoproteins'lipids: Sialic acids (Sia) (Ghazarian et al, 2011)

Such variations in the structure allow for such molecules to have a vast spectrum of activities apart from recognition and adhesion:

- Cell signaling
- Cell-cell recognition
- Contact inhibition
- Normal embryonic development
- Immune response
- Host-pathogen interaction
- Membrane integrity
- Intracellular trafficking

2.3.1 Carbohydrates and pathogens:

Glycans serve as ligands or receptors for several human pathogens. Examples include:

- *Pseudomonas aeruginosa* to fucose (L-Fuc)
- gp-120 glycoprotein of Human immunodeficiency virus (HIV) interacts with the CD4 receptors of CD4+ T cells
- *Neisseria gonorrhea* targets N-acetyllactosamine (Gal-β-4-GlcNAc, LacNAc)
- Influenza virus interacts with host sialic acids
- *Escherichia coli* (*E. coli*) binds host mannosides

2.3.2 Carbohydrates and cancer:

The transformation of normal cells into malignant ones involves metastasis which in turn depends on circumventing regulatory pathways. Few of these include altered telomerase activity to boost the ability to replicate indefinitely, not responding to inhibition by Rb and p53 (tumor suppressors) and changes to avoid detection by the immune system.

Research has shown that most malignant cells show changes in surface carbohydrates by changed glycosylation patterns. These altered patterns are due to altered glycosyltransferases and glycosidases. The molecules involved in the aberrations are glycoproteins. As far as glycolipids are concerned, the changes in cancer cells include:

- Synthesis of new carbohydrates
- Inhibiting the synthesis

Few changes are reported below:

- Several head cancers, neck cancer, and breast cancers show enrichment of gangliosides such as GD3 on their surface with the possible contribution of these sugars to vascular endothelial growth factor (VEGF) production by the tumor.
- N-linked β-1,6-GlcNAc oligosaccharides are enriched in cell surface carbohydrates of tumor cells.
- The enzymes α-1,6-fucosyltransferase (α-1,6-FT), β-1,6-N-acetylglucosominyltransferase V (GlcNAc-

- TV) and β-1,4-N-acetylglucosaminyltransferase III (GlcNAc-TIII) display higher activity in hepatocellular carcinoma (HCC).
- In central nervous system (CNS) tumors, cancers of thyroid, stomach, head, neck, etc. the levels of galectin-3 are elevated while these levels are lower in breast, uterus and ovary cancers.

2.3.3 Carbohydrates and treatment:

The Latin word *legere* meaning "to choose" gave rise to the word lectin that refers to "non-immune carbohydrate specific agglutinins". These carbohydrates show multiple roles ranging from:

- Apoptosis
- Cell adhesion
- Growth
- Fertilization
- Signaling
- Host-pathogen interactions
- Immune modulation
- Recognition
- Interactions between cells and the extracellular matrix (ECM)
- Folding of glycoproteins
- Homeostasis

The late 1980s saw the proposal of using specific drug delivery via lectins. Such drug delivery systems (DDS) referred to as targeted therapeutics lower the side effects while augmenting the efficacy of treatments. The use of lectins in DDS can be achieved via 2 approaches:

- Direct lectin targeting: endogenous cell surface lectins recognize carbohydrate moieties on the DDS
- Reverse lectin targeting: exogenous lectins on the DDS recognize glycolipids and glycoproteins

An example is the use of the overexpressed N-linked β-1,6-GlcNAc oligosaccharides on several cancer cells as targets for this approach (Ghazarian et al, 2011).

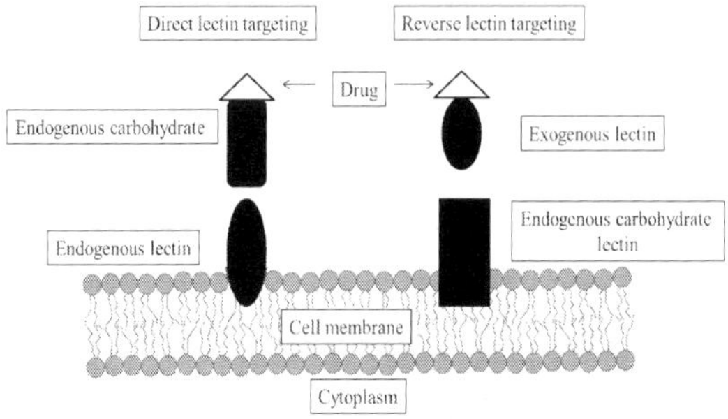

Figure 2.6 below summarizes the use of lectins as drug delivery systems:

While DDSs based on lectins can offer benefits in terms of avoiding the side effects of chemotherapy reagents, they can offer increased specificity in terms of targeting cancer cells. An example can be the use of *Phaseolus vulgaris* agglutinin (PHA-L) that is toxic to CCL-220 (human colon cancer cell line) while exerting a mitogenic effect on health colon cell lines. Studies have explored the toxic effects of ML-I, ML-II and ML-III: Mistletoe lectins on breast cancer.

The treatment of diseases caused by microbes can be addressed by carbohydrates that serve as receptors for pathogens. This anti-adhesion therapy aims at targeting pathogen lectin-host carbohydrate associations to possess advantages of environmental sensibility and lowered side effects. An example of this approach is seen naturally with the presence of Fuc-α-2-Gal-β-4-GlcNAc in breast milk-infants who received this milk were more immune to diarrhea caused by *Campylobacter jejuni*. Sialyloligosaccharides (SOS) could prevent the binding of cholera toxin to the GM1 receptor.

(Ghazarian et al, 2011)

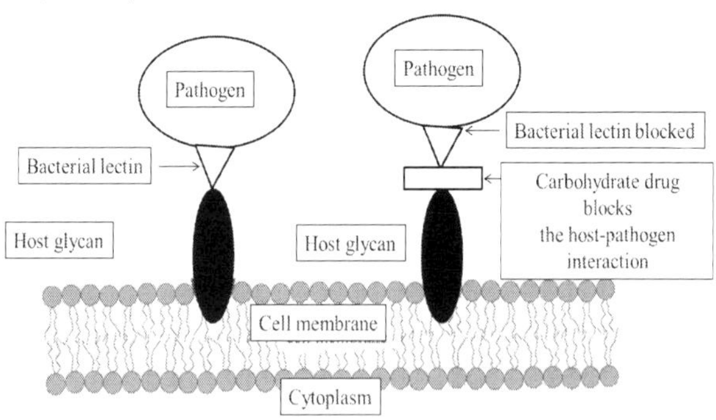

Figure 2.7 below summarizes the use of carbohydrates to treat microbial diseases:

2.4 STUDY: DIET AND CARBOHYDRATES (SEIDELMANN ET AL, 2018)

There are several dietary guidelines with regard to carbohydrates such as higher carbohydrate intake and lowered fat (30%) intake. Yet, there are practices of protein or fat taking the place of carbohydrates. While some cohort-based studies report that the risk of mortality is increased due to low carbohydrate intake, the 2017 Prospective Urban Rural Epidemiology (PURE) study reports the exact opposite!

Such contrary data called for revisions in the nutrition guidelines. However, these studies were found not to take into account the nature of the food source as animal-based versus plant-based. With these points in mind, a team published an article in 2018 about "a population-based study of overall carbohydrate consumption".

The researchers got 15,428 adults aged 45–64 years, in four US communities to complete a dietary questionnaire when they entered the Atherosclerosis Risk in Communities (ARIC) study. The caloric value of the subjects was: 600 kcal or >4200 kcal per day for men and <500 kcal or >3600 kcal per day for women. They filled a food frequency questionnaire (FFQ) or the frequency of consumption of particular foods and beverages in nine standard frequency categories from which nutrient intakes were derived using the Harvard Nutrient Database.

The primary outcome assessed was all-cause mortality using the percentage of energy from carbohydrate intake and all-cause mortality to include possible non-linear relationships. The results showed that those who consumed the lowest amount of carbohydrates consumed lesser plant protein and dietary fiber but higher animal fat and protein on an average. While the incidence of hypertension and weight gain was similar across carbohydrate quantiles, the lowest carbohydrate consumption, in both unadjusted and adjusted models was associated with the highest risk of mortality. There was a U-shaped association between carbohydrate consumption and risk of mortality; a carbohydrate consumption of 50–55% yielded the lowest risk.

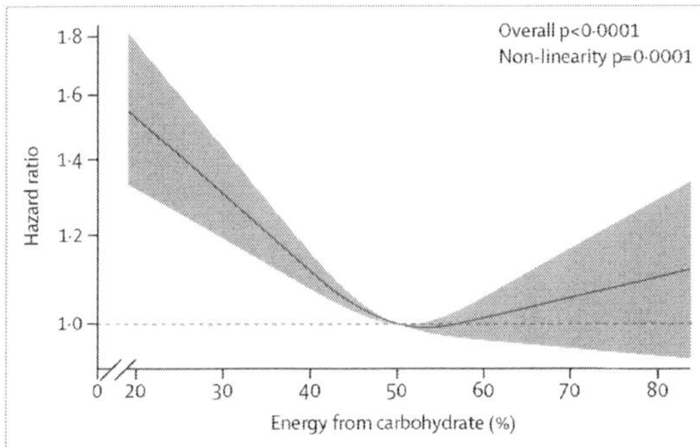

Figure 2.8 below summarizes the risk between carbohydrate consumption and mortality risk: (Seidelmann et al, 2018)

The association between high carbohydrate intake and mortality in the PURE study was within that of the confidence intervals of ARIC. A 2012 meta-analysis was updated with 2 more studies fitting the requirements added to the ARIC to finally amount to 432 179 participants in eight cohort studies.

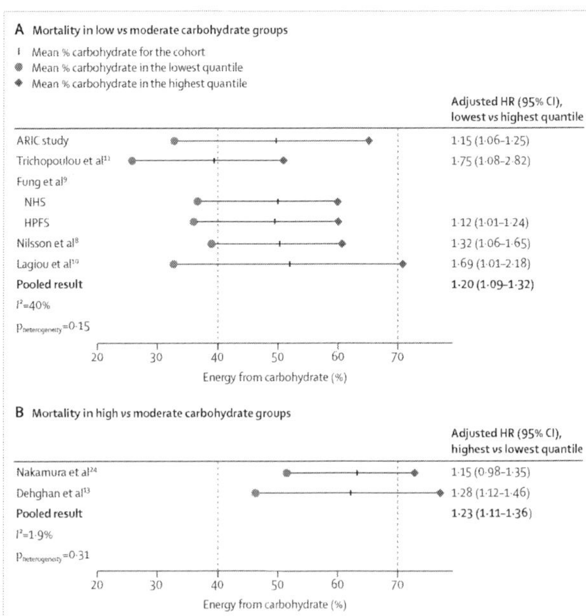

Figure 2.9: The all-cause mortality in participants with lower consumption of carbohydrates was observed for all studies even in the absence of the ARIC data as shown below. Dotted lines indicate cutoffs for low carbohydrate (<40%) and high carbohydrate (>70%). (A) Low carbohydrate vs. moderate carbohydrate

(40–70%) reference group. (B) High carbohydrate vs. moderate carbohydrate reference group. HR=hazard ratio; ARIC=Atherosclerosis Risk in Communities; NHS=Nurses' Health Study. HPFS=Health Professionals Follow-up Study (Seidelmann et al, 2018)

The meta-analysis, the association as mentioned earlier was found to be U-shaped with greater risk to mortality when the consumption was low carbohydrate consumption (<40%) and high carbohydrate (>70%). When animal-derived fat or protein (beef, chicken, lamb, pork) replaced carbohydrates, the risk increased but plant-based substitutions (whole grain bread, nuts, vegetables, peanut butter) lowered this risk.

To conclude, the authors state, "These data provide further evidence that animal-based low carbohydrate diets, which are more prevalent in North American and European populations, should be discouraged. Alternatively, if restricting carbohydrate intake is a chosen approach for weight loss or cardiometabolic risk reduction, replacement of carbohydrates with predominantly plant-based fats and proteins could be considered as a long-term approach to promote healthy ageing" (Seidelmann et al, 2018).

2.5 DISEASE AND CARBOHYDRATES

The share of the dietary energy is termed as "disproportionate" for the modern human's brain as against that of other primates. Initial food was animal food plus plant foods followed by cooked starchy foods that shaped the gastrointestinal tract to evolve as a smaller size. Changes in diet came along with increased gene copy number of salivary amylase (*AMY1*) and the digestion of milk lactose facilitated by lactase persistence even in adults. The development of an agricultural lifestyle allowed for grains as staples leading to lowered average height, nutrient deficiencies and dental caries.

Studies in populations are suggestive of more emphasis on the carbohydrate type than the percentage of carbohydrate of dietary energy in the risk of diseases. Based on the postprandial hyperglycemia and hyperinsulinemia, lowered disease risk is attributed to whole fruits, legumes and minimally processed grains while the risk is more with sugary - beverages, potatoes and refined grains.

There are 2 terms to classify food and blood glucose effects: glycaemic index (GI) that refers to the standard available carbohydrate while glycaemic load (GL) is considered as a better indicator and is defined as, "GI multiplied by the amount of carbohydrate in a typical serving".

The following **table 2.3** summarizes the content, GI and GL of select foods: (Ludwig et al, 2018).

Food	**Serving (g)**	**Available carbohy-drate (g)**	**GI**	**GL**
Boiled potato	150	20	78	16
Milk	250 mL	12	32	4
Boiled basmati Rice	120	30	57	17
Bread, wholemeal	40	13	54	7
Tropical fruit	120	16	58	9
Temperate fruit	120	14	42	6
Fruit juice	250 mL	24	50	12
Instant oat porridge	250	26	79	21
Nuts	30	7	25	2
White/brown Bread	40	19	75	14
Boiled legumes	150	22	31	7

As discussed in Chapter 1, metabolism and gastrointestinal absorption are influenced by fiber: that can be vicious or non-viscous. The former is present in psyllium, legumes, fruits and vegetable gums lower postprandial glycemia due to slowed -down digestion while insoluble fibers of wheat bran do not exert such metabolic effects.

The cardiovascular disease risk factors are increased by consuming more added sugars as shown by random trials. The levels of fat in the liver and other parts of the body were higher when the consumption of sugar-sweetened drinks was increased in other studies. The WHO guidelines of 2015 suggest that free sugars should be taken at below 10% of energy intake for both adults and children while the upper limit is 5% according to the Scientific Advisory Committee on Nutrition in the UK. According to the *2015-2020 Dietary Guidelines for Americans*, the consumption of added sugars is 10% of total energy (Ludwig et al, 2018).

2.5.1 Health and foods

- The risk of cancer, type 2 diabetes, cardiovascular disease and all-cause mortality is lowered in cohort studies when fruits are consumed regularly. While whole fruits of apple, grapes and blueberries offer protection from diabetes, the risk is increased with fruit juices.

- Apart from dietary carbohydrates, potatoes contain potassium, vitamin C and fiber. Cohort studies have shown that type 2 diabetes risk and weight gain were more when these vegetables were consumed at large amounts. The risk of hypertension was higher with potato consumption.
- The consumption of whole grain was found to be protective against several cancers, cardiovascular disease, type 2 diabetes, ischemic stroke and all-cause mortality.
- Legumes contain fiber, protein and low GI carbohydrate; the consumption of which offers lowered cholesterol and cardiovascular disease protection against non-consumers.

Overall, it is important to note that chronic disease is due to traditional starch-based diets plus higher body mass index and lack of exercise. Another vital point is the carbohydrate quality in health over that of the quantity. There is high variation among individuals for GI due to several factors such as genetic factors, glucose intolerance or insulin resistance.

As far as diabetes is concerned, early research is suggestive of lowered triglycerides and glycaemic control when diets with low amounts of foods increasing postprandial blood glucose are consumed. However, efficacy and safety remain to be addressed (Ludwig et al, 2018).

Points to be noted include:

- There are many grey areas as far as the metabolic effects of carbohydrates are concerned as data are mainly from observational studies that face challenges in approaches.
- Most of the randomized controlled trials are flawed due to the short time of study and lack of appropriate blinding of treatment intensity controls.
- A solution to achieve consensus is long term clinical trials using feeding studies and research based on observations along with aspects of economics and the environment.

2.6: CARBOHYDRATES AND FLAVOR

The use of dietary approaches can be utilized to manage the risk or issue of malnourishment. This includes the use of commercially available products, referred to as oral nutritional supplements (ONS) along with fortification or modification. These are formulated products that contain added ingredients inclusive of carbohydrates. Research shows that nutritional intake is augmented when ONS is used (Engfer and Green).

As general ONS fall under the category of "food"; a Nutrition Facts label is displayed on them. They are different from traditional foods as they can supply the total nutritional needs of an individual apart from supplementing the diet. They can supply nutrients under conditions of nutritional inadequacy until the regular intake is resumed.

The content of ONS is on the lines of the U.S. Dietary Reference Intakes (DRI):

- 45% to 65% of calories from carbohydrate,
- 10% to 35% from protein, and
- 20% to 35% from fat

The acceptance of an ONS product involves the critical role of taste and flavor. A 2012 published article of 46 studies involving 4328 patients showed that compliance towards ONS was an average of 78% (Hubbard et al, 2012). The taste in the case of humans is mainly due to the consumption of food that is nutritious and tasty. Metabolism has propelled humans to be innately attracted to sweet taste.

- In food and ONS, a major function of carbohydrates is to supply energy (4 kcal per gram). Unless under starvation, glucose meets the energy requirements of cells such as the brain, heart, skeletal muscle, renal cortex and neural tissues.
- Under conditions of the lack of available glucose, endogenous sources for energy are resorted to by the body. Lengthened periods of fasting boost the catabolism of lean body mass (LBM) and adipose tissue which can lead to maldigestion or malabsorption, respiratory and skeletal issues with time (over some days). The carbohydrates present in ONS or food circumvent this LBM/ protein catabolism to hence balance the energetics.
- Palatability and flavor are important aspects of food and ONS. The flavors of fruit and chocolate are augmented by sweeteners such as sucrose. Other nutrient tastes such as bitterness or saltiness are counteracted by the sweetness of sugars.

- Flavor congruency is contributed to by sugars and other carbohydrates. This refers to increased volatility of foods as seen with the taste- perception association of vanilla- its flavor and sweet taste of sugars.
- The mixing of sugar and water in ONS lower the freezing temperature while increasing the boiling temperature and viscosity to improve the "mouth feel".
- The number of hydroxyl groups dictates the viscosity of liquid foods; when there are no additional thickening agents, the presence of carbohydrates (contain hydroxyl groups) come into the picture.
- In terms of texture, the roles imparted by several synthetic compounds are taken care of by carbohydrates (Clemens et al, 2016).
- The aroma of some foods is augmented by sugar oxidation at high temperatures known as caramelization. Additional texture and browning are also the results of this process.
- Flavor development in food is also influenced by the Maillard reaction between amino acids and reducing sugars. The net result of this multi-step cascade of reactions is the formation of flavor compounds classified as toasted or roasted or nutty and brown colored melanoidins that add to the appearance of the food.
- Sugars also serve as precursors of the desired color and flavor in food.

(Reviewed by Kokkinidou et al, 2018)

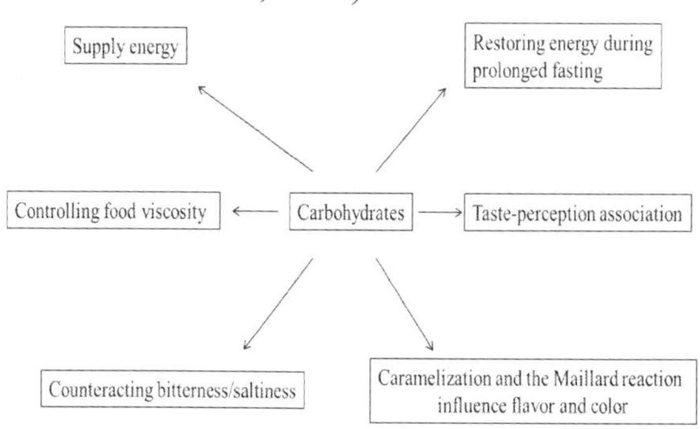

Figure 2.10 below summarizes the roles of carbohydrates in foods such as ONS:

2.7 REFERENCES

1. StubbsRJ,MazlanNandWhybrowS(Nov2001)Carbohydrates,Appetite and Feeding Behavior in Humans. J Nutr. 131(10):2775S-2781S.
2. Blundell JE and Stubbs RJ (1999) High and Low Carbohydrate and Fat Intakes: Limits Imposed by Appetite and Palatability and Their Implications for Energy Balance. Eur J Clin Nutr. 53:S148–S165.
3. Stubbs J, Ferres S and Horgan G (2000) Energy Density of Foods: Effects on Energy Intake. Crit Rev Food Sci Nutr. 40:481–515.
4. Hirsch E and Walsh M (1982) Effects of Limited Access to Sucrose on Overeating and Patterns of Feeding Physiol Behav. 29:129–134.
5. Harber GB, Heaton KW and Murphy D (1977) Depletion and Disruption of Dietary Fiber: Effects on Satiety, Plasma-Glucose, and Serum-Insulin. Lancet. 8040:679–682.
6. Burke LM, Hawley JA, Wong SHS and Jeukendrup AE (2011) Carbohydrates for Training and Competition. J Sports Sci. 29:sup1, S17-S27.
7. Ghazarian H, Idoni B and Oppenheimer SB (May 2011) A Glycobiology Review: Carbohydrates, Lectins, and Implications in Cancer Therapeutics. Acta Histochem. 113(3): 236–247. doi:10.1016/j.acthis.2010.02.004
8. Sharon N and Lis H (1989) Lectins as Cell Recognition Molecules. Science. 246:227–34.
9. Sharon N and Lis H (1993) Carbohydrates in cell recognition. Sci Am. 268:82–9.
10. Seidelmann SB, Claggett B, Cheng S, et al. (2018) Dietary Carbohydrate Intake and Mortality: A Prospective Cohort Study and Meta-Analysis. Lancet Public Health. 3:e419–28.
11. Ludwig DS, Hu FB, Tappy L and Brand-Miller J (Jun 2018) Dietary Carbohydrates: Role of Quality and Quantity in Chronic Disease. BMJ. 361:k2340. Published 2018 Jun 13. doi:10.1136/bmj.k2340.
12. Engfer M and Green C. Oral Nutritional Supplement to Tackle Malnutrition: Summary Booklet. Medical Nutrition International Industry. Available online: http://www.file:///C:/Users/bronsal/Downloads/MNIBookletFINAL2911%20(2).pdf.

13. Hubbard GP, Elia M, Holdoway A and Stratton RJ (2012) A Systematic Review of Compliance to Oral Nutritional Supplements. Clin Nutr Edinb Scotl. 31:293–312.
14. Clemens RA, Jones JM, Lee SY, Mayhew EJ, Slavin J and Zivanovic S (2016) Functionality of Sugars in Foods and Health. Compr Rev Food Sci Food Saf. 15:433–470.
15. Kokkinidou S, Peterson D, Bloch T and Bronston A (2018) The Important Role of Carbohydrates in the Flavor, Function, and Formulation of Oral Nutritional Supplements. Nutrients. 10(6):742. doi:10.3390/nu10060742.

3

Unified Theory of Bacterial Sialometabolism: How and Why Bacteria Metabolize Host Sialic Acids

Eric R. Vimr
Laboratory of Sialobiology, Department of Pathobiology, University of Illinois at Urbana-Champaign, Urbana, IL 61802, USA

ABSTRACT

Sialic acids are structurally diverse nine-carbon ketosugars found mostly in humans and other animals as the terminal units on carbohydrate chains linked to proteins or lipids. The sialic acids function in cell-cell and cell-molecule interactions necessary for organismic development and homeostasis. They not only pose a barrier to microorganisms inhabiting or invading an animal mucosal surface, but also present a source of potential carbon, nitrogen, and cell wall metabolites necessary for bacterial colonization, persistence, growth, and, occasionally, disease. The explosion of microbial genomic sequencing projects reveals remarkable diversity in bacterial sialic acid metabolic potential. How bacteria exploit host sialic acids includes a surprisingly complex array of metabolic and regulatory capabilities that is just now entering a mature research stage. This paper attempts to describe the

Citation: Unified theory of Bacterial Sialometabolism: How and why bacteria Metabolize Host Sialic Acids. Eric R. Vimr. Hindawi Publishing Corporation ISRN Microbiology Volume 2013, Article ID 816713, 26 pages http://dx.doi.org/10.1155/2013/816713

variety of bacterial sialometabolic systems by focusing on recent advances at the molecular and host-microbe-interaction levels. The hope is that this focus will provide a framework for further research that holds promise for better understanding of the metabolic interplay between bacterial growth and the host environment. An ability to modify or block this interplay has already yielded important new insights into potentially new therapeutic approaches for modifying or blocking bacterial colonization or infection.

INTRODUCTION

At least at some level common experience indicates to almost everyone that life is constrained by competition for limited resources. Formally trained biologists understand this competition as central to evolution, the only fundamental theory in biology. For some microorganisms competitive success in colonizing a mammalian or avian host depends upon specialized metabolism that may support growth in only certain niches. For example, Freter [1] has summarized his own and the work of others by describing the mechanisms of association of bacteria with mucosal surfaces. These mechanisms include "(a) chemotactic attraction of motile bacteria to the surface of the mucus [layer], (b) penetration and trapping within the mucus [layer], (c) adhesion to receptors…, (d) adhesion to epithelial cell surfaces, and (e) multiplication of the mucosa-associated bacteria." The combined set of traits or phenotypes expressed by a given bacterium defines its potential "virulence factors" or relative colonization success [1, 2]. In the current paper the final stage of the host-microbial interaction is exclusively focused upon multiplication of bacteria at mucosal surfaces. This focus further emphasizes *Escherichia coli* as the predominant facultative anaerobe in animal hosts and its metabolic uses of host sialic acids for nutrition or surface decoration. The narrative approach is intended to support a unified set of observations and hypotheses that could guide future research in the field designated microbial sialobiology [3].

By understanding the metabolic use of a single group of prevalent mucosal sugars, the sialic acids, it may be possible to at least partially identify factors controlling which bacteria colonize only certain areas of the gastrointestinal tract or other mucosal surfaces. This goal is central to understanding microbial colonization in disease and health of humans and livestock [2]. The gastrointestinal tract (GIT) is essentially an open tube containing a few valves located between mouth and anus and open to the environment at either end (Figure 1). Bacterial colonization begins during and after birth and may continue to change or become restructured over time as influenced by complicated factors such as diet, overall health, and even geographic location. Other mucosal or epithelial surfaces include the eyes and some sites not shown in Figure 1 like the nasopharynx, lungs, bladder, vagina, and urethra. Some of these sites are normally sterile, for example, eyes, lungs, and bladder unless colonized during an ongoing infectious disease process. Each of these extra-GIT sites expresses a variety of sialic acids that probably do not differ greatly from those found in the large intestine, though much less work has been done on this topic than on the mucus layer and epithelium of the animal large intestine. Regardless of the relative disparity in detailed information between datasets, information about the large intestine should facilitate generalizations to all mucosal sites in healthy and diseased states where microbial involvement is known or suspected. Note that listed in the legend to Figure 1 are bacteria that permanently colonize the large intestine. This group must be in constant competition thus separating the colon from normally sterile sites that usually remain uninfected or, when infected, it is usually by a single species resulting in either clearance by or death of the host.

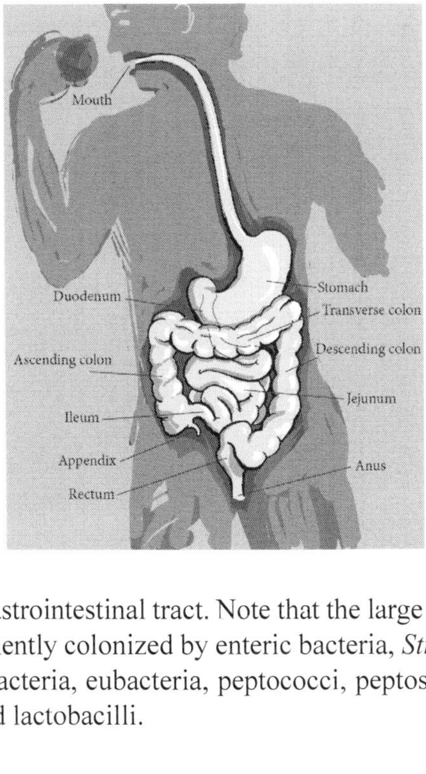

Figure 1. Human gastrointestinal tract. Note that the large intestine (colon with appendix) is permanently colonized by enteric bacteria, *Streptococcus faecalis*, bacteroides, bifidobacteria, eubacteria, peptococci, peptostreptococci, ruminococci, clostridia, and lactobacilli.

Most information about the pathogenic and commensal gut bacteria comes from standard (sometimes referred to as classical) methods of microbial culture and measurements of nutrient use. More current methods such as high-throughput sequencing for identifying both cultivable and noncultivable bacteria as well as nuclear magnetic resonance spectroscopy or mass spectrometry for identifying hundreds of small molecules in complex samples are generating datasets for statistical analyses [4]. However, when the exact identities or functions of important nutrients are unknown, or the metabolic pathways needed for their metabolism are not described, it is unclear how the more recent methods will offer many new insights until supported or refuted by direct experiments after the necessary basic pathways have been elucidated. Therefore, and at the risk of being repetitive, the goal of this current paper is to understand the metabolism of a remarkably distinct, chemically varied, and prevalent family of mucosal sugars that are known in some cases and hypothesized in others to influence in minor-to-major ways the capacity for bacterial niche specialization or disease potential. Some of the ways this information could be applied to specific practical (therapeutic) uses have been described [5].

STAGE DYNAMICS AND DRAMATIS PERSONAE: GIT MORPHOLOGY AND BACTERIAL INHABITANTS

Anatomic variation between GITs reflects the digestive needs of a given animal species. Carnivores such as cats and dogs, or human omnivores have a less developed cecum (the appendix) than monogastric herbivores, ruminants, or granivorous birds (Figure 2). Despite this and other anatomic differences most of the digestion and absorption of foodstuffs occurs in the animal small intestine such that carbohydrate, protein, and fat are all digested and mainly absorbed from this site before the undigested residuum empties into the colon [6]. The many factors limiting or selecting for bacterial diversity in most anatomical compartments is in stark contrast to the large intestinal microbiota in its richness and depth of both permanent and occasional inhabitants (Figure 1).

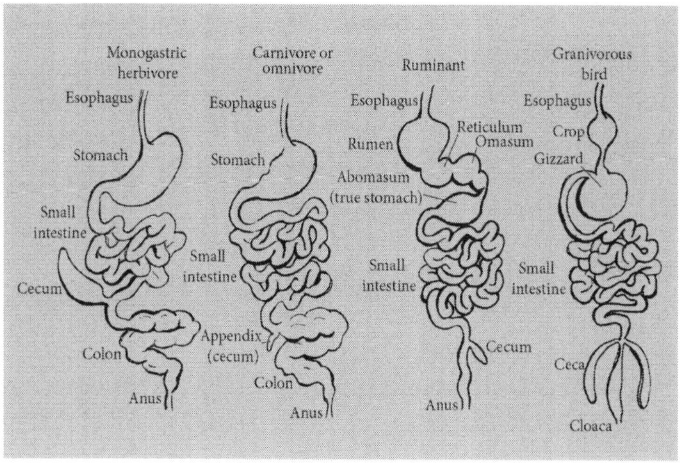

Figure 2. Intestinal structures reflecting different animal dietary types. Examples of monogastric herbivores are horse, rabbit, rat, and pig; carnivore or omnivores are cats, dogs, and man; ruminant examples are cow and sheep while granivorous bird examples are chickens or turkeys. The diagram is modified from reference [6].

Table 1 lists some of the bacterial species isolated from the healthy human intestine [6]. The genus/species designations given to some bacteria have changed over time, and other species unique to nonhumans (Table 2) expands the diversity of bacteria residing within animal GITs. Despite the enormous numbers of intestinal bacteria estimated at 10^{14} [7], and the wide species diversity of the colonic microbiota, two facts emerged from standard analyses of the major cultivable bacterial groups [6]. First, *E. coli* is 10 to 100 times more prevalent than clostridia, streptococci, or lactobacilli and one million times more common than yeasts, while 50 to 1000 times less prevalent than bacteroides in the normal human cecum or feces. Second, *E. coli* is found in the rumen and abomasum of cows and crop of chickens as well as the stomachs and entire small intestines of pigs, chickens, cats, and humans living in tropical environments [6]. A discussion of factors limiting *E. coli* to the terminal ileum and colon of healthy humans from temperate climates will not be attempted. Neither will an attempt be made to either support or refute metagenomic analyses that suggest many more, uncultivable bacterial species may exist than those species already identified by standard procedures (Tables 1 and 2). However, and for the purposes of this paper, it is essential to note the guiding principle suggested by the above data. Namely, bacteroides outcompetes *E. coli* by nutritionally exploiting residual foodstuffs not already absorbed by the host including carbohydrates that are undigestible by *E. coli*, while *E. coli* outcompetes all other enterics or other major bacterial groups by mechanism(s) unknown. This paper will address a hypothesis that could explain the evolutionary success of *E. coli*. Failing that goal, the present paper will at least provide a coherent assessment of recent data explaining how bacteria metabolize a major group of host-derived metabolites.

Table 1. Some of the bacterial species isolated from the healthy human intestine.

Microorganism (recent designation)	*E* value[a]	Maximum percent identity[b]
Gram-negative anaerobic rods		
Bacteroides asaccharolyticuslyitalics (Porphyromonas asaccharolyticus)	0.063	43% (46/106)
Bacteroides capillosus	4×10^{-6}	24% (43/180)
Bacteroides distasonis (Parabacteroides distasonis)	3×10^{-5}	25% (50/203)
Bacteroides eggerthii	3×10^{-5}	23% (70/311)
Bacteroides fragilis	8×10^{-6}	26% (58/222)
Bacteroides furcosus	No similarity found	
Bacteroides hypermegas (Megamonas hypermegale)	0.023	32% (23/93)
Bacteroides melaninogenicus subsp. *mel. (Prevotella melaninogenica)*	0.002	25% (30/120)
Bacteroides multiacidus (Mitsuokella multacida)	0.004	35% (16/46)
Bacteroides ovalis	No sequence available	
Bacteroides ovatus	2×10^{-29}	28% (83/294)
Bacteroides praeacutus (Tissierella praeacuta)	0.88	50% (7/14)
Bacteroides putredinis (Alistipes putredinis)	0.002	31% (36/117)
Bacteroides ruminicola subsp. *brevis*	No similarity found	
Bacteroides ruminicola subsp. *ruminicola (Prevotella ruminicola)*	1×10^{-27}	28% (80/287)

Bacteroides splanchinicus (Odoribacter splanchnicus)	5×10^{-4}	25% (33/131)
Bacteroides thetaiotaomicron	6×10^{-30}	28% (84/305)
Bacteroides uniformis	2×10^{-23}	26% (80/314)
Bacteroides vulgatus	1×10^{-28}	28% (82/297)
Desulfomonas pigra (Desulfovibrio piger)	0.001	31% (15/48)
Leptotrichia buccalis	0.002	27% (30/111)
Fusobacterium mortiferum	0.001	27% (33/124)
Fusobacterium naviforme	No similarity found	
Fusobacterium necrogenes	No similarity found	
Fusobacterium nucleatum	0.014	44% (11/25)
Fusobacterium plauti (F. plautii)	0.003	39% (17/44)
Fusobacterium prausnitzii (Faecalibacterium prausnitzii)	8×10^{-5}	22% (33/153)
Fusobacterium russi (F. russii)	None found	
Fusobacterium symbiosum	No sequence available	
Fusobacterium varium	8×10^{-79}	40% (124/310)
Butyrivibrio fibriosolvens	4×10^{-5}	22% (29/131)
Sucinimonas amylolytia	No sequence available	
Vibrio succinogenes	0.055	31% (25/81)
Gram-positive anaerobic rods		
Bifidobacterium adolescentis	0.018	26% (37/143)
Bifidobacterium angulatum	9×10^{-4}	56% (14/25)
Bifidobacterium bifidum	0.005	22% (25/114)
Bifidobacterium breve	0.012	50% (13/26)
Bifidobacterium catenulatum	0.084	43% (9/21)
Bifidobacterium cornutum	No sequence available	
Bifidobacterium dentium	0.016	29% (24/82)
Bifidobacterium infantis (B. longum subsp. *infantis)*	0.005	52% (13/25)
Bifidobacterium longum	0.006	52% (13/25)
Bifidobacterium pseudolongum (B. longum subsp. *longum)*	0.006	52% (13/25)
Clostridium bejerinkii (C. bejirinckii)	3×10^{-5}	20% (41/209)
Clostridium butyricum	1×10^{-76}	31% (123/402)

Clostridium cadaveris	3.3	50% (4/8)
Clostridium celatum	No similarity found	
Clostridium clostridiiforme	0.047	21% (17/80)
Clostridium difficile	0.023	28% (41/148)
Clostridium innocum	0.72	22% (11/50)
Clostridium leptum	0.026	25% (18/76)
Clostridium malenominatum	2.9	67% (4/6)
Clostridium nexile	1×10^{-4}	43% (20/47)
Clostridium paroputrificum	1.2	20% (15/77)
Clostridium perfringens	4×10^{-83}	43% (132/310)
Clostridium ramosum	0.11	28% (17/60)
Clostridium tertium	0.13	26% (19/72)
Eubacterium aerofaciens (Collinsella aerofaciens)	1×10^{-54}	34% (108/314)
Eubacterium contortum	No similarity found	
Eubacterium cylindroides	0.042	32% (13/41)
Eubacterium lentum (Eggerthella lentum)	0.064	32% (22/68)
Eubacterium limosum	0.003	32% (23/74)
Eubacterium rectale	0.018	26% (14/53)
Eubacterium ruminantium	2.5	36% (5/14)
Eubacterium tenue	No sequence available	
Eubacterium tortuosum	No similarity found	
Eubacterium ventriosum	1×10^{-4}	41% (22/54)
Lachnospira multiparus	1.6	60% (6/10)
Propionibacterium acnes	1×10^{-20}	25% (78/313)
Propionibacteriu granulosum	4.1	71% (5/7)
Propionibacterium jensenii	0.25	30% (17/57)
Anaerobic cocci		
Acidaminococcus fermantans (A. fermentans)	0.16	31% (18/58)
Megasphera elsdenii (Megasphaera elsdenii)	0.14	34% (11/32)
Peptococcus asaccharolyticus (Peptoniphilus asaccharolyticus)	0.023	49% (19/39)
Peptococcus magnus (Finegoldia magna)	6×10^{-4}	30% (17/57)

Peptococcus prevotii (Anaerococcus prevotii)	3×10^{-51}	34% (98/291)
Peptostreptococcus productus	No similarity found	
Ruminococcus albus	4×10^{-8}	28% (41/149)
Ruminococcus bromii	0.046	37% (14/38)
Ruminococcus flavefaciens	4×10^{-6}	26% (39/152)
Sarcina ventriculi	0.76	32% (8/25)
Streptococcus constellatus	0.039	28% (14/50)
Streptococcus intermedius	5×10^{-6}	23% (42/182)
Streptococcus morbillorium (Gemella morbillorum)	6×10^{-61}	35% (105/247)
Veillonella parvula	0.014	23% (28/120)
Coprococcus cutactus	No sequence available	
Coprococcus catus	0.022	29% (20/68)
Coprococcus comes	0.004	26% (16/61)
Gemiger formicilis	No sequence available	
Facultatively anaerobic bacteria		
Enteric bacteria		
Escherichia coli	9×10^{-6}	26% (35/137)
Citrobacter freundii	2×10^{-4}	23% (57/251)
Klebsiella pneumonia	8×10^{-5}	24% (34/140)
Enterobacter cloacae	6×10^{-4}	23% (30/133)
Enterobacter aerogenes	0.006	24% (60/249)
Proteus mirabilis	3×10^{-4}	42% (20/48)
Proteus morganii	0.15	41% (15/37)
Lactobacilli		
Lactobacillus acidophilus	6×10^{-4}	25% (36/144)
Lactobacillus brevis	0.012	26% (14/55)
Lactobacillus casei	0.038	26% (28/107)
Lactobacillus fermentum	0.011	22% (11/49)
Lactobacillus leichmannii (L. leichmanni)	3.9	29% (8/28)
Lactobacillus minutus	2.8	67% (6/9)
Lactobacillus plantarum	0.089	24% (29/121)
Lactobacillus rogosae	No similarity found	
Lactobacillus ruminis	4×10^{-13}	37% (29/78)

Lactobacillus salivarius	0.062	34% (16/47)
Streptococci		
Streptococcus faecalis (Enterococcus faecalis)	1×10^{-81}	43% (128/301)
Streptococcus faecium (Enterococcus faecium)	2×10^{-81}	43% (128/301)
Streptococcus bovis	2×10^{-99}	48% (152/319)
Streptococcus agalactiae	1×10^{-124}	55% (172/313)
Streptococcus anginosus	9×10^{-4}	43% (20/47)
Streptococcus avium (Enterococcus avium)	0.82	28% (10/36)
Streptococcus cremoris (Lactococcus lactis subsp. *cremoris)*	1×10^{-77}	40% (124/311)
Streptococcus equisimilius (S. dysgalactiae subsp. *equisimilis)*	0.004	29% (19/65)
Streptococcus lactis	2×10^{-78}	41% (125/307)
Streptococcus mitior (S. mitis)	0.0	94% (307/326)
Streptococcus mutans	0.001	26% (38/148)
Streptococcus salivarius	0.001	31% (37/121)
Streptococcus sanguis (S. sanguinis)	0.0	93% (303/326)
Staphylococcus epidermidis	0.111	22% (16/73)

[a]Sequenced microbial genomes of some of the species found in the healthy human intestine with recent designations given in parentheses [6]. Genera were queried for similarity to acetyl xylan esterase [8]. The lower the expected value (*E*) is reflects the likelihood that a match is not due to chance.

[b]Indicates the percentage of identical amino acids within the specified alignment length. The number of identical amino acids/the alignment length is given in parentheses.

Table 2. Additional bacterial species isolated from the healthy intestines of non-human animals.

Microorganism (recent designation)[a]	***E* value**	**Maximum percent identity**
Gram-negative anaerobic rods	0.018	27% (12/44)
Bacteroides succinogenes		
*(Fibrobacter succinogenes subsp. succinogenes)**		

*Bacteroides ruminicola (Prevotella ruminicola)**	1×10^{-27}	28% (80/287)
*Bacteroides amlylophilus (Ruminobacter amylophilus)**	1.3	25% (13/51)
Bacteroides termitidis (Sebaldella termitidis)	0.013	34% (21/61)
Bacteroides clostridiiformis (Clostridium clostridioforme)	0.047	21% (17/80)
Succinivibrio dextrinosolvens *	1	50% (5/10)
Selenomonas ruminatium (S. ruminantium subsp. *lactilytica)* *	5×10^{-4}	32% (19/59)
Desulfovibrio spp.	3×10^{-4}	35% (20/57)
Desulfotomaculum nigrificans	0.004	30% (13/44)
Desulfotomaculum ruminis *	0.004	30% (18/61)
Desulfotomaculum orientis	No similarity found	
Oscillospira guillermondii	No similarity found	
Gram-positive anaerobic rods		
Eubacterium cellulosolvens *	7×10^{-4}	50% (19/38)
Eubacterium nitrogenes (Eubacterium nitritogenes)	No similarity found	
Eubacterium helminthoides	No sequence available	
Cellulomonas flavigena	7×10^{-35}	29% (97/330)
Brevibacterium spp.	no similarity found	
Bifidobacterium asteroides	0.66	67% (8/12)
Bifidobacterium indicum	0.41	35% (9/26)
Bifidobacterium coryneforme	0.54	67% (8/12)
*Bifidobacterium pseudolongum**	0.63	34% (11/32)
*Bifidobacterium ruminale (Bifidobacterium thermophilum)**	1.3	75% (6/8)
*Lactobacillus ruminis**	4×10^{-13}	37% (29/78)
*Lactobacillus vitulinus**	0.6	42% (5/12)
*Clostridium lochheadii**	No sequence available	
*Clostridium longisporum**	0.97	43% (9/21)
*Clostridium cellobioparum**	2.2	64% (7/11)
Methanobacterium formicicum	0.53	55% (6/11)
Methanobacterium ruminantium (Methanobrevibacter ruminantium)	0.089	36% (10/28)

Methanobacterium mobile	No similarity found	
Anaerobic cocci		
Methanosarcina barkeri	1×10^{-7}	22% (40/181)
Lampropedia hyalina *	No similarity found	
Veillonella alcalescens	0.014	23% (28/120)
Facultatively anaerobic bacteria		
Salmonella spp.	4×10^{-5}	26% (35/137)
Hafnia alvei	0.016	27% (14/52)
Streptococcus equinus (Streptococcus bovis)	2×10^{-99}	48% (152/319)
Lactobacillus lactis (Lactobacillus delbrueckii subsp. *lactis)*	0.001	25% (42/165)
Lactobacillus buchneri	0.045	27% (20/73)
Lactobacillus cellobiosus (Lactobacillus fermentum)	0.011	22% (11/49)
Miscellaneous bacteria		
Borrelia spp.	0.017	45% (9/20)
Bacillus macerans (Paenibacillus macerans)	0.055	31% (12/39)
Acholeplasma bactoclasticum	No similarity found	
Gemmiger formicilis	No similarity found	
Alcaligenes faecalis	0.003	28% (13/47)
Fusosporus spp.	No similarity found	
Arthromitus spp. *(Candidatus arthromitus)*	6×10^{-4}	27% (24/88)
Anisomitus spp.	No sequence available	
Entomitus spp.	No sequence available	
Coleomitus spp.	No sequence available	
Bacillospira spp.	No sequence available	
Sporospirrillum spp.	No sequence available	
Metabacterium spp.	3×10^{-8}	25% (40/159)

[a]Bacteria isolated from nonhuman animal intestines with newer designations given in parentheses. Asterisks indicate bacteria isolated mainly or exclusively from the rumen [6]. Expect values (*E*) and maximum percentage identities are as described in Table 1.

If the central hypothesis introduced above is correct, which nutrient(s) does *E. coli* exploit for survival and outcompeting most of its rivals? Therefore, the conceit of this paper is that *E. coli* has evolved to efficiently exploit host-derived nutrients, and its success as the preeminent facultative large intestinal anaerobe is at least partly owed to an evolutionarily optimized use of host-derived sialic acids. Anyone interested in exploring the genesis of this paper's central hypothesis should read the delightful treatise by Koch [9]. In this work Koch argues, on the basis of biophysical data, at least a partial explanation for how *E. coli* became evolutionarily successful. Indeed, it is in this author›s opinion that Koch›s article is the single best of all possible articles on the subject of evolutionary success by suggesting that *E. coli* could be the most highly evolved species on the planet.

SIALIC ACIDS AND THEIR DISTRIBUTIONS IN THE LARGE INTESTINE

The sialic acids is a designation given to a group of over 40 naturally occurring nine-carbon keto acids found mainly in animals of the deuterostome embryonic lineage (starfish to humans). These sugars are synthesized rarely by bacteria, and then mostly by pathogens that use sialic acids to masquerade as immunological self, not at all in plants or protostomes except for perhaps a few larval insect stages, and probably not by fungi though the jury remains out in this case [3]. Therefore, when speaking of sialic acid metabolism (sialometabolism) the process is limited to mostly bacterial species that exist as animal commensals or pathogens [3, 10]. Faillard [11] covered the early history of sialic acids since their discovery in the 1930s to the modern era beginning around 1985. Chemists conducted most research during the initial stage of sialic acid discovery. However, Roland Schauer was an early proponent during this time of a different or at least more expansive view of sialic acids, as he clearly recognized that their unique chemical structures and skewed phylogenetic distribution was likely to be significant to diverse biological phenomena. His many insights helped lead to the modern field of sialobiology as a subset of glycobiology and ultimately to the current view of microbial sialobiology described in this paper. Indeed, Schauer was the first to show that a bacterium, *Clostridium perfringens*, appeared to have some mechanism for metabolizing sialic acid in the bacterial growth medium [12].

The most common sialic acid, 2-keto-3-deoxy-5-acetamido-D-*glycero*-D-*galacto*-nonulosonic acid, is abbreviated Neu5Ac reflecting the backbone neuraminic acid ring, the acetamido group at the carbon position 5, and the glycerol tail composed of carbons 7–9 (Figure 3(a)). Various chemical groups attached to the glycerol tail or ring carbon hydroxyl groups define most Neu5Ac derivatives. By far the most common derivatives bear O-acetyl groups at carbon positions 4, 7, 8, or 9. These additions are catalyzed by O-acetyl transferases in both bacteria and eukaryotes using acetyl-coenzyme A as acetyl donor. O-acetylated sialic acids are abbreviated Neu4(7,8,9),5$Ac_{2 \text{ or } 3}$ to reflect the position(s) of the acetyl ester(s). *N*-glycolylneuraminic acid (Neu5Gc), synthesized by a hydroxylase that adds a hydroxyl group to the carbon-5 acetamido of Neu5Ac (Figure 3(a)) is the other major form of sialic acid in most animals other than humans. Its absence in humans is due to a null mutation in the hydroxylase gene, indicating gene function was lost after the split of the human ancestor from that of the great apes. Ajit Varki and his colleagues have speculated about the biological consequences of Neu5Gc's absence in humans [13], but it is possible that the loss has no major consequence other than one less nutrient for bacteria to exploit in humans. Less common derivatives include an oxidized form of Neu5Ac, 4-(acetylamino)-2,4-dideoxy-D-*glycero*-D-*galacto*-octanoic acid (ADOA), a carbon position 1–7 lactone (Neu5Ac1,7L), and a 2-deoxy anhydro form, Neu5Ac2en (Figures 3(b)–3(d), resp.).

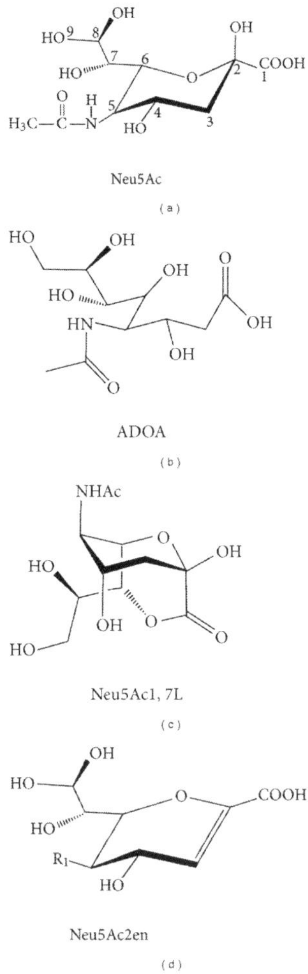

Figure 3. Sialic acid structural diversity. (a) Neu5Ac, the most common sialic acid. The 4 and 7–9 hydroxyls may all be substituted with acetyl groups or less commonly methy, lactyl, succinyl, or phosphate groups. A hydroxyl group on the C-5 acetamido yields Neu5Gc, which is common in all higher animals but humans. (b) Oxidized sialic acid. (c) Lactone detected in high amounts in humans. (d) Anhydro sialic acid and transition state analog of sialidases. Neu5Ac and its derivatives and Neu5Ac1,7L exist bound to other sugars on oligosaccharides of mucin and other glycoconjugates, while ADOA and Neu5Ac2en are free in solution and thus missed by most structural analyses.

The structural diversity of the sialic acids is matched by their *regio*-distributional differences along the length of the large intestine [14, 15], revealing an increasing sialic acid gradient from ileum to rectum. Figure

4 also shows that in humans Neu5Ac, Neu5Ac1,7L, and various O-acetylated forms are the most prevalent sialic acid derivatives. By contrast, the mouse as expected produces Neu5Gc [15], but a possibly lesser amount of the other derivatives found in humans (Figure 4). Except for Neu5Gc it is unclear whether differences between humans and mouse colonic sialic acids reflects true species diversity or artifacts of the sampling and analytical methods used for detection. If these differences were real, the mouse would be a poor model for investigating sialometabolism in humans.

Relative distribution of sialic acids in the human and murine gut.

Sialic acid	Human[a]	Mouse[c]
Neu5Ac	32–59	68
Neu5Ac8Me	1–4	7
Neu5Ac1, 7L	9–26	0
Neu5, 9Ac2	14–23	3
Neu	0–2	0
Neu5, 7Ac2	0–3	0
Neu5, 8Ac2	0-1	NR[d]
Neu5, 8, 9Ac3	0–2	NR
Neu5, 8Ac29Lt	0–2	NR
Neu5, 7, 9Ac3	0–13	NR
Neu4, 5Ac2	ND[b]	16
Neu5Gc	ND	6

[a]Range of sialic acid abundances in mucins isolated from adult human ileum, cecum, transverse and sigmoid colon, and rectum, shown at right

[b]ND: None detected

[c]Average sialic acid abundances in total adult mouse gut

[d]NR: None reported or analysis may not have been done

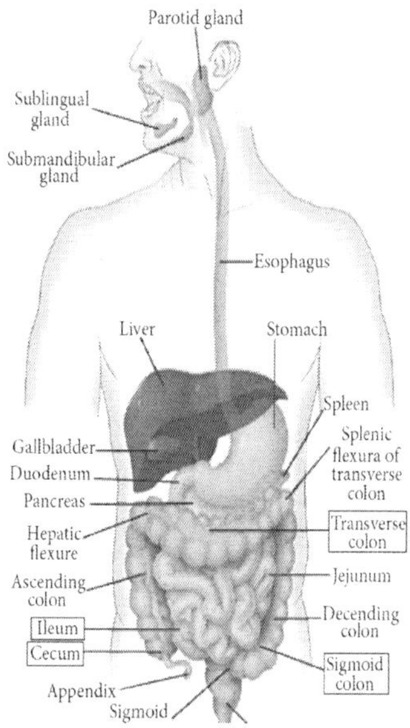

Figure 4. Relative distribution of sialic acids in the human and murine gut. Sialic acid abundances were determined for the human GIT compartments highlighted in rectangles [14]. Mouse values are for the small and large intestine [15].

Most sialic acid is linked to other sugars including other sialic acids and the di-, oligo-, or polysaccharides formed by these carbohydrate linkages are attached to lipids (forming glycolipids) or proteins (forming glycoproteins) comprising the group of molecules called glycoconjugates. Sialic acids are frequently, when present in a sugar chain, the terminal sugar linked to subterminal carbohydrate units through glycoketosidic bonds between the carbon-2 hydroxyl of the terminal sugar and subterminal hydroxyls at various positions depending on the acceptor. In the GIT as well as other mucosal surfaces sialic acids are a major component of mucins bathing the epithelial surfaces and the glycoconjugates comprising the epithelial glycocalyx including the glycolipids and glycoproteins bound to epithelial cell surfaces. The types of glycoconjugates and their interactions in health and disease have been recently reviewed [16–18]. Bound sialic acids by definition are unavailable to bacteria unless first released by sialidases (neuraminidase, E.C.3.2.1.18), which hydrolyze the linkages between terminal sialic acids and subterminal sugars. These hydrolases are produced by the host (endogenously) and by some bacterial species (exogenously). Bacterial sialidases come in a great variety of structures and may be multifunctional [19]. The combined actions of endogenous and exogenous (bacterially derived) sialidases is thought to be necessary for any further microbial utilization of host sialic acids for either synthetic or catabolic purposes [3]. The reader is directed to the original literature and reviews describing the molecular characterization of the bacterial sialidase superfamily [19–24], and a more recent review listing further examples of bacterial sialidases [25] for additional background information.

CANONICAL BACTERIAL PATHWAY FOR CATABOLISM (DISSIMILATION) OF HOST-DERIVED NEU5AC

Once sialic acids are released by hydrolysis they are available like most other sugars free in solution for transport into the cell and catabolic fermentation or oxidation. That the Neu5Ac catabolic pathway went undiscovered or, indeed, not even thought of until 1985 no doubt reflects the phylogenetic rarity of sialic acid and its commercial expense at the time as an available potential nutrient for experimental testing [26–29]. Figure 5 shows the canonical *E. coli* sialocatabolic operon (in color) and upstream repressor gene *nanR*. Induced catabolic genes, their encoded polypeptides, and functions where known are also depicted in the figure. Once transported into the cell by NanT the lyase encoded by *nanA* releases *N*-acetylmannosamine (ManNAc) and pyruvate; the latter enters the oxidative TCA pathway of energy production. NanK phosphorylates the ring form of ManNAc yielding the 6-phosphate derivative ManNAc-6-P. NanE converts ManNAc-6-P to *N*-acetylglucosamine-6-phosphate (GlcNAc-6-P), the inducer of the *nag* operon, with deacetylation and deamination by NagA and NagB, respectively, completing Neu5Ac dissimilation to fructose-6-phosphate. Neu5Ac thus serves as a carbon and nitrogen source, a source of energy, and a source of precursors for cell wall synthesis. With minor variation [30], the canonical biochemical pathway exists in all microorganisms known to catabolize Neu5Ac. Variations in gene organization are widespread in different species, but utilization of Neu5Ac for growth depends on some form of NanATEK and NagAB. The function of YhcH is unknown, but as discussed below it is hypothesized to function in dissimilation of sialic acids other than Neu5Ac, since deleting *yhcH* had no effect on growth of *E. coli* on Neu5Ac [31].

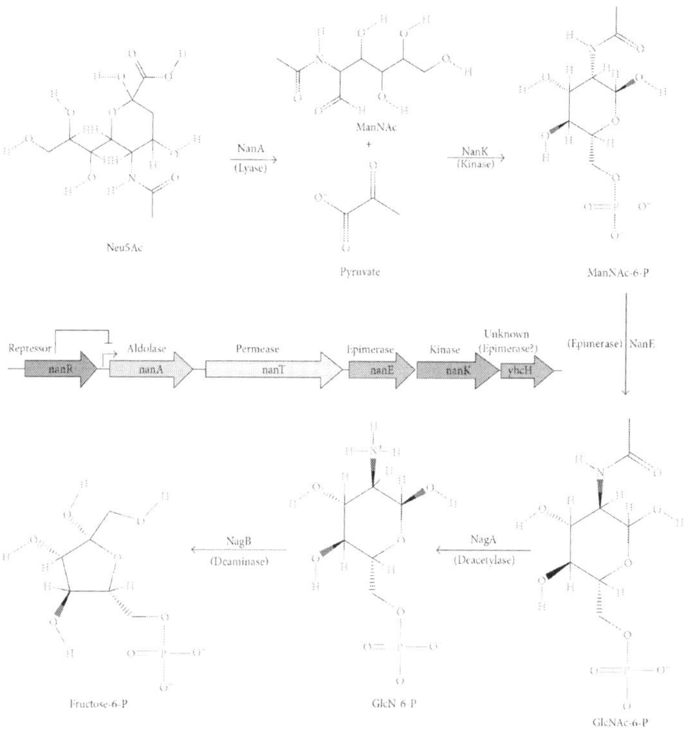

Figure 5. Canonical *N*-acylneuraminate (*nan*) dissimilatory pathway for metabolism of Neu5Ac by *E. coli*. Colored arrows indicate relative transcriptional directions and functions of genes involved in converting Neu5Ac to GlcNAc-6-P after transport of exogenous sialic acid by the permease, NanT (yellow): aldolase or lyase (blue), epimerase (green), kinase (purple), *yhcH* (orange). Expression of the structural genes of this operon are regulated by the repressor, NanR (red) located immediately upstream of the *nanA* start site. Depending on the bacterial species, NagA or NagB may be part of the canonical operon or, as in the case of *E. coli*, located in a separate operon.

THE GOOD, THE BAD, AND THE UGLY-EVIDENCE FOR OR AGAINST THE INVOLVEMENT OF SIALO-METABOLISM IN BACTERIAL COLONIZATION AND PATHOGENESIS

Bacteria such as *E. coli* serotype K1 and some neisserial serotypes synthesize sialic acids de novo and assemble them into cell surface glycolipids [32]. In some cases, the capsular polysaccharides synthesized by these bacteria mimic host molecules and thus fail to elicit a host immunological response [3]. In other sialo-positive serotypes the polysaccharides have no host analogs due to linkage differences and so form the basis of effective vaccines against bacterial meningitis. In all animal models tested loss of capsular sialic acids results in attenuation, indicating the essential role of the capsule in pathogenesis. The ineluctable conclusion is that bacteria presenting sialic acids at their surfaces do so as a mechanism of avoiding host immune surveillance or to otherwise mask the bacterial surface making it less susceptible to host defense mechanisms both innate and acquired. However, antibodies to capsules if present as a result of prior infection, passive transfer, or vaccination protect against disease, at least in the short term.

Another group of bacteria displaying surface sialic acid though lacking the de novo metabolic pathway include *Neisseria gonorrhoeae*, *Haemophilus influenzae* and other *Haemophilus* spp., and *Pasteurella multocida*. These bacteria acquire sialic acids from the host using either surface sialytransferase or hybrid synthetic-catabolic pathways including sialic acid transporters and a truncated synthetic system using only the activating enzyme and a sialyltransferase [33]. Experimental evidence in a natural *P. multocida* host, the cow, and a mouse model of invasive disease indicated that sialic acid transport was essential to pathogenesis [34]. Using substantially the same approaches, sialic acid uptake was independently confirmed to be essential in turkey pathogenesis, another natural host of this facultative pathogen [35]. A deletion of *nanA* did not affect *P. multocida* pathogenicity [34], suggesting that catabolism of host sialic acids was not essential to virulence. Similarly, *nanA* was unessential for colonization of the mouse nasopharynx by *H. influenzae* [36]. The combined results described above unambiguously support an essential function of bacterial sialic acid decoration for evading immune responses but provided little indication that an ability to catabolize these sugars was important to either colonization or disease. However, as discussed below, studies in other bacteria strongly suggest that host sialic acid catabolism has at least a minor role in pathogenesis in different species, and possibly a major role in colonization. None of the bacteria to be discussed below either synthesize sialic acid de novo or use a truncated catabolic-synthetic pathway for surface decoration. Therefore, the sole function of sialocatabolism in these bacteria must be growth at the expense of host mucosal sialic acids.

A variety of studies have suggested sialocatabolism is important to colonization or fitness in animal hosts. However, none of the studies has been independently confirmed, and some of the effects of deleting *nanA* or the sialate uptake system have shown less than dramatic effects on fitness. For example, deleting the *nanA* orthologs in *Vibrio cholerae* suggested a minor (<20-fold) decrease in competitive index when compared to wild type [37]. Single-infection experiments showed no significant difference with wild type, except at one early sampling interval [37]. A greater fitness effect (50–100 times less) was reported for a *nanA* deletion in *Vibrio vulnificus*, and a 500-fold increased LD_{50} after intraperitoneal injection in mice [38]. However, neither of the above studies rule out whether the effects were caused by an inability to metabolize sialic acids or toxicity resulting from intracellular sialic acid accumulation in the *nanA* mutants [26]. Jeong et al. [38] indicated there was no toxicity observed in vitro, but this statement was contradicted experimentally in a later study by some of the same authors [39], making the in vivo results suspect or at least requiring independent confirmation before they can be taken seriously. Furthermore, a *nanA* mutation in an uropathogenic strain of *E. coli* had no fitness defect in mouse bladder or kidneys, consistent with the effect of peptide or amino acid utilization in these extraintestinal sites [40]. However, some of the same authors later showed a 50-fold decreased fitness of an uropathogenic *E. coli nanA* mutant during bacteremia [41]. Again it is unclear whether this effect is due to sialic acid toxicity or lack of its contribution to growth under in vivo conditions. Much more work is needed before any conclusions can be drawn from these studies that do, however, at least point to either a nutritional or detoxifying effect of sialate aldolase (NanA) in bacterial-host interactions.

By constructing a double mutant defective in sialate uptake and aldolase, one can experimentally control for both the nutritional and antitoxicity functions of bacterial sialocatabolism. Using an *E. coli nanAT* double mutant in streptomycin-treated mice the mutant was 500–1000 times less able than wild type to colonize the host, consistent with a previously reported potential role of sialocatabolism in mucin utilization [42, 43]. Interestingly, enteropathogenic *E. coli* O157 did not appear to use sialic acid for colonization, which was one of the sugars used by commensal *E. coli*, suggesting sugars not used by the resident population support colonization of the pathogenic strain. In a recent study *E. coli* passage through the mouse intestine selected for derivatives with increased metabolic efficiency, including genes controlled by the NanR sialorepressor [44]. The problem with these otherwise elaborate studies [41–44] is that the mice are treated with streptomycin to reduce the normal intestinal microbiota to allow a better chance of introduced strains to colonize. In terms of sialic acid utilization, this treatment means that all or most of the free sialic acid for growth must come from endogenous (host) sialidases, and any residual sialidase-positive bacteria remaining after the drug treatment. This conclusion follows from repeated observations that *E. coli* lacks sialidase and must rely on other providers for free sialic acids in vivo. Uncompromised studies are needed before any firm conclusions can be drawn about the role of sialocatabolism in *E. coli* gut colonization.

A seemingly more convincing study suggesting the role of sialocatabolism in *Streptococcus agalactiae* (group B streptococci, GBS) was recently published [45]. GBS are a leading cause of neonatal meningitis in human newborns and a common inhabitant of the vagina mucosal surface. Except for glucose there are few obvious carbohydrates that GBS can use for energy production. Because GBS lacks sialidase, any source of free sialic acid must come from endogenous or other microbial sialidase activities in the vagina. The authors showed that exogenous addition of sialic acid in a mouse model increased wild type growth in the vagina and had, as expected, no effect on a sialate transport-defective mutant [45]. These results add to the overall hypothesis of the current and earlier paper [3] by suggesting that host-derived sialic acids are important to colonization and disease potential.

A study similar to that described above for GBS [45] was previously carried out in *S. pneumoniae* [46]. By contrast to GBS, inactivating the *S. pneumoniae* sialate uptake system had only a 50-fold decreased fitness. However, the dramatic in vivo effects seen with GBS were only observed when exogenous sialic acid was injected into the animal model, which is problematic unless the results are compared to the expected general increase in all coresident species utilizing sialic acids in the nares, lungs, or vagina. In other words, the sialouptake defect in GBS had little or no effect on colonization in any of these sites relative to wild type unless exogenous sialic acid were coadministered, which is the expected result essentially making the mouse an unnecessary "furry test tube." Both streptococcal studies [45, 46] also can be criticized on the basis of genomic comparisons of sialocatabolic loci in *S. pneumoniae*. Figure 6 shows the known or predicted sialocatabolic genes in three sequenced strains: D39, one of the original Avery isolates, ATCC700669, and TIGR4. Despite a few differences in overall gene arrangement the gene duplications or triplication of *nan* orthologs *nanA* (lyase, blue), *nanE* (epimerase, green), *yhcH* (unknown, orange), *yjhC* (unknown, grey), and *nanK* (kinase, purple) point to past recombination events in the streptococcal sialocatabolism regions of these strains (Figure 6). Of note from this analysis is the *nanA* orthologs of strain D39 bear identical point mutations early in the sequence resulting in an inability to catabolize exogenous sialic acids. Despite this defect D39 is as pathogenic for mice as TIGR4 or other wild type streptococcal strains indicating that a natural sialocatabolic-defective mutant might be unaffected for colonization or disease potential. Work is in progress in my laboratory to resolve the contradictory evidence, which includes one other study claiming *S. pneumoniae* D39 uses sialic acid derived from hog gastric mucin for growth [47].

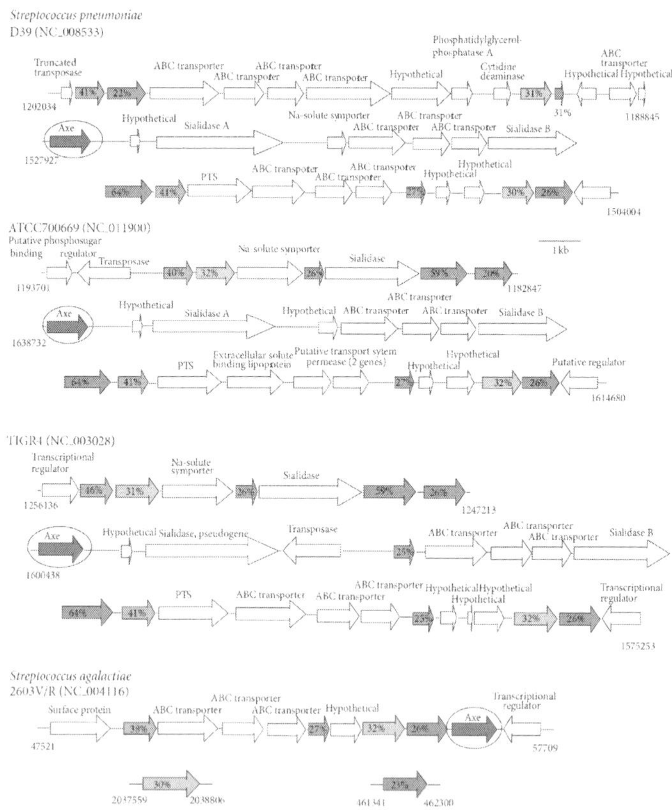

Figure 6. Proposed organization of *nan* gene clusters in selected streptococci. Canonical nan genes in *S. pneumoniae* strains D39, ATCC700669 and TIGR4, and GBS *S. agalactiae* have the same color designations as given in the legend

to Figure 5, with the addition of Axe (magenta), YjhC (grey), and YjhB (gold) based on orthologs of *E. coli* genes described in the text. The known or proposed functions of other genes in the various clusters are listed above the open arrows. The numbers below at left or right of the first or last gene in the cluster gives the beginning and ending nucleotide positions of each gene segment.

As discussed above, *H. influenzae* and *P. multocida* catabolize sialic acids and sialic acid transport is essential for virulence while use of sialic acid as energy source is not [33–36]. These findings were confirmed and extended in vivo for nontypeable *H. influenzae* (NTHi), an important agent of middle ear (otitis media) infections especially in children [48, 49]. Thus, unlike *E. coli* and possibly GBS and *S. pneumoniae*, an ability to catabolize host-derived sialic acids might not necessarily correlate with colonization or pathogenesis. The regulatory mechanism controlling *H. influenzae* sialic acid uptake and catabolism is similar to that described previously for *E. coli* [31, 50–52]. However, the importance ascribed to this regulatory system [50–52] has been recently challenged [49]. Regardless of the discrepancies, another area where host-derived sialic acid may be important to NTHi is biofilm formation under both in vivo and in vitro conditions [53–56]. While these studies support a role for sialic acid in biofilm formation in vitro, the entire concept of NTHi biofilms in the middle ear and by extension the role of host-derived sialic acids in otitis media has been challenged [57]. The discrepancies between groups investigating substantially identical phenomena using similar methodologies warrants caution when extrapolating in vitro results to in vivo conditions. Even in vivo results may be misleading when the relevance of the animal model might be flawed.

Other bacteria where biofilms and sialic acids might be important to infection include *Pseudomonas aeruginosa*, an environmental opportunist, and *S. pneumoniae*, an important cause of ear infections, meningitis, septicemia, and pneumonia in especially young, old or immunocompromised human beings. Both microorganisms express sialidase(s) at their surfaces, although the *P. aeruginosa* enzyme seems to cleave sialic acid-like molecules (pseudaminic acids) found on a variety of bacterial species including *P. aeruginosa* [58], but not animals of the deuterostome lineage [59]. Furthermore, and unlike *S. penumoniae*, *P. aeruginosa* lacks the catabolic genes to transport or metabolize sialic acids. However, in both bacterial species sialidase seems to be required for biofilm formation in vivo [60–62]. Competitive sialidase inhibitors that bind to the respective enzyme's active sites appeared to reduce biofilm formation and in vivo fitness, suggesting that these inhibitors, normally prescribed for viral influenza infections, may be useful clinically for treating pseudomonad and streptococcal infections. Similar to biofilm formation in NTHi, where host-derived sialic acid presumably influences biofilm formation by incorporation into bacterial surface structures, the pseudomonad sialidase might modulate pseudaminic acid levels on bacterial surface structures thereby promoting or inhibiting biofilm formation. The streptococcal situation is much more complicated, not least by conflicting evidence showing an effect of Neu5Ac but not Neu5Gc on biofilm formation when contaminating amounts of Neu5Ac in the Neu5Gc used was probably in excess of the effective Neu5Ac concentration [62]. Furthermore, *S. pneumoniae* expresses up to three sialidases each producing a different hydrolytic product [63]. More work is obviously needed to confirm the potentially exciting findings, especially when competitive sialidase inhibitors might form the basis of a useful therapeutic approach. For example, the major sialidase expressed by all strains of *S. pneumoniae* has been linked to phase-variation during infection and modification of the leukocyte inflammatory response [64–66], supporting the possibility of a general approach aimed at blocking sialidase activity.

MECHANISMS OF BACTERIAL ACQUISITION OR SCAVENGING OF HOST SIALIC ACIDS

As indicated throughout the current paper sialic acids are present in free form at low amounts presumably resulting from the actions of endogenous sialidases. At least four forms of human sialidase have been identified with one located at the plasma membrane [67]. In principle any one of the endogenous sialidases could gain access to mucosal sialoglycoconjugates and release free sialic acid product. In complex microbial communities like those at mucosal surfaces, bacteria express a wide variety of sialidases that can either be excreted, surface-associated, intracellular, or periplasmically located. For example, Mizan et al. [68] showed that *P. multocida* uses its two different surface sialidases to grow on different sialoglycoconjugates by releasing free sialic acid for transport and catabolism by products of the sialocatabolic operon [34]. However, the complexity of sialometabolism at mucosal surfaces is likely to be greater than a simple scavenging model might otherwise indicate.

Consider in addition to simple scavenging of free sialic acids (Figure 7(a)) two models with distinctly different outcomes but both involving sialidases unique to the bacterial species. Figure 7(b) shows an example of a "spitter" in which terminal sialic acid, as part of a glycoconjugate, is cleaved by a periplasmic sialidase but by a strain otherwise lacking all other sialocatabolic functions [69]. The outcome is a sort of acid reflux whereby the released sialic acid enters the extracellular milieu while the subterminal sugars are subsequently hydrolyzed by specific glycosidases, then transported and used for cell growth. Clearly, this is a growth strategy that sacrifices the sialic acid in turn to gain access to subterminal sugars on carbohydrate chains, underscoring the previous conclusion that the diversity of bacterial sialocatabolic pathways evolved in response to the apotheosis if not emergence of sialic acids in the deuterostome lineage [3]. Therefore, the only difference between a spitter and a "swallower" (Figure 3(c)) is that the latter has a sialocatabolic system to exploit the full richness of carbohydrate substrates. By definition, swallowers like *Bacteroides* spp [30, 70] should not be in competition with *E. coli* unless their sialo uptake systems are so inefficient that they allow sialic acid reflux and consequent scavenging by *E. coli* or other sialidase-negative species. Koch [9] explained how *E. coli* carbohydrate transport systems are evolutionarily optimized for scavenging sugars.

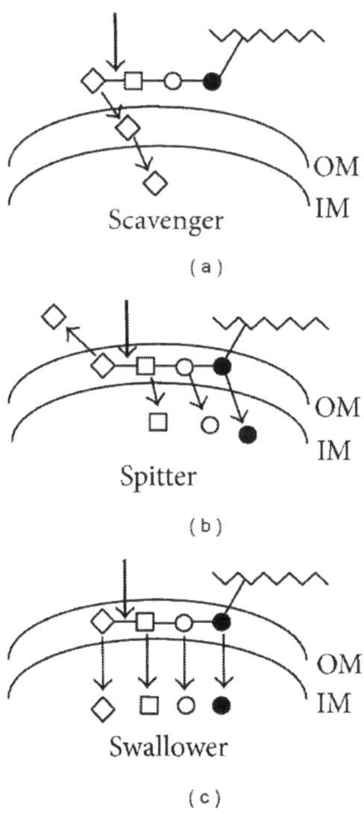

Figure 7. Modes of acquisition of host sialic acids by typical Gram-negative bacterial species. (a) Bacteria with a scavenger mode of sialic acid acquisi-

tion depend on either their own or another source of sialidase (bold arrows) to release free sialic acid (diamond) from carbohydrate chains linked to host substrates (jagged lines). Released sialic acids diffuse into the periplasm between the outer and inner membranes (OM and IM, resp.,) for transport by specific permease(s) into the cytoplasm. (b) The spitter mode of acquisition involves sialidase release but inability to further metabolize sialic acid. These bacteria then sequentially release GlcNAc (open squares), galactose (Gal, open circles), and *N*-acetylgalactosamine (GalNAc, bold circles) from the idealized oligosaccharide for subsequent dissimilatory pathways. Note that the entire oligosaccharide chain may be degraded within the periplasm. (c) The swallower mode is identical to that of the spitter, except that swallowers catabolize the released sialic acid(s). Note that the scavenger and spitter modes are available to Gram-positive bacteria that lack a periplasmic space.

THE SALAD BAR-SUBSTRATE AVAILABILITY AT MUCOSAL SURFACES

That some gut bacteria utilize host glycoconjugates for nutrition is strongly supported by an ingenious study looking at the microbiological consequences of feeding neonatal pigs a normal diet parenterally or intravenously (i.v.) bypassing the GIT. Sixty-two percent of the i.v.-fed ileal microbiota were mucolytic species compared to 33% of the species detected from piglets fed parenterally [71]. This result points to the nutritional foraging by GIT bacteria of carbohydrate substrates abundant at the mucosal surface when other food sources are absent. The conclusion derived from this in vivo study is consistent with the extensive carbohydrate-utilization systems in *Bacteroides thetaiotaomicron*, which is dedicated to foraging mucosal surfaces as the bacterium searches for alternative energy sources [70]. The foraging system requires surface-associated glycosidases, outer membrane oligosaccharide transporter, and periplasmic glycosidases to release monosaccharides, inner membrane transporters, and the intracellular metabolic functions to produce energy from the imported sugars. In other words, *B. thetaiotaomicron* is an example of a spitter or a swallower (Figures 7(b) and 7(c), resp.), depending on its ability to metabolize sialic acid. While the above studies identify the nutritional use of enteral carbohydrates for bacterial nutrition, they contribute directly nothing to understanding the metabolism of specific mucosal sialoglycoconjugates.

Some investigators have demonstrated metabolism of sialylated mucins isolated from various mucosal surfaces. In one study mucins from germ-free rats were incubated with total cecal microbiota from conventionally raised rats. Sialylated mucins were degraded more rapidly than the neutral or sulfated forms suggesting an overall optimized use of sialic acids by intestinal bacteria [72]. Although ocular fluid from many humans is sterile, some studies have shown that other people maintain a commensal bacterial population without incident. These commensals were shown to degrade sialylated ocular mucins indicating the primary carbon and energy sources for these bacteria are carbohydrates found at ocular mucosal sites [73]. Similarly, Burnaugh and colleagues showed that in vitro growth of *S. pneumoniae* on human glycoconjugates relied on the sequential action of several different surface-bound glycosidases, including the major sialidase-A [74]. However, a mutant defective for this sialidase was still able to colonize the mouse lung, suggesting either free sialic acid is not essential to the host-microbe interaction or that other sources of this sugar are to be found in the lung [75]. Alternatively, the contribution of the sialidase to disease might be host species-specific, underscoring the potential pitfall when extrapolating too freely between in vitro and in vivo results.

Probably the best commercially available source of chemically characterized sialomucin for experimental investigation is bovine submaxillary gland mucin (SGM). Sensitive fluorometric methods exist to identify Neu5Ac, all of its O-acetylated derivatives, and Neu5Gc or its derivatives [76]. However, SGM cannot adequately represent the vast variety of carbohydrates detected in human mucins. For example, using electrospray ionization quadrupole time-of-flight mass spectrometry, 46 neutral, and 50 acidic carbohydrate chains were detected from mucin oligosaccharides isolated from the ileum, cecum, transverse colon, sigmoid colon, and rectum [77]. Neutral oligosaccharides do not contain Neu5Ac or sulfate residues while acidic chains included Neu5Ac, sulfate, or both Neu5Ac and sulfate [76]. Anthony Corfield and his colleagues were the first to show that some mucosal bacteria synthesized sialidase, glycosulfatase, and sialate O-acetyl esterase, supporting the idea that acidic sugars are a nutritional source for bacteria residing in the GIT [78]. As these authors noted [78], because sulfated carbohydrates and O-acetylated sialic acids reduce glycosidase activity, bacteria evolved mechanisms to remove the modifications so that the "released" carbohydrates could become more readily available for nutrition. Some in vivo experimental results support this conclusion.

Research with *Capnocytophaga canimorsus*, a member of the *Bacteroidaceae* family, underscores how bacteria feeding off mammalian cell surface glycoconguates gain competitive advantage [69]. However, the authors failed to cite an earlier publication by Michael Malamy and his associates demonstrating essentially the same phenomenon with *Bacteroides fragilis* [79]. These investigators showed that a *B. fragilis* sialidase-negative mutant could not compete against wild type when growing in tissue culture or a rat-pouch model of human abscess. Bacterial growth in both models was equivalent until the time glucose was exhausted, suggesting that the wild type exploited sialoglycoconjugates that were unavailable to the mutant [79]. Both studies [69, 79] focus on the need for increased attention to bacterial sialidase substrate specificities because the variety of sialic acids and their linkages to subterminal sugars is so diverse. For example, using a novel system of chemoselective labeling, Parker et al. [80] showed that the minor *S. pneumoniae* sialidase-C strongly preferred Neu5Ac to Neu5Gc. The paucity of Neu5Gc in humans may in part explain why *S. pneumoniae* is such a successful human pathogen while not generally a problem in other animals. These observations concerning substrate availability further suggest that animal models of human infectious diseases may not accurately report reliable information. The unavoidable conclusion is that testing therapeutics aimed at inhibiting sialometabolism could require human volunteers.

While investigations of sialomucin and other sialoglycoconjugate substrates will continue to expand understanding of bacterial sialometabolism, it seems essential to have a unified theory for at least one bacterium. This theory would include all known and putative sialocatabolic functions thus allowing directed approaches aimed at understanding metabolic pathways while facilitating extrapolation to other sialo-capable bacterial species. *E. coli* remains the best model organism for developing a unified theory of sialometabolism.

IDENTIFICATION OF THE *E. COLI* SIALOREGULON

The *E. coli* sialocatabolic system is regulated by repressor protein, NanR, whose structural gene is located immediately upstream of the *nanATEK-yhcH* operon (Figure 8). NanR binds to a unique operator with three GGTATA repeats separated by two or three nucleotides [31]. The *nan* operon responds to exogenous sialic acid with *nanA* induction up to 1000-fold [26], indicating the important function of the lyase for both nutritional use of sialic acids and detoxification [26, 27]. Except for the unknown function of YhcH, the canonical *nan* operon is dedicated to catabolism of Neu5Ac [26–29, 31]. However, when transcriptome analysis of a *nanR* mutant was compared to wild type, or when wild type bacteria were grown with Neu5Ac or glycerol as sole carbon source [3], two additional NanR coregulated operons were identified by their increased message production representing five additional genes (Figure 8). Both *nanCMS* and *yjhBC* include NanR operators upstream of the putative transcriptional start sites for each operon. The functions of three of the five coregulated genes is known or at least supported by some experimental evidence.

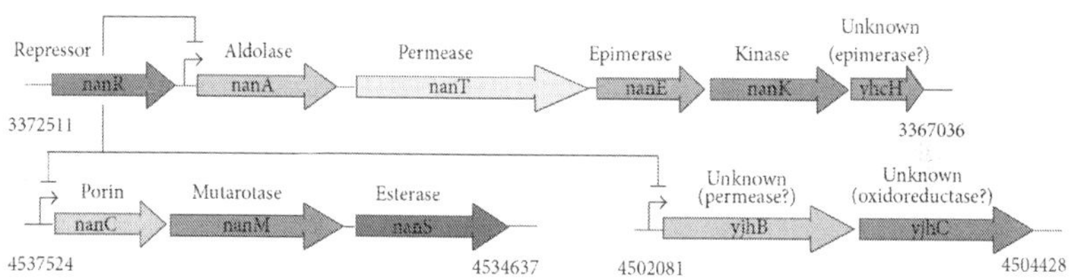

Figure 8. The *E. coli* sialoregulon. In addition to the canonical *nan* operon (with colored arrows having the same designations as given in Figure 6), other genes regulated by NanR include *nanC* (pink), *nanM* (teal), and *nanS* (magenta), which is homologous to the *axe* genes shown for streptococci in Figure 6. Another coregulated operon is composed of a putative permease, YjhB (gold) and oxidoreductase, YjhC (grey).

The *nanCMS* operon is composed of genes encoding an outer membrane porin (*nanC*), sialate mutarotase (*nanM*), and sialate O-acetyl esterase (*nanS*). The porin is not required for growth of *E. coli* on Neu5Ac unless outer membrane porins OmpF and OmpC are absent [81]. The recently solved crystal structure of NanC confirms its similarity to porins with presumed selectivity for acidic oligonucleotides [82]. The NanC 12-stranded β-barrel tertiary structure defines an open pore with average radius of 3.3 Å lined by two strings of basic amino acid residues apposed across the pore. The alignment of basic residues is conserved within a family of diffusion channels that likely facilitates the entry of acidic oligosaccharides [82]. The similarity of NanC to this family of diffusion channels was thought to indicate preferential uptake of sialooligomers [82]. However, there is little indication that such oligomers would exist outside of polysialic acid in the central nervous system [3], nor any known periplasmic or intracellular *E. coli* sialidase that could convert oligomers to free sialic acids. A recent transcriptome analysis of *E. coli* indicated that *nanC* was induced when *E. coli* is growing in biofilms [83]. Given that *nanC* is part of an operon controlled by NanR, it is difficult to see how induction could occur unless the operon was under control of some regulator other than NanR. Interestingly, *nanC* was one of the genes identified by a targeted mutagenesis approach in *Salmonella enterica* serovar Typhimurium strain ATCC14028 as having decreased fitness during competitive mouse infection [84]. The combined results of crystallography, transcriptome, and animal studies strongly suggest that NanC is a porin that is important to host colonization and disease. That it is part of the sialoregulon further suggests it somehow facilitates utilization of host sialoglycoconjugates or at least their released sialic acids.

Like most sugars in solution, the pyranose Neu5Ac ring continuously rotates by opening and closing between the thermodynamically more stable β-anomer with axially directed hydroxyl at the carbon-2 position (Figure 1(a)) and the α-anomeric form (<10% of the total Neu5Ac in an equilibrium solution) with hydroxyl directed equatorially. The mutarotation time to equilibrium starting from a pure solution of the α-anomer is on the order of an hour, such that at equilibrium the mixture contains >90% β-anomer [85]. By contrast to this equilibrium mixture, all Neu5Ac or derivatives attached to glycoconjugates are in α-glycoketosidic linkages [86]. Because mammalian and bacterial sialidases are retaining hydrolases, the α-isomer is exclusively released from substrates after enzyme cleavage. Since spontaneous rotation is slow, and if as seems logical bacterial sialate transporters recognize the thermodynamically predominant sialate in solution, bacterial mutarotase encoded by *nanM* catalyzing the α- to β-isomeric sialate conversion may enhance competitive success at mucosal surfaces. Thus, NanM could increase the scavenging potential for sialates in an animal host where bacteria rely at least in part on sugars released by endogenous or exogenous sialidases for growth. This is an attractive idea with some supporting evidence [87]. Mutarotation from α- to β-Neu5Ac is easy to entertain when the enzyme is located in the periplasm. However, some bacteria have more than one copy of *nanM* suggesting a cytoplasmic location for at least some Neu5Ac mutarotases [87]. A cytoplasmic location for mutarotase is problematic because the lyase encoded by *nanA* requires α-Neu5Ac substrate (Figure 3(a)). It is conceivable that NanA pulls the β-anomeric form, presumably the form transported by NanT, in the direction of the α-isomer. However, at best, there would seem to be competition between cytoplasmic NanM and NanA. Therefore, because Neu5Ac accumulation in the cytoplasm is potentially toxic [26], perhaps NanM functions primarily as a detoxifying enzyme in the event that α-Neu5Ac is the toxic form. In any case, NanM and its predicted orthologs are found in many but by no means all bacterial species with known or predicted canonical Neu5Ac dissimilatory pathways

(Figure 5), suggesting the mutarotase is not an essential component of sialocatabolism. Indeed, an *E. coli nanM* mutant had at most a 20% reduction in growth rate relative to wild type under experimental conditions favoring overabundance of the α-anomeric form [87], as might exist while bacteria scavenge Neu5Ac in their natural hosts (Figure 7(a)). This relatively modest growth defect might be, however, a significant factor helping to explain part of the overall puzzle why or how *E. coli* became the preeminent facultative anaerobe in the GIT. One obvious test would be to construct an *E. coli nanM* mutant and compare its fitness to wild type in an appropriate animal model. Unfortunately, as discussed above, it is not entirely clear what an appropriate model would be unless the phenotypic effect in, say the mouse, were a dramatic one.

The third and last gene of the *nanCMS* operon, *nanS*, was previously thought to be a conditionally essential gene of *E. coli* for growth on glycerol as sole carbon source [88]. However, Steenbergen et al. [89] published evidence that NanS is a sialate O-acetyl esterase, indicating that the glycerol-growth defect previously reported [88] was almost certainly caused by an uncharacterized secondary mutation in the test strain. In other words, growth of newly constructed *nanS* mutants on Neu5,9Ac$_2$ was eliminated while the mutant grew normally with glycerol [89]. Discerning the true function of NanS was made possible by two key observations: a commercially available source of Neu5,9Ac$_2$ and a bioinformatics survey of *nanS* against the microbial genomic database which identified weak similarity to an acetyl xylan esterase (*axe*) [8]. Because esterases frequently share conserved primary structural similarities including active site residues [90], it was logical that NanS might be a sialate O-acetyl esterase because it mapped within a NanR-coregulated operon and was at least partly similar to Axe [88]. Remarkably, when NanS is screened against its close bacterial relatives none has a discernable copy of *nanS* despite the presence of genetic information known to or to potentially encode and regulate the canonical Neu5Ac dissimilatory pathway (Figures 8 and 9). Some of the species shown in Figure 9 that are related to *E. coli* include orthologs of *nanC*, *namM*, or *yjhBC* though none has a copy of *nanS* regardless of whether the database is screened against *nanS* or *axe* other than *Shigella dysenteriae* (see below). By contrast to the absence of nanS or axe orthologs in enteric bacteria closely related to E. coli, potential axe orthologs abound in GIT bacterial species (Tables and 2), suggesting that an ability to metabolize O-acetylated sialic acids is a common phenotype of bacteria living on or at a mucosal surface [89]. One drawback working with commercially available

O-acetylated sialic acids is their relative lack of purity such that preparations of Neu5,9Ac$_2$ or Neu4,5Ac$_2$ contain impurities including Neu5,(7,8) Ac$_2$ contaminants [89]. Clarke et al. [91] recently reported the chemical synthesis of Neu2,5Ac$_2$, Neu4,5Ac$_2$, and Neu5(7,8)Ac$_2$ derivatives in pure form. Relatively straightforward chemical synthetic methods for preparing O-acetylated sialic acids should facilitate future research on these interesting and phylogenetically widespread Neu5Ac derivatives. The identity of NanS as an O-acetyl esterase was recently confirmed by Rangarajan et al. [92], who presented a crystal structure of the NanS homolog from *E. coli* O157:H7. While there is nothing remarkable about the structure partial characterization of the NanS active site residues suggested NanS is the founding member of a subfamily of esterase [92].

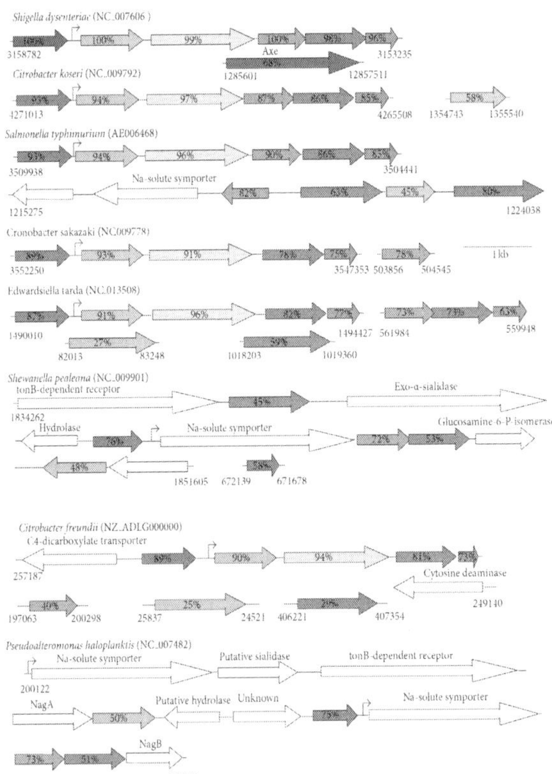

Figure 9. Bacteria with putative *nan* genes coregulated by NanR. Colored and open arrows have their same designations as given in Figures Figures66 and and8.8.

Unlike *E. coli* with its three coregulated *nan* operons all known close relatives containing predicted NanR orthologs include only one or in the case of *P. haloplanktis* two *nan* operators (Figure 9). These observations predict a general lack of coordinated *nan* expression in species related to *E. coli*, and that only *E. coli* is capable of metabolizing O-acetylated sialic acids within this related bacterial group. Evidence that the latter conclusion is true came from an analysis of wild type *S. enterica* var Typhimurium (*S. typhimurium*) grown on Neu5Ac or Neu5,9Ac_2, while *S. typhimurium* wild type grew as expected with Neu5Ac as sole carbon source, a result supported by the inability of a *nanA* mutant to grow under the same condition, the wild type did not grow when the O-acetylated sialic acid was provided as sole carbon source [89]. This last result is consistent with the predicted absence of *nanS* in *S. typhimurium* [89]. The clear implication of these results is that with the possible exception of *S. dysenteriae*, the *nan* regions of species shown in Figure 9 lack the genetic information to encode esterase or the ability to metabolize O-acetylated sialic acids. *S. dysenteriae*, the causative agent of dysentery, includes a gene with an internal domain paralogous to *nanS* and two predicted domains of unknown function at the N- or C-termini flanking the *nanS* paralog (Figure 9). Interestingly, the *nanS* paralog is located in the *S. dysenteriae* prophage that encodes shiga toxin. Indeed, the prophage copy of *nanS* immediately follows in the same transcriptional direction as the two genes encoding subunits of the holotoxin. One possibility for the close association of *nanS* with toxin genes is that the epithelial toxin receptor somehow involves the need to convert O-acetylated sialic acid(s) to Neu5Ac. Other prophage copies of *nanS* exist in E. coli O157 strains and other serotypes causing hemorrhagic disease (Table 3).

Table 3. Distribution of *nanS* and *stx* in sequenced *E. coli* strain.

Strain	**Copies of *nanS***	**Copies of *stx***
Enterohemorrhagic *E. coli* (EHEC)		
O23:H11 11368	1 short, 10 long, 1 broken into 2 pieces of 108 aa and 513 aa	1
O26:H11	1 short, 11 long	1
O103:H2 12009	1 partial, 1 short, 1 med., 7 long	2
O104:H4	1 med., 4 long	2
O111:H$^-$ 11128	1 short, 10 long	2
O145:H28	1 long	1

O157:H7 EDL933	1 short, 5 partials, 7 long	2
Enteroaggregative *E. coli* (EAEC)		
55989	3 long	0
Extraintestinal pathogen-ic *E. coli* (ExPEC)		
CFT073 (K2)	2 short	0
NA114	1short, 1 very long	0
O7:K1 CE10	1 short, 2 long	0
UMN026 (K1)	2 short	0
Enteropathogenic *E. coli* (EPEC)		
E110019	1 short, 2 long	0
B171	1 short, 3 long	0
O55:H7 CB9615	1 short, 6 long	0
O127:H6 E2348/69	1 short, 3 long	0
Enterotoxigenic *E. coli* (ETEC)		
H10407	1 short	0
UMNF18	1 short	1
Other *E. coli*		
E22	1 short, 3 long	0
B088	2 short, 1 long	0
83972	2 short, 1 long	0
S88	1 short, 1 long	0
ED1a	2 short, 6 long	0
NA114	1 short, 1 long	0
101-1	1 short	0
536	1 short	0
AA86	1 short	0
ABU83972	1 short	0
APEC01	1 short	0
B185	1 short	0
B354	1 short	0
B7A	1 short	0
Bl21	1 short	0
BW2952	1 short	0
E24377A	1 short	0

F11	1 short	0
FVEC1412	1 short	0
HS	1 short	0
IAI1	1 short	0
IAI39 (O7:K1)	1 short	0
IHE3034	1 short	0
K-12 DH10B	1 short	0
K-12 MC1655	1 short	0
K-12 W3110	1 short	0
KO11FL	1 short	0
LF82	1 short	0
IO83:H1 NRG 857C	1 short	0
REL606	1 short	0
SE15	1 short	0
SMS-3-5	1 short	0
UM146	1 short	0
UT189	1 short	0
ATCC8739	0	0
SE11	0	0
UMNK88	0	0

The prophage carrying shiga-like toxin in most EHEC strains is similar to the *S. dysenteriae* phage as are the encoded toxin (Stx or Stx-like) subunits. As shown in Table 3, some *stx*-positive bacteria are predicted to express a variable number of *nanS* paralogs, where short refers just to the *E. coli* K-12 homolog (Figure 8), long to *nanS* with N- and C-terminal domains, and partial to *nanS* plus one or the other flanking domain. Other strains of pathogenic *E. coli* from EAEC, ExPEC, and EPEC groups lack *stx* but may have multiple copies of *nanS* that are invariably associated with prophage remnants. Remarkably, one EHEC appears to lack even the NanR-regulated copy of *nanS* whereas 24 other sequenced strains, like *E. coli* K-12, lack *stx,* and *nanS* paralogs (Table 3) and 3 strains, ATCC8739, SE11, and UMNK88 lack any versions of *nanS*. This bioinformatics survey beggars many questions warranting future investigation. Do *nanS* paralogs have O-acetyl esterase activity? If so, why are seemingly redundant copies of *nanS* located in prophage or prophage remnants? Is expression of *nanS* essential for dysentery or hemorrhagic diseases; if so, why do some strains lack even the otherwise common *nanS* copy? Indeed, one strain lacks even the canonical NanR-regulated *nanATEK-yhcH* operon

(Table 3). In other words, so many *E. coli* strains have already been or are being sequenced that it is possible to find nearly every conceivable variant of *nan* organization. Does this variation mean that some or all *nan* genes are nonessential to the *E. coli* lifestyle, or more likely that variants might be on their way to extinction or have partially different lifestyle than the majority of *E. coli* strains? What are the functions if any of the N- and C-terminal domains flanking *nanS* paralogs? Why is *nanS* absent in some bacteria with otherwise intact *nan* systems? Finally, is *E. coli nanS* really essential for the evolutionary success of this bacterium as a human and animal commensal, facultative, and sometimes frank pathogen? Determining the answers to some of these questions will surely increase understanding of sialometabolism and have the potential to suggest new ways of manipulating mucosal bacterial physiology in general.

As shown in Tables 1 and 2 and Figure 6, many bacterial GIT-inhabitants with predicted sialocatabolic systems include a copy of *nanS* (*axe*). This finding is consistent with a potentially important role of NanS in supporting the commensal lifestyle involving utilization of host-derived sialic acids other than Neu5Ac. For example, it is unclear why pneumococcal strains have distinct *nan* genetic organizations whereas all strains examined, like GBS, include one copy of *nanA* in their genomes (Figure 6). Unpublished data from the author›s laboratory has shown that the *nanS* homologs in streptococci encode functional Neu5,9Ac$_2$ O-acetyl esterases. The obvious experimental approach to extend these findings is to eliminate streptococcal esterase(s) and determine the effects on host colonization or disease. However, because the role of sialocatabolism in pneumococcal infection is suspect, the best candidate organism for the proposed studies is GBS, which seem to have a clearer dependency on sialocatabolism for colonization than pneumococci [45].

Compared to the at least partially characterized functions of NanATEK and NanCMS, little is known about *yhcH* or *yjhBC* except that these genes are coregulated by NanR in *E. coli* K-12 (Figure 8). Species closely related to *E. coli* have one or in the case of *E. tarda*, two *yhcH* copies, whereas *yjhB* and *yjhC* are infrequently detected. The YhcH ortholog in *H. influenzae* was purified and its crystal structure solved [93]. The resulting conjecture that it might function in catabolism of Neu5Gc was not supported when an *E. coli yhcH* null mutant was shown to grow as well as wild type on Neu5Gc [31]. However, solving the crystal structure of YhcH does support a possible epimerase activity [93]. Despite the absence of positive data, the similarity of YhcH to an epimerase, YjhB primary structure being similar to NanT, and YjhC primary structure suggesting it is a possible oxidoreductase (Figure 8), strongly suggests that like NanS, genes coregulated as part of the sialoregulon function in metabolism of sialates other than Neu5Ac or O-acetylated sialates. Note that *S. pealeana* and *P. haloplanktis* lack *yhcH* and *yjhB* (Figure 9), suggesting that the spectrum of sialates metabolized by these bacteria might be less than for most *E. coli* strains. Were a panel of all likely sialic acids present at mucosal surfaces available, it would be straightforward to determine all those derivatives of Neu5Ac metabolized by *E.coli* but not by *S. pealeana* or *P. haloplanktis*. Indeed, since some *E. coli* lack certain genes of the sialoregulon (Table 3), these strains alone might suffice to determine the functions of *yhcH* and *yjhBC*. Therefore, instead of waiting for chemical methods that would probably be available only to a few laboratories, simply isolating all sialates from selected mucosae and exposing them to *E. coli* and relevant mutants or natural mutant phenocopies could facilitate identification of all currently unknown gene functions, as long as the results are combined with simple chemical detection methods [5, 76].

Bacteria have evolved diverse sialate transport systems including symporters, ABC- and TRAP-transporters [3, 94]. NanT is a proton symporter with 14 instead of the usual 12 membrane spanning domains [3, 94]. By contrast, YjhB though similar to NanT lacks the central hydrophilic domain found in NanT [95]. This domain is thought to be essential for uptake of Neu5Ac, Neu5Gc, and certain other sialates [3, 26, 89]. Therefore, the presumed sialate(s) transported by YjhB should be structurally distinct from more common sialates and might have specificity for less common forms like ADOA or Neu5Ac1,7L (Figures 3(b) and 3(c), resp.). As shown in Figure 4, Neu5Ac1,7L seems to be a relatively common sialate in the large intestine, suggesting it could be a potentially important source of bacterial nutrition. Chemical synthesis of sialyl lactones has been reported [96], suggesting simple experiments to determine its utilization by *E. coli* and possibly identify the function of *yjhB*. ADOA is an oxidized form of Neu5Ac that may serve as an essential hydroxyl free radical scavenger in tissues [97, 98].

S. typhimurium is closely related to *E. coli* but has only one predicted operon regulated by NanR (Figure 9). However, immediately downstream of a duplicated copy of *nanE* (ManNAc-6-P to GlcNAc-6-P epimerase) is a predicted sodium-solute symporter that was shown to complement an *E. coli nanT* mutant for growth on Neu5Ac in trans [99]. This result suggests that *S. typhimurium* spends at least some of its time in an environment with at least physiological levels (*c.* 140-mM) of sodium, concentrations found commonly in all human or other animal hosts. The problem with the complementation study is no evidence was presented showing the sodium-solute sialate symporter (here designated *nanV*) in fact functions as such in *S. typhimurium* [99]. Figure 10 shows the results of an auxanographic analysis of *S. typhimurium* strain 14028 wild type, *nanT*, *nanV* and *nanTV* double mutant growth on Neu5Ac as sole carbon source. Auxanography is a common procedure where bacteria suspended in soft (0.7%) agar are plated on top of 1.5% bottom agar, both lacking at least one essential growth factor [100, 101]. The analysis can be carried out qualitatively by sprinkling about 1 mg of substrate at one point of the plate, or semiquantitatively by applying a precise amount either in a small liquid volume or onto a paper disk [26]. The results as expected show growth of the wild type (WT) on Neu5Ac and none by the *nanT* mutant. However, whereas growth was observed for the WT on a plate where the top agar was supplemented with 100 mM sodium, similar growth was observed for the *nanT* mutant demonstrating the sodium-dependency of another sialate uptake system. The sodium-dependent phenotype of the *nanT* mutant was lost when a *nanTV* double mutant was plated in the presence or absence of sodium (Figure 10). However, some few colonies observed in the double mutant with sodium suggest another sodium-sialate transporter remains to be identified. It will be interesting to test this isogenic mutant series for fitness defects in animal models of salmonellosis. These studies are in progress.

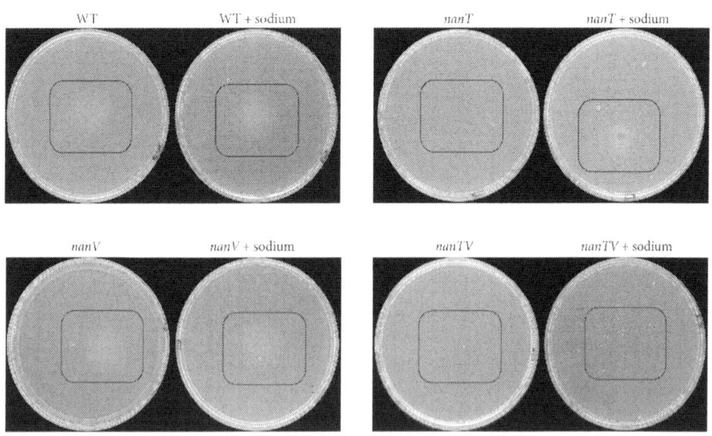

Figure 10. Auxanographic analysis of Neu5Ac utilization by *S. typhimurium nanT* and *nanV* mutants. The indicated strains were grown in minimal medium with glycerol as sole carbon source and plated in top agar with no carbon source and with or without 100 mM sodium chloride. Black rectangles indicate areas

where Neu5Ac was added, with growth shown by the hazy zones or individual colonies.

HOW BACTERIA CATABOLIZE NEU5GC AND METHYLATED SIALIC ACID (NEU5ACME)

It has been known since 1985 that *E. coli* uses Neu5Gc as a sole carbon source [26]. Although as discussed above this sialic acid is not synthesized by humans it is found in most other animal hosts where it could serve an important nutritional source for bacterial colonization. Because Neu5Gc differs from Neu5Ac by a single hydroxyl group at the carbon-5 position of Neu5Ac (Figure 3(a)), there might be an enzyme that first removes the group before or after transport of Neu5Gc by NanT. However, no such enzyme is known to exist in nature indicating that Neu5Gc metabolism probably begins with cleavage by NanA to release pyruvate and ManNGc. This activity of the sialate lyase has been demonstrated biochemically for the mammalian homolog of NanA [102]. Since we know Neu5Gc serves as a sole carbon source for *E. coli* and is cleaved by NanA, how *E. coli* handles the resulting hydroxylated ManNAc derivative, ManNGc, should define the pathway for catabolism of Neu5Gc in *E. coli* and probably all other bacteria that catabolize this substrate. In a preliminary experiment from the author's laboratory the expected accumulation of Neu5Gc by an *E. coli nanA* mutant was confirmed by previously described chemical methods [76], demonstrating that the hydroxyl, as expected, is stable after uptake, that is, there is no dehydroxylase in the cell. Figure 11 shows that once Neu5G is cleaved by NanA, the resulting ManNGc is likely to be phosphorylated by NanK and epimerized by NanE to yield *N*-glycolylglucosamine-6-phosphate (GlcNGc-6-P). If the NagA deacetylase can remove the glycolyl group, yielding GlcN-6-P, then NagB would complete the pathway by converting GlcN-6-P to fructose-6-P just as it does in the canonical Neu5Ac pathway (Figure 5). The remaining glycolic acid would then be a substrate for the *glc* system, which is induced by glycolate, yielding glyoxylate [103]. Glyoxylate can then either be condensed with acetyl coenzyme-A by malate synthase G, or two molecules acted upon by glyoxylate carboligase, which simultaneously decarboxylates the condensation product, tartronic semialdehyde, and reduces it to glycerate that is then phosphorylated to glycerate-3-phosphate. These reactions constitute what is known as the glycerate pathway [103]. The *E. coli* pathway proposed for metabolism of Neu5Gc in Figure 11 is straightforward to verify as it involves readily available methods of bacterial mutagenesis of known target genes. If Neu5Gc is an important carbon source in vivo, a mutant defective in glycolate oxidase might have an interesting phenotype that should be easy to determine. A variant of the pathway proposed in Figure 11 has been speculated upon in mammalian cellular metabolism of Neu5Gc [104].

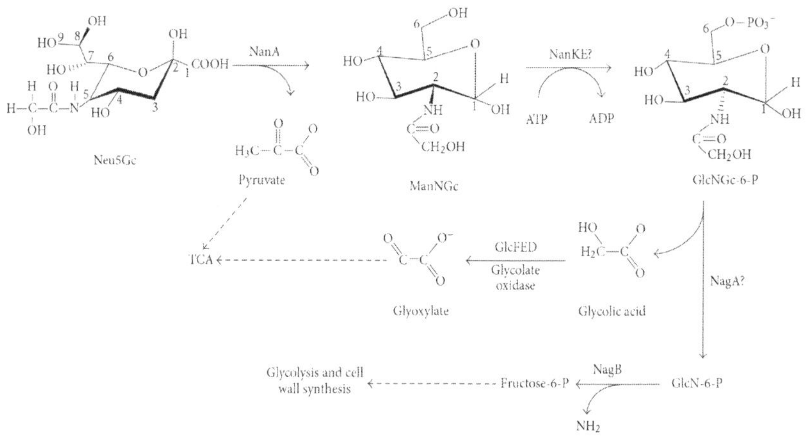

Figure 11. Proposed pathway for utilization of Neu5Gc by *E. coli*. After transport by NanT, Neu5Gc is degraded to pyruvate and *N*-glycolylmannosamine (ManNGc) by NanA. It is proposed that the combined actions of NanK and NanE convert ManNGc to *N*-glycolylglucosamine-6-phosphate (GclNGc-6-P). The rest of the pathway combines the actions of NagAB with the glyoxylate shunt to complete catabolism of Neu5Gc. This pathway implies that the hydroxyl group of Neu5Gc does not impede already known enzymatic activities as described in the text.

Neu5AcMe is sialic acid with a methyl group attached to the carbon-1 carboxylate group (Figure 3(a)). *E. coli* uses this sugar as sole carbon source despite the inability of NanA to cleave methylated sialic acid [26]. On the basis of pervious studies with NanS [89], it is reasonable to conclude that an as yet unidentified methyl esterase(s) exists in the periplasm of *E. coli* to convert the methylated sialate to Neu5Ac. A pattern of metabolism similar to that of *N*-glycolyl or methyl group removal is envisioned for other sialates with lactyl or other simple chemical substitutions found in the GIT (Figure 4). The picture emerging from the admittedly still limited number of studies concerning the sialoregulon is that bacteria and especially *E. coli* have evolved metabolic functions to funnel the diversity of host-derived sialates to Neu5Ac or other readily digestible forms of this sugar.

This dataset suggests a simple model of sialocatabolism for at least some of the sialoregulon parts that should be universally true for all bacteria with homologous sialocatabolic functions.

MODEL OF KNOWN OR PROPOSED SIALOCATABOLIC PATHWAYS IN *E. COLI*

The model is subdivided into five parts specifying the various *E. coli* cellular compartments: extracellular space, outer membrane (OM), periplasm or periplasmic space, inner membrane (IM), and cytoplasm (Figure 12). The various components of the sialocatabolon are then either substrates, porins allowing ingress of substrates to the periplasm where modifiers convert sialate derivatives into forms transportable by NanT or YjhB (permeases) located in the IM, and further conversion cytoplasmically as needed to Neu5Ac, followed by the actions of the canonical metabolic pathway (Figure 5).

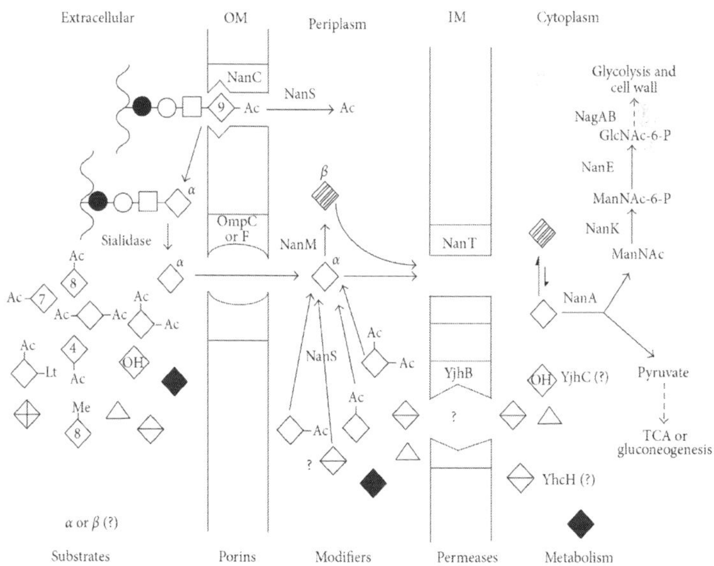

Figure 12. Known or proposed functions of the *E. coli* sialoregulon. Wavy lines indicate mucin peptide backbones with idealized O-linked (serine/threonine) oligosaccharides composed of GalNAc (bold circle); GlcNAc (open squares); Gal (open circles); Neu5Ac (open diamonds, indicating the alpha anomer while hatched diamonds represent the thermodynamically favored beta anomer). Neu5Ac derivatives: acetyl (Ac) groups indicated by their linkages to differ-

ent Neu5Ac carbon position (numbers in diamonds); Neu5,9Ac$_2$, Neu5,7Ac$_2$, Neu5,8Ac$_2$, Neu4,5Ac$_2$, Neu5,7,9Ac$_3$, Neu5,8,9Ac$_3$, and Neu5Gc (diamond with OH) is Neu5Ac with hydroxyl at position 5 (see Figure 1), Lt (lactyl), Me (methy). Bold diamond is neuraminic acid (Neu, which lacks the carbon-5 acetamido group). Triangle represents ADOA while diamond with horizontal line represents Neu5Ac1,7L (see Figures 3(b) and 3(c)). Diamond with cross lines represents Neu5Ac2en (see Figure 3(d)). Other abbreviations are as given in the legend to Figure 7.

Sialates released from sialoglycoconjugates by endogenous or exogenous sialidase(s) are immediately available for passive diffusion into the periplasm through outer membrane pores (porins) OmpC, OmpF, and NanC with molecular weight exclusion sizes of about a disaccharide. Though NanC is unessential for growth on Neu5Ac it might allow certain oligosaccharides with O-acetylated sialic acids to gain access to NanS in the periplasm, which would release the acetyl group(s) thus facilitating conversion in the extracellular milieu to free sialic acids. This idea is easy to test because NanS modifies glycoketosidically linked O-acetylated sialic acids as well as the free sugars. Therefore, a strain with a fully induced sialoregulon when exposed to SGM would allow one to determine if NanC facilitates oligosaccharide deacetylation. Alternatively, NanC might simply facilitate ingress of one or more of the minor sialates not recognized well by the major porins. Once in the periplasm NanM, NanS, and other unidentified modifying enzymes facilitate conversion of sialates to form(s) recognized by NanT and, possibly, YjhB. After transport to the cytoplasm other reactions presumably occur that convert any remaining sialates to Neu5Ac for subsequent cleavage to pyruvate and ManNAc by NanA. The resulting ManNAc, pyruvate, and glycolate in the case of Neu5Gc are then available for final conversion to carbon, nitrogen, and energy sources or cell wall precursors. Sialocatabolism thus is capable of fulfilling all cellular metabolic needs, consistent with its widespread occurrence and diversity in mucosal bacteria.

CONCLUSIONS

Any new or original idea goes through at least three stages: first many say it is not true, then they say it is true but not interesting; finally, it is deemed true and interesting but not new (paraphrased from "anonymous"). It is only when a new scientific field reaches the third ideation that it has any chance of attracting adherents and the subsequent funding necessary for expansion.

The field of microbial sialobiology began in 1985 with discovery of a sialic acid catabolic system in *E. coli* [26, 27]. At that time most microbiologists had no idea what sialic acids were, and when explained to them they said they are "just another sugar" of no particular interest or importance. Those few researchers who knew something about sialic acid structural and phylogenetic diversity either thought the sugars were either too unstable or otherwise inaccessible to have any special relevance to microbial growth and overall bacterial physiology, and so the field has had a long gestation. The current paper focused on advances in microbial sialobiology since the field was last reviewed in 2004 [3]. Major advances since then have been the expanding knowledge of the sialoregulon and tantalizing in vivo experiments supporting minor to definitive roles of sialometabolism in diverse host-microbe interactions. These recent findings are quite separate from the well-known functions of host sialic acids as microbial or toxin adhesins or regulators of innate immunity, knowledge of which has had little success generating practical advances in biomedicine. By contrast, targeting *E. coli* K1, certain neisserial serotypes, *Haemophilus* spp, and *P. multocida* synthetic or hybrid catabolic systems of sialic acid surface decoration are already known to have therapeutic potential [3, 5, 34, 35, 48, 51, 105]. On the basis of these practical advances and the basic theoretical and experimental underpinnings the pace of research in microbial sialobiology is likely to increase. Therefore, the point of the current paper is to increase optimism by presenting a coherent unified theory of bacterial sialometabolism. Certainly some ideas will not withstand detailed scrutiny. Indeed, for reasons described some of the in vivo studies are not even likely to be reproducible. All the likely failures should stimulate rather than impede attempts to sharpen experimental approaches. It is hoped that the basic scientific findings presented in this paper will stimulate the proper scrutiny and help guide the field during its mature stage.

ACKNOWLEDGMENTS

The work from the author's laboratory was supported by Grants from the National Institutes of Health Institute of Allergy and Infectious Disease. The author thanks Eric Sixmister for the helpful discussions. Susan Steenbergen assisted with editing and formatting the paper. Kerry Helms was responsible for the excellent artwork that increased the overall appearance of the paper.

REFERENCES

1. Freter R. Mechanisms of association of bacteria with mucosal surfaces. *Ciba Foundation Symposium*. 1981;80:36–55.
2. Freter R. Control mechanisms of the large-intestinal microflora and its influence on the host. *Acta Gastroenterologica Latinoamericana*. 1989;19(4):197–217.
3. Vimr ER, Kalivoda KA, Deszo EL, Steenbergen SM. Diversity of microbial sialic acid metabolism. *Microbiology and Molecular Biology Reviews*. 2004;68(1):132–153.
4. Li JV, Marchesi JR. Gut microbe-host metabolic interactions in health and disease: exploring host and gut microbiota connections could uncover the mechanisms of various diseases along with targets for drugs with which to treat them. *Microbe*. 2012;7(7):310–318.
5. Vimr ER, Steenbergen SM. Targeting microbial sialic acid metabolism for new drug development. In: Bewley CA, editor. *Protein-Carbohydrate Interactions in Infectious Diseases*. London, UK: RCS; 2006.
6. Drasar BS, Barrow PA. Intestinal microbiology. *Aspects of Microbiology*. 1985;10
7. Lepage P, Leclerc MC, Joossens M, et al. A metagenomic insight into our gut's microbiome. *Gut*, 2013. In press.
8. Glöckner FO, Kube M, Bauer M, et al. Complete genome sequence of the marine planctomycete *Pirellula* sp. strain 1. *Proceedings of the National Academy of Science*. 2003;100:8298–8303.
9. Koch AL. How bacteria face depression, recession, and depression. *Perspectives in Biology and Medicine*. 1976;20(1):44–63.
10. Almagro-Moreno S, Boyd EF. Insights into the evolution of sialic acid catabolism among bacteria. *BMC Evolutionary Biology*. 2009;9(1, article 118)
11. Faillard H. The early history of sialic acids. *Trends in Biochemical Sciences*. 1989;14(6):237–241.
12. Nees S, Schauer R. Induction of neuraminidase from *Clostridium perfringens* and the cooperation of this enzyme with acylneuraminate pyruvate lyase. *Behring Institute Mitteilungen*. 1974;55:68–78.
13. Davies LRL, Pearce OMT, Tessier MB, et al. Metabolism of vertebrate amino sugars with *N*-glycolyl groups: resistance of α2-8-

linked *N*-glycolylneuraminic acid to enzymatic cleavage. *The Journal of Biological Chemistry*. 2012;287(34):28917–28931.

14. Robbe C, Capon C, Maes E, et al. Evidence of regio-specific glycosylation in human intestinal mucins: presence of an acidic gradient along the intestinal tract. *The Journal of Biological Chemistry*. 2003;278(47):46337–46348.
15. Rinninger A, Richet C, Pons A, et al. Localisation and distribution of *O*-acetylated *N*-acetylneuraminic acids, the endogenous substrates of the hemagglutinin-esterases of murine coronaviruses, in mouse tissue. *Glycoconjugate Journal*. 2006;23(1-2):73–84.
16. Moran AP, Gupta A, Joshi L. Sweet-talk: role of host glycosylation in bacterial pathogenesis of the gastrointestinal tract. *Gut*. 2011;60:1412–1425.
17. Johansson MEV, Ambort D, Pelaseyed T, et al. Composition and functional role of the mucus layers in the intestine. *Cellular and Molecular Life Sciences*. 2011;68(22):3635–3641.
18. McGuckin MA, Lindén SK, Sutton P, Florin TH. Mucin dynamics and enteric pathogens. *Nature Reviews Microbiology*. 2011;9(4):265–278.
19. Vimr ER. Microbial sialidases: does bigger always mean better? *Trends in Microbiology*. 1994;2(8):271–277.
20. Vimr ER, Lawrisuk L, Galen J, Kaper JB. Cloning and expression of the *Vibrio cholerae* neuraminidase gene *nanH* in *Escherichia coli* . *Journal of Bacteriology*. 1988;170(4):1495–1504.
21. Hoyer LL, Hamilton AC, Steenbergen SM, Vimr ER. Cloning, sequencing and distribution of the *Salmonella typhimurium* LT2 sialidase gene, *nanH*, provides evidence for interspecies gene transfer. *Molecular Microbiology*. 1992;6(7):873–884.
22. Hoyer LL, Roggentin P, Schauer R, Vimr ER. Purification and properties of cloned *Salmonella typhimurium* LT2 sialidase with virus-typical kinetic preference for sialyl $\alpha 2 \rightarrow 3$ linkages. *Journal of Biochemistry*. 1991;110(3):462–467.
23. Crennell SJ, Garman EF, Laver WG, Vimr ER, Taylor GL. Crystal structure of a bacterial sialidase (from *Salmonella typhimurium* LT2) shows the same fold as an influenza virus neuraminidase. *Proceedings of the National Academy of Sciences of the United States of America*. 1993;90(21):9852–9856.
24. Roggentin P, Schauer R, Hoyer LL, Vimr ER. The sialidase

superfamily and its spread by horizontal gene transfer. *Molecular Microbiology*. 1993;9(5):915–921.

25. Lewis AL, Lewis WG. Host sialoglycans and bacterial sialidases: a mucosal perspective. *Cellular Microbiology*. 2012;14(8):1174–1182.
26. Vimr ER, Troy FA. Identification of an inducible catabolic system for sialic acids (nan) in *Escherichia coli* . *Journal of Bacteriology*. 1985;164(2):845–853.
27. Vimr ER, Troy FA. Regulation of sialic acid metabolism in *Escherichia coli*: role of *N*-acylneuraminate pyruvate-lyase. *Journal of Bacteriology*. 1985;164(2):854–860.
28. Plumbridge J, Vimr E. Convergent pathways for utilization of the amino sugars *N*-acetylglucosamine, *N*-acetylmannosamine, and *N*-acetylneuraminic acid by *Escherichia coli* . *Journal of Bacteriology*. 1999;181(1):47–54.
29. Ringenberg MA, Steenbergen SM, Vimr ER. The first committed step in the biosynthesis of sialic acid by *Escherichia coliK1* does not involve a phosphorylated *N*-acetylmannosamine intermediate. *Molecular Microbiology*. 2003;50(3):961–975.
30. Brigham C, Caughlan R, Gallegos R, Dallas MB, Godoy VG, Malamy MH. Sialic acid (N-acetyl neuraminic acid) utilization by *Bacteroides fragilis* requires a novel *N*-acetyl mannosamine epimerase. *Journal of Bacteriology*. 2009;191(11):3629–3638.
31. Kalivoda KA, Steenbergen SM, Vimr ER, Plumbridge J. Regulation of sialic acid catabolism by the DNA binding protein NanR in *Escherichia coli* . *Journal of Bacteriology*. 2003;185(16):4806–4815.
32. Vimr E, Steenbergen S, Cieslewicz M. Biosynthesis of the polysialic acid capsule in *Escherichia coli* K1. *Journal of Industrial Microbiology*. 1995;15(4):352–360.
33. Vimr E, Lichtensteiger C. To sialylate, or not to sialylate: that is the question. *Trends in Microbiology*. 2002;10(6):254–257.
34. Steenbergen SM, Lichtensteiger CA, Caughlan R, Garfinkle J, Fuller TE, Vimr ER. Sialic acid metabolism and systemic pasteurellosis. *Infection and Immunity*. 2005;73(3):1284–1294.
35. Tatum FM, Tabatabai LB, Briggs RE. Sialic acid uptake is necessary for virulence of *Pasteurella multocida* in turkeys. *Microbial Pathogenesis*. 2009;46(6):337–344.
36. Vimr E, Lichtensteiger C, Steenbergen S. Sialic acid metabolism's

dual function in *Haemophilus influenzae* . *Molecular Microbiology*. 2000;36(5):1113–1123.

37. Almagro-Moreno S, Boyd EF. Sialic acid catabolism confers a competitive advantage to pathogenic *Vibrio cholerae* in the mouse intestine. *Infection and Immunity*. 2009;77(9):3807–3816.
38. Jeong HG, Oh MH, Kim BS, Lee MY, Han HJ, Choi SH. The capability of catabolic utilization of *N*-acetylneuraminic acid, a sialic acid, is essential for *Vibrio vulnificus* pathogenesis. *Infection and Immunity*. 2009;77(8):3209–3217.
39. Kim BS, Hwang J, Kim MH, Choi SH. Cooperative regulation of the *Vibrio vulnificus nan* gene cluster by NanR protein, cAMP receptor protein, and *N*-acetylmannosamine 6-phosphate. *The Journal of Biological Chemistry*. 2011;286(47):40889–40899.
40. Alteri CJ, Smith SN, Mobley HLT. Fitness of *Escherichia coli* during urinary tract infection requires gluconeogenesis and the TCA cycle. *PLoS Pathogens*. 2009;5(5)e1000448
41. Smith SN, Hagan EC, Lane MC, Mobley HLT. Dissemination and systemic colonization of uropathogenic *Escherichia coli* in a murine model of bacteremia. *mBio*. 2010;1(5)e00262-10
42. Chang DE, Smalley DJ, Tucker DL, et al. Carbon nutrition of *Escherichia coli* in the mouse intestine. *Proceedings of the National Academy of Sciences of the United States of America*. 2004;101(19):7427–7432.
43. Fabich AJ, Jones SA, Chowdhury FZ, et al. Comparison of carbon nutrition for pathogenic and commensal *Escherichia coli* strains in the mouse intestine. *Infection and Immunity*. 2008;76(3):1143–1152.
44. Fabich AJ, Leatham MP, Grissom JE, et al. Genotype and phenotypes of an intestine-adapted *Escherichia coli* K-12 mutant selected by animal passage for superior colonization. *Infection and Immunity*. 2011;79(6):2430–2439.
45. Pezzicoli A, Ruggiero P, Amerighi F, Telford JL, Soriani M. Exogenous sialic acid transport contributes to group B *Streptococcus* infection of mucosal surfaces. *Journal of Infectious Diseases*. 2012;206(6):924–931.
46. Marion C, Burnaugh AM, Woodiga SA, King SJ. Sialic acid transport contributes to pneumococcal colonization. *Infection and Immunity*. 2011;79(3):1262–1269.
47. Yesilkaya H, Manco S, Kadioglu A, Terra VS, Andrew PW. The

ability to utilize mucin affects the regulation of virulence gene expression in *Streptococcus pneumoniae* . *FEMS Microbiology Letters*. 2008;278(2):231–235.

48. Bouchet V, Hood DW, Li J, et al. Host-derived sialic acid is incorporated into *Haemophilus influenzae* lipopolysaccharide and is a major virulence factor in experimental otitis media. *Proceedings of the National Academy of Sciences of the United States of America*. 2003;100(15):8898–8903.
49. Jenkins GA, Figueira M, Kumar GA, et al. Sialic acid mediated transcriptional modulation of a highly conserved sialometabolism gene cluster in *Haemophilus influenzae* and its effect on virulence. *BMC Microbiology*. 2010;10, article 48
50. Johnston JW, Zaleski A, Allen S, et al. Regulation of sialic acid transport and catabolism in *Haemophilus influenzae* . *Molecular Microbiology*. 2007;66(1):26–39.
51. Johnston JW, Apicella MA. Sialic acid metabolism and regulation by *Haemophilus influenzae*: potential novel antimicrobial therapies. *Current Infectious Disease Reports*. 2008;10(2):83–84.
52. Johnston JW, Shamsulddin H, Miller AF, Apicella MA. Sialic acid transport and catabolism are cooperatively regulated by SiaR and CRP in nontypeable *Haemophilus influenzae* . *BMC Microbiology*. 2010;10, article 240
53. Greiner LL, Watanabe H, Phillips NJ, et al. Nontypeable *Haemophilus influenzae* strain 2019 produces a biofilm containing *N*-acetylneuraminic acid that may mimic sialylated *O*-linked glycans. *Infection and Immunity*. 2004;72(7):4249–4260.
54. Swords WE, Moore ML, Godzicki L, Bukofzer G, Mitten MJ, VonCannon J. Sialylation of Lipooligosaccharides Promotes Biofilm Formation by Nontypeable *Haemophilus influenzae* . *Infection and Immunity*. 2004;72(1):106–113.
55. Jurcisek J, Greiner L, Watanabe H, Zaleski A, Apicella MA, Bakaletz LO. Role of sialic acid and complex carbohydrate biosynthesis in biofilm formation by nontypeable *Haemophilus influenzae* in the chinchilla middle ear. *Infection and Immunity*. 2005;73(6):3210–3218.
56. Sandal I, Inzana TJ, Molinaro A, et al. Identification, structure, and characterization of an exopolysaccharide produced by *Histophilus somni* during biofilm formation. *BMC Microbiology*. 2011;11, article

186

57. Moxon ER, Sweetman WA, Deadman ME, Ferguson DJP, Hood DW. *Haemophilus influenzae* biofilms: hypothesis or fact? *Trends in Microbiology*. 2008;16(3):95–100.
58. Xu G, Ryan C, Kiefel MJ, Wilson JC, Taylor GL. Structural Studies on the *Pseudomonas aeruginosa* Sialidase-Like Enzyme PA2794 Suggest Substrate and Mechanistic Variations. *Journal of Molecular Biology*. 2009;386(3):828–840.
59. Lewis AL, Desa N, Hansen EE, et al. Innovations in host and microbial sialic acid biosynthesis revealed by phylogenomic prediction of nonulosonic acid structure. *Proceedings of the National Academy of Sciences of the United States of America*. 2009;106(32):13552–13557.
60. Soong G, Muir A, Gomez MI, et al. Bacterial neuraminidase facilitates mucosal infection by participating in biofilm production. *The Journal of Clinical Investigation*. 2006;116(8):2297–2305.
61. Parker D, Soong G, Planet P, Brower J, Ratner AJ, Prince A. The NanA neuraminidase of *Streptococcus pneumoniae* is involved in biofilm formation. *Infection and Immunity*. 2009;77(9):3722–3730.
62. Trappetti C, Kadioglu A, Carter M, et al. Sialic acid: a preventable signal for pneumococcal biofilm formation, colonization, and invasion of the host. *Journal of Infectious Diseases*. 2009;199(10):1497–1505.
63. Xu G, Kiefel MJ, Wilson JC, Andrew PW, Oggioni MR, Taylor GL. Three *Streptococcus pneumoniae* sialidases: three different products. *Journal of the American Chemical Society*. 2011;133(6):1718–1721.
64. King SJ, Hippe KR, Gould JM, et al. Phase variable desialylation of host proteins that bind to *Streptococcus pneumoniaein vivo* and protect the airway. *Molecular Microbiology*. 2004;54(1):159–171.
65. Li-Korotky HS, Lo CY, Banks JM. Interaction of pneumococcal phase variation, host and pressure/gas composition: virulence expression of NanA, HylA, PspA and CbpA in simulated otitis media. *Microbial Pathogenesis*. 2010;49(4):204–210.
66. Chang Y-C, Uchiyama S, Varki A, Nizet V. Leukocyte inflammatory responses provoked by pneumococcal sialidase. *mBio*. 2012;3(1) e00220-11
67. Bigi A, Morosi L, Pozzi C, et al. Human sialidase NEU4 long and short are extrinsic proteins bound to outer mitochondrial membrane and the

endoplasmic reticulum, respectively. *Glycobiology*. 2009;20(2):148–157.

68. Mizan S, Henk A, Stallings A, Maier M, Lee MD. Cloning and characterization of sialidases with 2-6′ and 2-3′ sialyl lactose specificity from *Pasteurella multocida* . *Journal of Bacteriology*. 2000;182(24):6874–6883.
69. Mally M, Shin H, Paroz C, Landmann R, Cornelis GR. *Capnocytophaga canimorsus* a human pathogen feeding at the surface of epithelial cells and phagocytes. *PLoS Pathogens*. 2008;4(9)e1000164
70. Martens EC, Chiang HC, Gordon JI. Mucosal glycan foraging enhances fitness and transmission of a saccharolytic human gut bacterial symbiont. *Cell Host and Microbe*. 2008;4(5):447–457.
71. Deplancke B, Vidal O, Ganessunker D, Donovan SM, Mackie RI, Gaskins HR. Selective growth of mucolytic bacteria including *Clostridium perfringens* in a neonatal piglet model of total parenteral nutrition. *American Journal of Clinical Nutrition*. 2002;76(5):1117–1125.
72. Fontaine N, Meslin JC, Doré J. Selective *in vitro* degradation of the sialylated fraction of germ-free rat mucins by the caecal flora of the rat. *Reproduction Nutrition Development*. 1998;38(3):289–296.
73. Berry M, Harris A, Lumb R, Powell K. Commensal ocular bacteria degrade mucins. *British Journal of Ophthalmology*. 2002;86(12):1412–1416.
74. Burnaugh AM, Frantz LJ, King SJ. Growth of *Streptococcus pneumoniae* on human glycoconjugates is dependent upon the sequential activity of bacterial exoglycosidases. *Journal of Bacteriology*. 2008;190(1):221–230.
75. King SJ, Hippe KR, Weiser JN. Deglycosylation of human glycoconjugates by the sequential activities of exoglycosidases expressed by *Streptococcus pneumoniae* . *Molecular Microbiology*. 2006;59(3):961–974.
76. Vimr ER, Steenbergen SM. Chromatographic analysis of the *Escherichia coli* polysialic acid capsule. In: Anne Delcour H, editor. *Methods in Molecular Biology-the Bacterial Cell Surface, Methods and Protocols*. 2013. pp. 109–120.
77. Robbe C, Capon C, Coddeville B, Michalski JC. Structural diversity and specific distribution of *O*-glycans in normal human mucins along

the intestinal tract. *Biochemical Journal*. 2004;384(2):307–316.

78. Corfield AP, Wagner SA, O'Donnell LJD, Durdey P, Mountford RA, Clamp JR. The roles of enteric bacterial sialidase, sialate *O*-acetyl esterase and glycosulfatase in the degradation of human colonic mucin. *Glycoconjugate Journal*. 1993;10(1):72–81.
79. Godoy VG, Dallas MM, Russo TA, Malamy MH. A role for *Bacteroides fragilis* neuraminidase in bacterial growth in two model systems. *Infection and Immunity*. 1993;61(10):4415–4426.
80. Parker RB, McCombs JE, Kohler JJ. Sialidase specificity determined by chemoselective modification of complex sialylated glycans. *ACS Chemical Biology*. 2012;9:1509–1514.
81. Condemine G, Berrier C, Plumbridge J, Ghazi A. Function and expression of an *N*-acetylneuraminic acid-inducible outer membrane channel in *Escherichia coli* . *Journal of Bacteriology*. 2005;187(6):1959–1965.
82. Wirth C, Condemine G, Boiteux C, Bernèche S, Schirmer T, Peneff CM. NanC crystal structure, a model for outer-membrane channels of the acidic sugar specific KdgM porin family. *Journal of Molecular Biology*. 2009;394(4):718–731.
83. Domka J, Lee J, Bansal T, Wood TK. Temporal gene-expression in *Escherichia coli* K-12 biofilms. *Environmental Microbiology*. 2007;9(2):332–346.
84. Santiviago CA, Reynolds MM, Porwollik S, et al. Analysis of pools of targeted Salmonella deletion mutants identifies novel genes affecting fitness during competitive infection in mice. *PLoS Pathogens*. 2009;5(7) e1000477
85. Moustafa I, Connaris H, Taylor M, et al. Sialic acid recognition by *Vibrio cholerae* neuraminidase. *The Journal of Biological Chemistry*. 2004;279(39):40819–40826.
86. Angata T, Varki A. Chemical diversity in the sialic acids and related α-keto acids: an evolutionary perspective. *Chemical Reviews*. 2002;102(2):439–469.
87. Severi E, Müller A, Potts JR, et al. Sialic acid mutarotation is catalyzed by the *Escherichia coli*β-propeller protein YjhT. *The Journal of Biological Chemistry*. 2008;283(8):4841–4849.
88. Joyce AR, Reed JL, White A, et al. Experimental and computational assessment of conditionally essential genes in *Escherichia coli* . *Journal of Bacteriology*. 2006;188(23):8259–8271.

89. Steenbergen SM, Jirik JL, Vimr ER. YjhS (NanS) is required for *Escherichia coli* to grow on 9-*O*-acetylated *N*-acetylneuraminic acid. *Journal of Bacteriology*. 2009;191(22):7134–7139.
90. Steenbergen SM, Lee YC, Vann WF, Vionnet J, Wright LF, Vimr ER. Separate pathways for *O* acetylation of polymeric and monomeric sialic acids and identification of sialyl *O*-acetyl esterase in *Escherichia coli* K1. *Journal of Bacteriology*. 2006;188(17):6195–6206.
91. Clarke PA, Mistry N, Thomas GH. Synthesis of the complete series of mono acetates of *N*-acetyl-d-neuraminic acid. *Organic and Biomolecular Chemistry*. 2012;10(3):529–535.
92. Rangarajan ES, Ruane KM, Proteau A, et al. Structural and enzymatic characterization of NanS (YjhS), a 9-*O*-acetyl *N*-acetylneuraminic acid esterase from *Escherichia coli* O157:H7. *Protein Science*. 2011;20(7):1208–1219.
93. Teplyakov A, Obmolova G, Toedt J, Galperin MY, Gilliland GL. Crystal structure of the bacterial YhcH protein indicates a role in sialic acid catabolism. *Journal of Bacteriology*. 2005;187(16):5520–5527.
94. Roy S, Douglas CW, Stafford GP. A novel sialic acid utilization and uptake system in the periodontal pathogen *Tannerella forsythia* . *Journal of Bacteriology*. 2010;192:2285–2293.
95. Martinez J, Steenbergen S, Vimr E. Derived structure of the putative sialic acid transporter from *Escherichia coli* predicts a novel sugar permease domain. *Journal of Bacteriology*. 1995;177(20):6005–6010.
96. Allevi P, Rota P, Scaringi R, Colombo R, Anastasia M. Chemoselective synthesis of sialic acid 1,7-lactones. *Journal of Organic Chemistry*. 2010;75(16):5542–5548.
97. Ogasawara Y, Namai T, Yoshino F, Lee MCI, Ishii K. Sialic acid is an essential moiety of mucin as a hydroxyl radical scavenger. *FEBS Letters*. 2007;581(13):2473–2477.
98. Iijima R, Takahashi H, Ikegami S, Yamazaki M. Characterization of the reaction between sialic acid (*N*-acetylneuraminic acid) and hydrogen peroxide. *Biological and Pharmaceutical Bulletin*. 2007;30(3):580–582.
99. Severi E, Hosie AHF, Hawkhead JA, Thomas GH. Characterization of a novel sialic acid transporter of the sodium solute symporter (SSS) family and *in vivo* comparison with known bacterial sialic acid transporters. *FEMS Microbiology Letters*. 2010;304(1):47–54.

100. Vimr ER, Miller CG. Dipeptidyl carboxypeptidase-deficient mutants of *Salmonella typhimurium*. *Journal of Bacteriology*. 1983;153(3):1252–1258.

101. Vimr ER, Green L, Miller CG. Oligopeptidase-deficient mutants of *Salmonella typhimurium*. *Journal of Bacteriology*. 1983;153(3):1259–1265.

102. Bulai T, Bratosin D, Artenie V, Montreuil J. Characterization of a sialate pyruvate-lyase in the cytosol of human erythrocytes. *Biochimie*. 2002;84(7):655–660.

103. Pellicer MT, Badía J, Aguilar J, Baldomà L. *glc* Locus of *Escherichia coli*: characterization of genes encoding the subunits of glycolate oxidase and the *glc* regulator protein. *Journal of Bacteriology*. 1996;178(7):2051–2059.

104. Bergfeld AK, Pearce OMT, Diaz SL, et al. Metabolism of vertebrate amino sugars with *N*-glycolyl groups: incorporation of *N*-glycolylhexosamines into mammalian glycans by feeding *N*- glycolylgalactosamine. *The Journal of Biological Chemistry*. 2012;287(34):28898–28916.

105. Goller CC, Seed PC. High-throughput identification of chemical inhibitors of *E. coli* group 2 capsule biogenesis as anti-virulence agents. *PLoS ONE*. 2010;5(7)e11642

4

Glycosylation-Based Serum Biomarkers for Cancer Diagnostics and Prognostics

Alan Kirwan[1], Marta Utratna[1], Michael E. O'Dwyer[2], Lokesh Joshi[1] and Michelle Kilcoyne[1,3]

[1]Glycoscience Group, National Centre for Biomedical Engineering Science, National University of Ireland Galway, Galway, Ireland
[2]Department of Hematology, National University of Ireland Galway, Galway, Ireland
[3]Carbohydrate Signalling Group, Microbiology, School of Natural Sciences, National University of Ireland Galway, Galway, Ireland

ABSTRACT

Cancer is the second most common cause of death in developed countries with approximately 14 million newly diagnosed individuals and over 6 million cancer-related deaths in 2012. Many cancers are discovered at a more advanced stage but better survival rates are correlated with earlier detection. Current clinically approved cancer biomarkers are most effective when applied to patients with widespread cancer. Single biomarkers with satisfactory sensitivity and specificity have not been identified for the most common cancers and some biomarkers are ineffective for the detection

Citation: Glycosylation-Based Serum Biomarkers for Cancer Diagnostics and Prognostics. Alan Kirwan, Marta Utratna, Michael E. O'Dwyer, Lokesh Joshi, and Michelle Kilcoyne. Hindawi Publishing Corporation BioMed Research International Volume 2015, Article ID 490531, 16 pages http://dx.doi.org/10.1155/2015/490531

of early stage cancers. Thus, novel biomarkers with better diagnostic and prognostic performance are required. Aberrant protein glycosylation is well known hallmark of cancer and represents a promising source of potential biomarkers. Glycoproteins enter circulation from tissues or blood cells through active secretion or leakage and patient serum is an attractive option as a source for biomarkers from a clinical and diagnostic perspective. A plethora of technical approaches have been developed to address the challenges of glycosylation structure detection and determination. This review summarises currently utilised glycoprotein biomarkers and novel glycosylation-based biomarkers from the serum glycoproteome under investigation as cancer diagnostics and for monitoring and prognostics and includes details of recent high throughput and other emerging glycoanalytical techniques.

INTRODUCTION

Cancer is the second most common cause of death in developed countries. According to a survey of worldwide cancer rates, there were approximately 14 million newly diagnosed cases and estimated 6,234,000 cancer-related deaths in 2012 [1]. The most commonly diagnosed and leading causes of cancer-related deaths worldwide are malignancies of the lung, bronchus, and trachea in males and breast cancers in females (Figure 1).

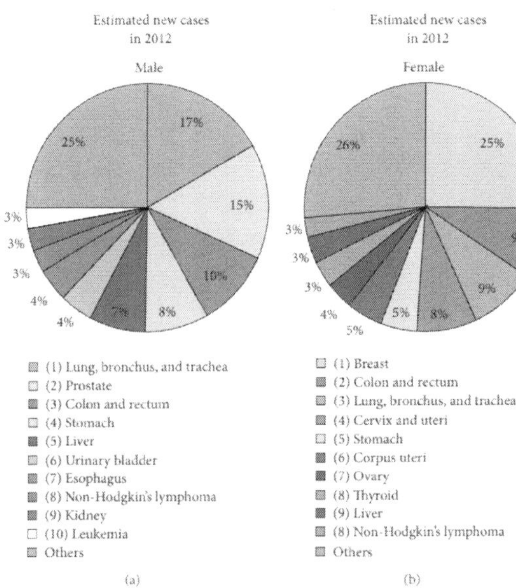

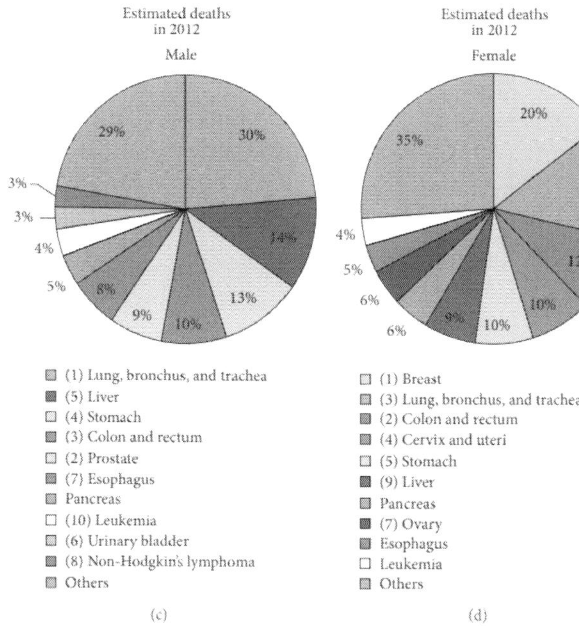

Figure 1. Global cancer statistics. Based on data for 2012 from Torre et al., 2015 [1]. (a) and (b) depict the top 10 most frequently diagnosed types of cancer as a percentage of all detected ones. (c) and (d) represent the top 10 causes of death with each type as a percentage of all cancer-related deaths.

Due to a lack of early symptoms and a hesitation to seek medical investigation, many cancer cases are discovered late, when the disease is at a relatively advanced stage. Survival rate is strongly correlated with the stage at which the disease is diagnosed. The early detection of the disease and the development of minimally invasive screening methods that have wide patient acceptability is the most promising approach for improving the long-term survival of cancer patients.

Recent advances in molecular biology tools and computational methods have enabled the identification of novel cancer biomarkers. Biomarkers are currently used as a complementary strategy to imaging or histopathology techniques and aim to provide minimally invasive and source-effective information which can be prognostic and predictive [2].

The current clinically approved cancer biomarkers have greatest value when applied to patients with widespread cancer. However, despite years of effort and a plethora of publications suggesting novel screening tools, single biomarkers with satisfactory sensitivity (ability to detect individuals with the disease) and specificity (ability to distinguish individuals with the disease from those that are either normal or have some other condition) have not been identified for the most common cancers [3]. This is possibly due to the molecular heterogeneity of tumours from patient to patient and the fact that an individual organ can contain a tumour of several stages in the same tissue [4]. Moreover, the majority of cancer biomarkers are elevated in benign diseases, and some biomarkers are undetectable in early stage cancers. However, in most cases extremely abnormal biomarker concentrations correlate to a poor prognosis and inform clinicians that a more aggressive treatment method is required [3]. Thus, despite their limitations, a variety of biomarkers are routinely used in clinical laboratories (Table 1) [5]. Increasing clinical technical capabilities and better characterization of existing biomarkers might contribute to the introduction of multimarker combinations with better diagnostic, monitoring, and prognostic performance and to the discovery of new candidate biomarkers.

Table 1 . List of FDA-approved cancer biomarkers currently used in clinical practice.

Marker	Full name	Cancer types	Detection type	Clinical applications	Year of FDA approval
AFP	α-Fetoprotein	Liver	Protein concentrations and core fucosylation (for AFP-L3)	Diagnosis, staging, detecting recurrence, and monitoring therapy	1992/2008
PSA, Pro2PSA	Prostate-specific antigen	Prostate	Protein concentrations	Screening, discriminating cancer from benign disease	1986/1994/2012
CA125 (MUC16)	Cancer antigen 125	Ovarian	Protein concentrations	Monitoring therapy, detecting recurrence	1997/2011
HE4 (WFDC2)	Human epididymis protein 4	Ovarian	Protein concentrations	Monitoring therapy, detecting recurrence	2008
OVA1 test (multiple proteins)	-2 Microglobulin + CA 125II (up), apolipoprotein A1 + prealbumin + transferrin (down)	Ovarian	Protein concentrations	Prediction	2009
ROMA test	HE4 + CA125	Ovarian	Protein concentrations	Prediction	2011

CA15-3 (MUC1)	Cancer antigen 15-3	Breast	Sialylated *O*-linked oligosaccharide on MUC1	Monitoring therapy	1997
CA27-29	Cancer antigen 27-29	Breast	MUC1 protein levels	Monitoring therapy	2002
CA19-9	Carbohydrate antigen 19-9 or cancer antigen 19-9	Pancreatic, ovarian	SLe^a on mucin glycoproteins and gangliosides	Monitoring therapy	2002
CEA	Carcinoembryonic antigen	Colon, gastric, pancreatic, lung, and breast	Protein concentrations	Monitoring therapy, detecting recurrence	1985
HER2/neu	Human epidermal growth factor receptor 2	Breast	Protein concentrations	Therapy choice	1998
Tg	Thyroglobulin	Thyroid	Protein concentrations	Monitoring therapy	1997
hCG	Human chorionic gonadotropin	Testicular, ovarian	Protein concentrations	Diagnosis, staging, detecting recurrence, and monitoring therapy	Not approved

Aberrant glycosylation of proteins is a well-known hallmark of cancer and represents a valuable source of information [6, 7]. However, in contrast to proteins and nucleic acids, biosynthesis of oligosaccharides in mammals is not template driven [8]. The structural complexity of carbohydrates underpins their wide range of biological roles and involvement in many cell-cell and cell-matrix interactions related to cancer through modulation of adhesion and cell trafficking [9]. Interestingly, the majority of the human serum proteome is made up of glycoproteins [10]. Proteins enter the circulatory system from tissues or blood cells through active secretion or leakage, including necrotic and apoptotic processes. Thus, carbohydrate structures of great complexity fluctuate in response to multiple stimuli reflecting the physiological and pathological state of the organism. Serum, with its ease of accessibility from the peripheral blood and reduced risk to the patient due to the minimally invasive nature of harvesting, is an attractive option from a clinical and diagnostic perspective [2].

Many technical approaches have been undertaken to describe glycosylation changes associated with cellular conditions and to address the challenges of carbohydrate structure detection and determination [11–16]. In many cases, specific cancer-associated carbohydrate alterations (reviewed elsewhere [6, 17]) can be detected using the separation of oligosaccharides released from glycoproteins by hydrophilic-interaction chromatography (HILIC) high performance liquid chromatography (HPLC), capillary electrophoresis (CE), and mass spectrometry (MS). Monoclonal antibodies and lectins, carbohydrate-binding proteins which are highly specific for various carbohydrate moieties [18], are also commonly employed for the detection of abnormal structures and the proportion of alterations can be quantified [14, 19–21].

The aim of this review is to summarise the current status and the potential for contribution of the serum glycoproteome to cancer diagnostics, monitoring, and prognostics. Glycosylation-based cancer biomarkers which have crossed the boundary from the laboratory into routine clinical use (Table 1) and those which are under development (Table 2), together with the most recent advantages of high throughput (HTP) and other emerging analytical techniques, are described.

Table 2. Clinical trials using blood/plasma or serum carbohydrate analysis to diagnose and monitor cancer. Information on recent clinical trials (https://clinicaltrials.gov/) that involve analysis of glycosylation-based biomarkers in blood components to monitor and diagnose various cancers. *Status of trials was correct at time of submission (April 2015).

Trial title	Description of trial	Statu	Clinicaltrials.gov identifier
Glycoprotein and Glycan in Patients with Stage I, Stage II, and Stage III or Stage IV Cervical Cancer Undergoing Surgery to Remove Pelvic and Abdominal Lymph Nodes	Studying samples of tumor tissue and blood from patients to identify cancer biomarkers; the current primary objectives of this study are to detect the presence of T-synthase or COSMIC. Measuring the level of staining for Tn and STn antigens as well as measuring the differences in expression of 50 different genes on a customized glycogen array and differences in 10 carbohydrate structures using a customized glycan array.	Study is ongoing but not recruiting	NCT00460356

The Association between Alpha 1 Acid Glycoprotein Level and Outcome Metastatic Cancer Treated with Docetaxel	The association between the baseline plasma level of alpha 1 acid glycoprotein and progression-free survival of docetaxel based therapies in patients with metastatic nonsmall cell lung carcinoma, breast cancer, gastric cancer, prostate cancer, and bladder cancer.	Study is not yet open for participant recruitment	NCT00897962
Blood Glycan Biomarkers in Women with Stage IV Breast Cancer	Profiling serum glycan biomarkers in patients with metastatic breast cancer, healthy controls, and patients with noncancer medical illness.	Study is active but no longer recruiting	NCT00897962
Glycan Analysis in Diagnosing Cancer in Women with Ovarian Epithelial Cancer and in Healthy Female Analysis	Comparison of a new assay to the standard CA125 assay.	Study is currently recruiting participants	NCT00628654

CLINICALLY APPROVED BIOMARKERS

α-Fetoprotein (AFP)

The presence of α-fetoprotein (AFP), a glycoprotein of approximately 70 kDa, was initially reported in the serum of the human fetus in 1956 [22]. AFP is mainly produced by the yolk sac and the fetal liver, reaching its maximum concentration of 3–5 × $10^6\mu$g/L at the end of the first trimester [23]. The concentration of AFP in fetal serum decreases to approximately 1–20 × $10^5\mu$g/L at term and rapidly declines after birth to adult reference values (0.5–15 μg/L) reached at 2 years of age. Elevated concentrations of AFP also appear in maternal serum during pregnancy and peak at about weeks 30–32 of gestation (200–300 μg/L). Under certain pathological conditions, the expression of AFP is elevated and high serum concentrations are usually an indication of underlying diseases, including hepatocellular carcinoma (HCC), pancreatic and gastrointestinal carcinomas, germ cell tumours of the testis, and brain tumours [24]. Despite its low specificity for individual cancer types, AFP is the best-studied serological biomarker for HCC which is the most common type of liver cancer. Liver cancer is the fifth most frequent type of cancer diagnosed in males worldwide and the second most common in terms of number of cancer-related deaths in males (Figure 1). In a systematic review which evaluated AFP concentrations at all stages of HCC [25], sensitivities of 41–65% and specificities of 80–94% were reported for a cut-off of 20 ng/mL. AFP concentrations are correlated with increased HCC tumour size but have poor sensitivity at early stages, which is insufficient for early detection of cancer [26]. However, a sensitivity of 66% was reported for early stage HCC using a lower cut-off of 10.9 ng/mL [27].

AFP has a single *N*-linked oligosaccharide with a biantennary complex-type structure which has altered terminal sialylation and core fucosylation during cancer (Figure 2). This fucosylation is detectable by the lectin *Lens culinaris* agglutinin (LCA) and increased fucosylation can be correlated with HCC progression [28]. Due to the limitation of AFP concentration for early detection of HCC, the proportion of the LCA-reactive fraction of AFP (AFP-L3) compared to total AFP has been proposed as an improved biomarker [12, 29]. With a 10% cut-off for AFP-L3/AFP, a specificity of 90% and sensitivity of 60% for this biomarker were achieved for all stages of HCC, for those patients with AFP concentrations exceeding 10 ng/mL, including the early disease stages. The United States (U.S.) Food and Drug Administration (FDA) approved a laboratory test for AFP-L3 in 2006 for determining the risk of developing liver cancer [20]. The development of a highly sensitive assay for AFP-L3 enabled measurement in individuals with AFP concentrations as low as 2 ng/mL and the accuracy of this biomarker is under further investigation [30, 31]. The activity of α-(1,6)-fucosyltransferase was also correlated with HCC progression [32]. The addition of measuring the enzymatic activity of α-fucosidase, which specifically removes the fucose residue from the *N*-linked oligosaccharide of AFP-L3, can further increase the specificity and sensitivity for the early detection of primary HCC [33]. Coupling the use of the AFP cut-off concentration of 20 ng/mL to classify patients as AFP-positive or AFP-negative [21] with the ratio of fucosylated paraoxonase 1 to paraoxonase can be used to distinguish between HCC and liver cirrhosis (LC) with a sensitivity of 90% and a specificity of 75% in AFP-negative patients. These results were confirmed in a small cohort of patients (20 HCC, 20 LC) in which 17 patients were correctly diagnosed with HCC, providing support for the use of multiple biomarkers as a means of diagnosing early stage cancers [21].

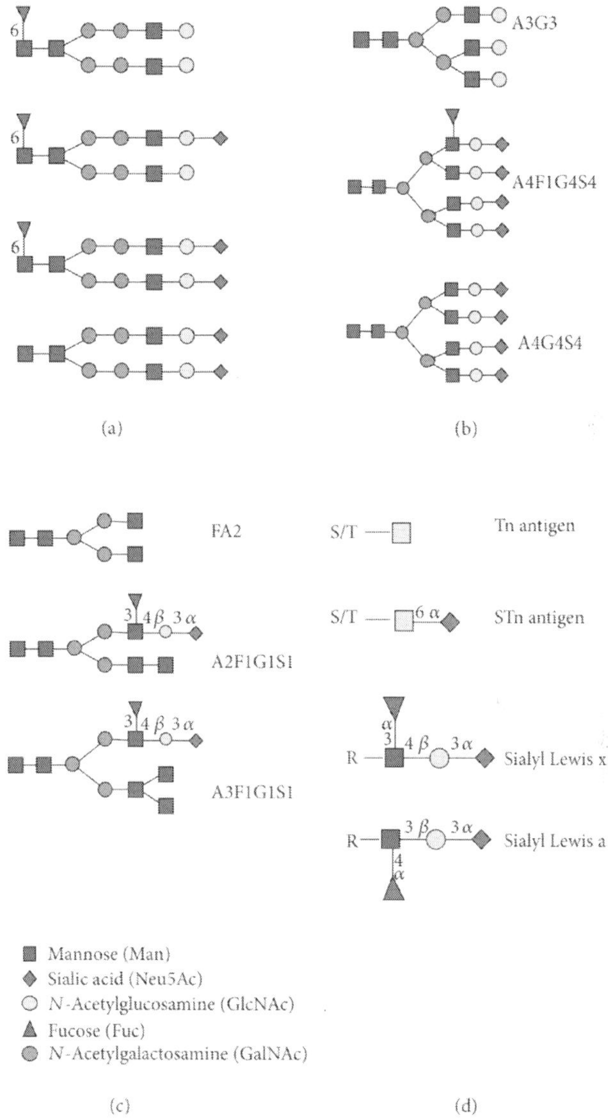

Figure 2. Altered carbohydrate structures expressed in various cancers. (a) N-linked oligosaccharides expressed on AFP in HCC patients, the majority of which have core fucosylation based on Johnson et al. [160]. (b) N-linked oligosaccharide structures which change in abundance as the cancer progresses according to Saldova et al. [41]. (c) N-linked oligosaccharide structures that are upregulated in lymph node metastasis positive breast cancers based on Pierce et al. [159]. (d) Tumour associated carbohydrate structures.

Prostate-Specific Antigen (PSA)

Prostate-specific antigen (PSA), also known as gamma-seminoprotein, kallikrein-3, and KLK3, has been widely used to screen for prostate cancer in men. Prostate cancer is the second most commonly diagnosed cancer and the fifth leading cause of cancer-related deaths in men (Figure 1). PSA is member of the kallikrein family of peptidases and is secreted by the prostate epithelium and periurethral glands. It is a 28.4 kDa glycoprotein with one *N*-linked glycosylation site and is further subcategorized into glycosylated (gp28, gp22, gp18, and gp12) or nonglycosylated (p26-full length nonglycosylated PSA, p20, p16, p10, and p6) peptides [34]. The function of PSA is to liquefy semen in the seminal coagulum to enable sperm to swim in the ejaculate.

Disruption of the prostatic epithelium in inflammation and prostate disorders, including benign prostatic hyperplasia (BPH) and prostate cancer, causes diffusion of PSA into the tissue around the epithelium and leads to elevated concentrations of circulating PSA in these conditions. PSA is present in small quantities in the serum of men with healthy prostates (up to 2.5 ng/mL before their 40 s and around 6.5 ng/mL after 70 years of age) but concentrations above 4 ng/mL are considered indicative of prostate cancer or BPH [35]. PSA as a diagnostic by itself currently has a low specificity and has led to extensive overdiagnosis, overtreatment, and potential harm, especially from unnecessary biopsies. However, serum PSA screening in conjunction with a digital rectal exam (DRE) and Gleason scoring of prostate biopsy samples has been approved by the FDA for the early detection of prostate cancer [36, 37]. Recent approaches for improving the specificity and sensitivity of the serum PSA test include research into the individual molecular forms of PSA (proPSA, benign PSA, and intact PSA), kallikreins other than PSA, calculating the proportion of total PSA complexed with α1-chymotrypsin and α2-macroglobulin (tPSA) compared to free PSA (fPSA), comparing PSA with other markers such as prostate cancer antigen 3 (PCA3) [38] and examining PSA modifications such as glycosylation [39–41].

Several studies have reported altered fucosylation and sialylation in PSA and other proteins isolated from the serum of prostate cancer patients [42, 43]. Serum PSA contains an additional α-(2,3)-linked sialic acid to the terminal galactose residue on N-linked oligosaccharides in prostate cancer when compared to healthy individuals [39, 44] (Figure 2). The binding of prostate cancer-associated PSA to the α-(2,3)-linked sialic acid-recognizing lectin *Maackia amurensis* agglutinin (MAA) was more intense compared to PSA from a healthy individual [13]. Analysis of PSA by 2D electrophoresis identified five PSA glycoforms (F1, F2, F3, F4, and F5) in prostate cancer and BPH sera [40]. The F5 glycoform was nonglycosylated and the F4 glycoform had a lower degree of sialylation compared to the F1–F3 glycoforms. The N-linked oligosaccharides on the most abundant PSA glycoform F3 had a greater proportion of α-(2,3)-linked sialic acid and a decrease in core fucosylation in prostate cancer (Figure 2). The relative percentage of F3 (%F3) compared to all glycoforms (F1–F5) negatively correlated with the stage of prostate cancer while the relative percentages of the F4 (%F4) glycoform, which contained monosialylated N-linked oligosaccharides (Figure 2), were increased in prostate cancer patients [40].

Li et al. [45] showed that fucosylated PSA had better predictive power to differentiate between aggressive and nonaggressive forms of prostate cancer compared to total PSA. Yoneyama et al. [46] used a magnetic microbead-based immunoassay and a free serum PSA glycoform that terminates in α-(2,3)-linked sialic acid to develop a more sensitive diagnostic PSA assay. The novel assay has a sensitivity of 90.2% and a specificity of 64.2% when used on a cohort of patients with (n = 138) and without (n = 176) prostate cancer. This method was more sensitive and accurate than either PSA alone or percentage of fPSA in diagnosing prostate cancer in these patients.

Routine differentiation between prostate cancer and BPH is far from clear-cut and on-going research concentrates on the altered microheterogeneity of each PSA glycoform to distinguish between the two conditions [47, 48].

MUC16 (CA125)

MUC16, initially named cancer antigen 125 (CA125), was first described as a biomarker in a screen of monoclonal antibodies developed against the OVCA433 ovarian cancer cell line [49]. MUC16 is a membrane-spanning mucin and the largest mucin known to date. It has a molecular mass as high as 2×10^6 Da [50, 51]. MUC16 is expressed by the various normal epithelial cells of the human body, including bronchial, endometrial, ovarian, and corneal. MUC16 protects the cells and sheds its extracellular portion into the bloodstream. Soon after its discovery, MUC16 was established as a serum biomarker for diagnosing and monitoring stability or progression in ovarian cancer [52]. However, observations of other conditions including nongynecological cancers and benign conditions such as endometriosis, as well as individuals during menstruation and pregnancy, reported elevated MUC16 serum concentrations [53]. Despite being nonspecific and unreliable for diagnosing early stage ovarian cancer, monitoring serum MUC16 together with ultrasonography is a standard procedure for detection of ovarian malignancies [54].

Several studies have reported attempts to use MUC16 glycoforms to discriminate between endometriosis and ovarian cancer and to evaluate the clinical stage, cytological grade, and histological type of ovarian cancer [55–57]. Varying concentrations of sialyl-Tn antigen (STn, Neu5Ac-α-(2,6)-GalNAc-α-*O*-Ser/Thr) were expressed in MUC16-enriched fractions from the peritoneal fluid of patients with endometriosis and ovarian cancer [55]. A lectin microarray analysis of selected carbohydrate structures, including STn and Tn (GalNAc-α-*O*-Ser/Thr) (Figure 2), on MUC16 and MUC1 (CA15-3) was able to distinguish benign ovarian neoplasms from invasive epithelial ovarian/tubule cancer with a specificity of 61.1% and 90% sensitivity [56]. This HTP method is a promising approach for differential diagnosis and requires further investigation in other cancers.

Human Epididymis Protein 4 (HE4)

Human epididymis protein 4 (HE4, also known as WFDC2) was first identified in differential cDNA screening of human epididymal tissue [58, 59]. HE4 is a small (23–27 kDa), secretory protein with hydrophobic amino acids at the N-terminus consistent with a signal peptide which cleaves to yield a mature secretory polypeptide with a consensus site for N-linked glycosylation at amino acid position 15. HE4 contains two whey acidic protein (WAP) domains characterized by a four-disulfide core arrangement of 50 amino acids, including eight cysteines. Based on gene expression data, the HE4 gene is one of the most frequently upregulated genes in epithelial ovarian carcinomas [60, 61]. HE4 has also been shown to be expressed and secreted as a glycoprotein by ovarian carcinoma cells [62]. Moreover, HE4 expression is lower than MUC16 in benign gynecological conditions and low-malignant potential tumours and HE4 is found in a fraction of endometrial and ovarian cancers which are deficient for MUC16 expression.

In June 2008, the HE4 enzyme immunoassay (EIA) test kit (Fujirebio Diagnostics, Sweden) and, in March 2010, the ARCHITECT HE4 automated version (Abbott Diagnostics, UK) were approved by the FDA as substantially equivalent to a MUC16 assay for ovarian cancer. The HE4 EIA is a solid-phase, noncompetitive immunoassay based on the direct sandwich technique, which measures concentrations between 15 pM and 900 pM [63]. The most recent review of the performance of the HE4 and MUC16 in multiple studies concluded that HE4 exhibits a significantly higher specificity than MUC16 (93% *versus* 78%, resp.) and outperforms MUC16 in identifying patients with early stage ovarian cancer [64]. In September 2011, the FDA approved the combination of the HE4 test with the MUC16 test in the Risk of Ovarian Malignancy Algorithm (ROMA) test, to determine the likelihood of finding malignancy at surgery in premenopausal or postmenopausal women presenting with an ovarian adnexal mass [65, 66]. A study involving 349 female patients with pelvic masses and with different menopausal status confirmed that the ROMA test outperforms the individual biomarkers in their ability to detect both early and late stage ovarian cancers, and this reached statistical significance in postmenopausal women [67].

Despite the fact that HE4 was shown to be glycosylated [62, 68], there has only been limited studies addressing the role of glycosylation for HE4 function [69] and none on the diagnostic or prognostic capability of different glycoforms. Further studies of HE4 glycoforms may lead to insights in to the occurrence, development, or migration of cancerous cells and facilitate early diagnosis or improve the therapeutic options in ovarian cancer.

MUC1 (CA15-3/CA27.29)

MUC1, also known as cancer antigen 15-3 (CA15-3), MAM6, milk mucin antigen, and CA27.29 [70], is a transmembrane mucin expressed by most glandular epithelial cells as a high molecular mass glycoprotein which is heavily substituted with *O*-linked oligosaccharides. It was first identified in human milk, where it is shed from lactating mammary epithelial cells which surround the fat globules [71, 72]. MUC1 was identified on the surface of many types of cancer cells, for example, breast and ovarian, lung, pancreatic, and prostate cancers [70]. It is shed into the blood stream where it can be found in the serum of cancer patients in considerable amounts by certain therapeutic antibodies. To date, multiple monoclonal antibodies, recognizing different portions of the molecule, have been developed against the mucinous antigens of MUC1 [73–75]. Thus, in many publications, the terms CA15-3- and CA27.29-targeting epitopes of MUC1 protein are used interchangeably. Despite the lack of specificity, MUC1-directed assays in combination with other serum biomarkers are routinely used in the complex diagnosis of breast cancer [76]. The anti-CA27.29 monoclonal antibody developed against one of the MUC-1-associated epitopes binds to an eight-amino-acid sequence that partially overlaps the antigen binding site for the

DF3 antibody [77]. Thus, it provides comparable results to the first results reported in MUC1 tests assessed by anti-CA15-3 radioimmunoassay [74, 77]. However, serum monitoring with the CA27.29 antibody cannot distinguish stage I from stage II patients and CA27.29 monitoring is primarily used in metastatic breast cancer to detect treatment failure in the absence of readily measurable disease [74].

The altered glycosylation of serum MUC1 in breast cancer is another possibility for the early diagnosis of breast cancer. MUC1 in breast malignancies is more heavily glycosylated in comparison to MUC1 from a healthy tissue and MUC1 peptide fragments bearing aberrant *O*-linked oligosaccharides are secreted from epithelial cell surfaces to serum [78]. The *O*-linked oligosaccharides on the MUC1 shed in to serum of an advanced breast cancer (ABC) patients were analysed by HPLC [79]. Mucin type core 1 *O*-linked oligosaccharide structures dominated (83%) over core 2 structures (17%) and the majority of structures had high levels of sialylation. Additionally, truncated structures of MUC1 are observed on tumor cells with short, often prematurely sialylated side chains of oligosaccharides, including the Thomsen-Friedenreich antigen (T antigen), its precursor (Tn antigen), and their respective sialylated derivatives STn and α-(2,6)-sialylated T antigen (Figure 2). Because MUC1 antigen is abundantly expressed and aberrantly glycosylated in carcinomas, the tumor associated glycopeptides and epitopes which are masked in normal cells are considered an attractive target in cancer immunotherapy and immunodiagnostics and have been the subject of intensive research efforts [80–82].

Human Epidermal Growth Factor Receptor 2 (HER2)

The human epidermal growth factor receptor 2 (HER2) is encoded by the ERBB2 gene and is also known as cluster of differentiation 340 (CD340) or protooncogene Neu. It is a 185 kDa glycoprotein consisting of three domains; a 105 kDa extracellular domain (ECD), a transmembrane lipophilic segment, and an intracellular domain with tyrosine kinase activity. The ECD portion can be released by cleavage from the HER2 receptor and shed into serum [83]. Overexpression of HER2 is observed in 20–30% of breast cancers, resulting in an aggressive tumour phenotype, reduced survival, and possible treatment eligibility with the monoclonal antibody trastuzumab or other therapies targeted against the HER2 receptor protein [84, 85]. The prognostic value of HER2 ECD combined with MUC1 in early breast cancer was shown to be valuable in identifying high-risk breast cancer patients. These two independent indicators of a worse disease-free survival are used to identify patients in need of more aggressive therapies and intensified surveillance [86]. HER2 ECD has also been shown to be a potential diagnostic and prognostic biomarker in HER2-positive gastric cancer. Not only was there a direct correlation between serum and tumour HER2 concentration, there was also a correlation between serum HER2 concentration and patient responses to chemotherapy [87].

Although the ECD of HER2 contains several potential N-linked glycosylation sites, studies of its glycoforms have been limited to an examination of how HER2 glycosylation affected the specificities of a panel of anti-HER2 antibodies [88]. HER2 oligosaccharide structures have not yet been elucidated.

Carcinoembryonic Antigen (CEA)

Carcinoembryonic antigen (CEA) has molecular weight of approximately 180 kDa, belongs to the immunoglobulin superfamily, and is a glycosylphosphatidylinositol-anchored cell surface glycoprotein. CEA is normally produced by mucosal cells in gastrointestinal tissue during fetal development and the expression decreases before birth, with the highest concentrations in the second trimester 80–100 ng/mL in amniotic fluid at week 19, reducing to 50 ng/mL at full term [89]. CEA is not elevated in maternal serum during pregnancy since it does not cross the placenta and is present only at very low concentrations in healthy adult serum of both genders (less than 2.5 ng/mL). However, CEA serum concentrations are elevated for heavy smokers, who express up to 5 ng/mL, and under certain pathological conditions, including colorectal, gastric, pancreatic, nonsmall cell lung, and breast carcinomas [90]. CEA is the primary biomarker used for the staging of colorectal carcinoma and monitoring the recurrence or spread of colon cancer after surgical resection, as rising concentrations of CEA precede other clinical indicators by several months [91, 92].

N-linked oligosaccharides account for more than 50% of the molecular mass of CEA and it is hypothesized that the reduction in mass of human colonic CEA to 170 kDa is as a result of alterations in glycosylation [93, 94]. CEA expressed by CD44-double knockdown LS174T colon carcinoma cells is more densely substituted with sialylated and fucosylated epitopes than CEA on wild-type LS174T cells [95]. The avidity of the altered glycoforms of CEA for selectins was increased when compared to glycoforms from the wild-type cells, which may contribute to metastatic dissemination [95]. However, further studies are required for CEA glycosylation and the role of this glycosylation and to investigate whether these potentially altered glycoforms can enhance the diagnostic ability of CEA.

Carbohydrate Antigen (CA19-9)

Carbohydrate antigen 19-9 (CA19-9), or cancer antigen 19-9, is the sialyl Le^a (s Le^a) blood group structure which is recognised by the antibody N-19-9 [96] (Figure 2). CA19-9 is used primarily in combination with other biomarkers (e.g., CEA) for the monitoring and management of pancreatic cancer [90]. CA19-9 is also currently recognised as one of the most common tumour markers for colorectal, gastric, and hepatocellular cancer [97]. The latter three types of cancer contribute to 28% and 16% of cancer-associated deaths in males and females, respectively (Figure 1). The biggest disadvantage of using the CA19-9 testing is that s Le^a structure is neither exclusively expressed for a specific tumour type nor is it expressed in cancer only, but it is expressed at a lower concentration in tissue and serum of healthy individuals of appropriate blood types. Additionally, patients who are genotypically negative for the Le^a antigen cannot produce CA19-9, even when affected by cancer [98].

Increased expression of CA19-9 is used to indicate the presence of pancreatic cancer before any evidence of disease is obtained with other methods [99] and strictly correlates with the clinical response after pancreatectomy. Thus, it is used for the monitoring of disease recurrence [100]. Similarly, CA19-9 testing in combination with other biomarkers was recommended in multiple studies for estimating the relapse of gastric carcinoma after surgery [101, 102]. Recently, CA19-9 has been used as a prognostic biomarker for HCC and postoperative cholangiocarcinoma patients. In HCC, patients serum concentrations in excess of 100 U/mL independently predicts poorer overall survival while, in cholangiocarcinoma patients, serum CA19-9 concentrations in excess of 150 U/mL were associated with a worse overall survival [97, 103].

Thyroglobulin (Tg)

Thyroglobulin (Tg) is a 660 kDa dimeric glycoprotein with 20 potential N-linked glycosylation sites, of which 16 sites were shown to be glycosylated in the mature protein [104]. Tg is produced by the follicular cells of the thyroid and is used by the thyroid gland as a substrate for the synthesis of thyroxine and triiodothyronine and for the storage of the inactive forms of thyroid hormone and iodine. Serum Tg concentration is a biomarker for monitoring postoperative thyroid cancer recurrence [105, 106]. However, the usefulness of preoperative Tg measurements (partly related to difficulties with antibody interference and nonspecific recognition) remains unclear [107–110]. The glycosylation of Tg is well known and carbohydrate structures correlated with Tg function playing a role in the secretion of Tg, transportation of Tg to cell compartments, iodination, hormone synthesis, and immunoreactivity [104].

Structure elucidation of Tg glycosylation in cancer has not been performed to date but may be useful for thyroid cancer diagnostics. Preliminary studies showed that the interaction of lectin LCA with Tg from thyroid carcinoma was significantly lower than that in normal thyroid tissue and in patients with benign thyroid tumor [111, 112]. The percentage of LCA-reactive Tg could discriminate between benign and malignant lesions [113]. It was also found that the percentage of LCA-reactive Tg was significantly decreased in thyroid carcinoma patients who were positive for lymph node metastasis compared to thyroid carcinoma patients who were negative for lymph node metastasis [112].

POTENTIAL NOVEL BIOMARKERS

The translation of biomarkers from discovery to clinical practice is still ongoing for hundreds of potential biomarkers which have been identified and published. The validation process of a putative biomarker requires time, hundreds of specimens, and large cohorts of patients to be shown reproducibly. Examples of promising biomarkers routinely checked in clinical practice but not approved for specific cancer due to low specificity or sensitivity are described below.

Human Chorionic Gonadotropin (hCG)

Human chorionic gonadotropin (hCG) is a heterodimeric glycoprotein hormone produced by the placenta and comprises an α-subunit and a β-subunit that can vary in glycosylation [114]. The α-subunit structure is common to luteinizing hormone, follicle stimulating hormone, and thyroid stimulating hormone while the β-subunits of the aforementioned hormones display various degrees of homology with each other, conferring the distinct biological activity of each heterodimer. In addition, two variants of hCG, regular and hyperglycosylated, have independent activities. The regular form maintains the arteries and the vascular supply of the placenta during the full course of pregnancy while the hyperglycosylated hCG (hCG with *O*-linked oligosaccharides) is responsible for embryo implantation during pregnancy [115].

The hyperglycosylated form of hCG is also expressed by several tumours, including male germ cell tumours (GCTs) and choriocarcinomas [114], and has been suggested to play a central role in cancer invasion [116]. High concentrations of hCG are usually indicative of adverse prognosis for cancer progression [114, 117]. More complex carbohydrate structures were reported for cancer-related hCG when compared to hCG expressed during pregnancy [118, 119]. However, the relative proportion of hCG isoforms may vary among healthy and diseased states and false positive hCG results are a major problem in the management of gestational trophoblastic disease and cancer [120]. While hCG is well-known indicator of tumours, it has not been approved for this application by the FDA.

α-1-Antitrypsin (A1AT)

A1AT is a 52 kDa serine protease inhibitor with three potential glycosylation sites which is produced mainly by hepatocytes and is upregulated in the serum of lung cancer patients [121–124]. A1AT is present in various different glycoforms which can be used to distinguish between various subtypes of lung cancer and benign pulmonary diseases (BPDs) [124]. The galactosylated A1AT and fucosylated A1AT glycoforms can both distinguish nonsmall cell lung carcinoma (NSCLC) ($n = 23$) from BPD ($n = 25$) with identical degrees of accuracy (AUC = 0.834). Fucosylated A1AT can also efficiently distinguish adenocarcinoma ($n = 28$) from BPD (AUC 0.919). The poly- *N*-acetyllactosamine (polyLacNac) A1AT glycoform can distinguish between small cell lung carcinoma and BPD with a high degree of accuracy (AUC = 0.905) [124]. While the preliminary data is promising, these biomarkers were only examined on 81 patients and need to be investigated in a larger cohort of patients.

Fucosylated Haptoglobin (Fuc-Hpt)

Haptoglobin is a 40 kDa glycoprotein that is produced mainly in the liver and has a low proportion of fucosylation in healthy individuals [125, 126]. Highly fucosylated haptoglobin (Fuc-Hpt) was identified as a potential biomarker in pancreatic cancer upon *Aleuria aurantia* lectin (AAL) blot analysis of the serum of pancreatic cancer patients [127]. Fuc-Hpt has also been shown to be upregulated in the serum of pancreatic cancer patients, with increased branching and fucosylation of the antennae of the N-linked oligosaccharides on the beta chain of Hpt [128]. Fuc-Hpt of pancreatic cancer patients had more intense binding to AAL compared to the healthy controls [125]. The concentrations of Fuc-Hpt in 300 pancreatic cancer patients and 315 healthy volunteers were analysed using lectin-based ELISAs. Fuc-Hpt concentrations were significantly higher in the pancreatic cancer patients ($P < 0.01$) and the ELISA had an AUC of 0.91, a sensitivity of 85.1%, and a specificity of 82.3% [129]. Fuc-Hpt was also elevated in certain colorectal cancer patients, in relation to the proximity of the tumour to the liver and distance metastasis. When Fuc-Hpt was combined with CEA, it was shown that it had the potential to be a novel prognostic marker in colorectal cancer [125, 130]. Fuc-Hpt could also be a potential prognostic biomarker in prostate cancer, as it significantly correlated with Gleason scores and biochemical recurrence after radical prostatectomy. PSA also correlated with overall and progression-free survival and the clinical stage of prostate cancer [131].

YKL-40

YKL-40, also known as chitinase-3-like 1 (CHI3L1) or human cartilage glycoprotein-39, is a 40 kDa secreted glycoprotein with two potential N-linked glycosylation sites which has been proposed as a biomarker in a variety of cancers but has not received FDA approval [132]. High serum concentrations of YKL-40 have previously been associated with high risk disease and increased bone destruction [133, 134]. YKL-40 was investigated as a prognostic marker in multiple myeloma (MM) [135]. A study carried out in 230 MM patients showed that age-corrected serum YKL-40 concentration is an independent prognostic biomarker in MM and indicates a quicker progression to the first skeletal related complications (e.g., bone lesions) [135]. The data shown is promising but a larger multicentre clinical trial is required before YKL-40 can be accepted as a prognostic marker in MM.

CARBOHYDRATES AS POTENTIAL SERUM BIOMARKERS

Advances in HTP glycoanalytical methods have led to investigation of the carbohydrate structures present on glycoproteins in the serum of cancer patients and healthy controls. Many research groups have evaluated whether the variation in structure and/or abundance of these carbohydrates can distinguish between cancer patients and healthy controls [136–139]. This section presents the most recent publications (from 2010 to the present) on alterations in carbohydrate structures on serum glycoproteins of cancer patients and their potential utility as clinical biomarkers.

Ovarian Cancer

The biomarker currently used to diagnose ovarian cancer, MUC16, can only detect late stage ovarian cancer and cannot distinguish between ovarian cancers and benign ovarian diseases (BOD). Enzyme-released N-linked oligosaccharides from the serum of patients with ovarian cancer and BOD were analyzed by MS [138, 140]. MS analysis revealed a panel of N-linked oligosaccharides which could accurately distinguish between ovarian cancer and BOD with greater sensitivity (81–84%) and specificity (83%) than MUC16 (sensitivity = 78%) when tested on a small cohort of patients (37 ovarian patients and 23 healthy controls [140] and 20 ovarian cancer patients, 20 BOD patients, and 33 healthy controls [138]). The use of carbohydrates as improved biomarkers for diagnosing ovarian cancer compared to MUC16 is currently being investigated in a clinical trial (NCT00628654, Table 2).

Increased sialylation is a common glycosylation alteration in various cancer types and sialylation has been investigated as a possible cancer biomarker [141–143]. Measuring the alteration in the serum concentrations of both sialic acid and hydroxyproline distinguishes between ovarian cancer and healthy controls [144]. This assay outperformed the MUC16 and HE4 assays in the diagnosis of ovarian cancer [44]. However, this assay was not significantly better than the ROMA test, which currently remains the best method for diagnosing and monitoring ovarian cancer [66].

Gastric Cancer

Gastric cancer is the second most common cause of cancer-related death (Table 1). Studies have shown that the infections with Helicobacter pylori which cause gastritis can progress to gastric adenocarcinoma [145– 148]. H. pylori infection is associated with a sixfold increased risk of gastric cancer [149]. H. pylori infection also causes peptic ulcer disease but, unlike gastritis, it is inversely correlated to gastric cancer. However, there are currently no methods for the early stage detection of gastric cancer and most cases present with advanced or metastatic disease [150]. MS was used to investigate whether alterations in the structure or abundance of N-linked oligosaccharides in the serum of gastric cancer patients (n = 36) could distinguish them from patients with gastritis (n = 18) or duodenal ulcers (n =

18) [139]. Gastric cancer patients had altered serum *N*-linked glycosylation when compared to patients with gastritis. Gastric cancer patients showed reductions in high-mannose type *N*-linked oligosaccharides, those with one complex-type antenna and bigalactosylated biantennary structures and increased levels of nongalactosylated biantennary *N*-linked oligosaccharides [139]. Significant differences in *N*-linked oligosaccharides only existed between gastric cancer and gastritis patients. While these results will need to be confirmed in a larger cohort of patients, it does support the use of serum glycosylation as potential diagnostic biomarkers in gastric cancer.

Pancreatic Cancer

Alpha-1-acid glycoprotein (AGP), a 40 kDa acute phase serum glycoprotein with five complextype *N*-linked oligosaccharides attached to the polypeptide backbone, shows variations in abundance and glycosylation in various different cancers [151]. The structures of the *N*- linked oligosaccharides on AGP from patients with pancreatic cancer (n=6) and patients with chronic pancreatitis (n=2) were analysed using LC-MS to investigate their potential as diagnostic biomarkers [152]. There was an increase in fucosylated triantennary trisialylated and fucosylated tetra-antennary trisialylated *N*-linked oligosaccharides in the pancreatic patients when compared to the pancreatitis patients. The increased abundance of these *N*- linked oligosaccharides also differed between the various stages of pancreatic cancer and could be potentially used as prognostic biomarkers [152]. While the sample size used in this study was too small for a statistical analysis to be carried out, a larger cohort of patients can confirm whether these glycosylation alterations can be used as diagnostic and prognostic markers.

Current clinical interest in AGP is related to its abundance in the serum of cancer patients. The serum concentration of AGP affects the pharmacokinetics and dynamics of the chemotherapeutic drug docetaxel and may predict a patient's reaction to the therapy [153]. The effect of AGP on docetaxel therapy is currently being examined in a large scale clinical trial (Table 2).

Another serum glycoprotein that is abnormally glycosylated in pancreatic cancer is ceruloplasmin. Ceruloplasmin is an acute-phase protein that is produced by the liver and secreted into the plasma. Ceruloplasmin has four *N*-linked glycosylation sites with complex type, bi-, tri-, and tetraantennary structures, fucosylated and sialylated, containing the sialyl Lewis x (sLe^x) epitope (Figure 2) [154]. Analysis of the *N*-linked oligosaccharides on ceruloplasmin using MS showed that it had a trend towards higher proportions of sLe^x in pancreatic patients ($n = 20$), when compared to healthy controls ($n = 13$) and patients with chronic pancreatitis ($n = 14$) [155]. A larger sample size is required before it can be confirmed whether the trend of higher sLe^x expression on ceruloplasmin can be used as biomarker for pancreatic cancer diagnosis and progression.

Colon Cancer

The expression of the cancer-related epitopes sLe^x and sLe^a on glycoproteins present in the serum of colon cancer patients was analysed using a novel antibody microarray [156]. A panel of five serum glycoproteins were identified that could distinguish between stage 3 and stage 4 colon cancer patients and healthy controls with an AUC of 90%. Although the glycoproteins were not named, they may represent novel biomarkers that could improve the sensitivity of current tests for colorectal cancer.

Oesophageal Cancer

LC-MS was used to determine the site specific alterations in *N*-linked oligosaccharides in oesophageal cancers [137]. This novel method was applied to serum isolated from patients with oesophageal cancer ($n = 15$) and disease-free controls ($n = 15$). The study also included patients with diseases that can develop into oesophageal cancer, for example, high grade dysplasia ($n = 12$) and Barrett's disease ($n=7$) [137, 157, 158]. Significant alterations in site-specific glycosylation were successfully identified on the serum proteins vitronectin, ceruloplasmin, alpha-2-macroglobulin, and complement factor 1 between oesophageal cancer and control patients. These findings will have to be verified in a larger cohort of patients before any definitive conclusions can be made.

Breast Cancer

Increased sialylation, changes in fucosylation, and higher proportions of sialyl Lewis x were reported in *N*-linked oligosaccharide structures in serum from breast cancer patients [19]. The abundance of sLe^x containing *N*-linked oligosaccharides was investigated in the serum from 52 breast cancer patients and 134 patients with benign breast disease using exoglycosidase digestion and HPLC analysis to determine whether it could be used as a diagnostic/prognostic tool. While there was no significant difference in serum glycosylation between early stage breast cancer and benign breast disease, there were differences in serum glycosylation between breast cancer patients with lymph-node positive and lymph-node negative breast cancer. Patients with lymph-node positive breast cancer showed increased proportions of biantennary (FA2) and terminally sialylated N-linked oligosaccharides (A3F1G1S1 and A2F1G1S1) containing the sLe^x structure in their serum when compared to lymph node-negative patients with early breast cancer [159]. These results need to be confirmed in a larger cohort of patients to verify the prognostic utility of serum glycan analysis in breast cancer.

Prostate Cancer

Analysis of serum glycosylation using HPLC and exoglycosidase digestion showed taht there were differences in fucosylation and sialylation between prostate cancer patients and patients with BPH. Serum form prostate cancer patients had increased core fucosylation, as well as increased expression of α-(2,3)-linked sialic acid when compared to serum BPH patients [41]. These alterations in serum glycosylation could also distinguish between different stages of prostate cancer. Triantennary trigalactosylated (A3G3) and tetra-antennary tetrasialylated N-linked oligosaccharides with outer arm fucose (A4FS4) (Figure 2) were significantly decreased on serum PSA from patients with a Gleason score of 7 (more aggressive cancer and a higher chance of relapse) compared to a Gleason score of 5. In contrast, tetra-antennary tetrasialylated N-linked oligosaccharides (A4S4) (Figure 2) were increased in the serum of PSA patients with a Gleason score of 7. The serum glycome analysis was better than PSA at distinguishing between BPH and prostate cancer and at distinguishing between patients with a Gleason score of 7 and patients with a Gleason score of 5 [41]. While the results of this study are promising, they must be confirmed in a larger cohort of patients.

CONCLUSIONS

Research into cancer-specific alterations in glycosylation of serum glycoproteins has provided a promising source of novel biomarkers. Various groups have reported that altered glycoforms of serum glycoproteins can be used to diagnose and monitor various cancers with greater sensitivity and specificity than the currently used biomarkers [21, 44, 156]. Preliminary data has shown that serum glycome analysis is potentially a very sensitive method of discriminating between cancer and control patients or patients with related benign conditions and can detect cancers at a much earlier stage than the currently used biomarkers [139, 140]. Given the possible diagnostic power of glycoproteins and serum glycome analysis, glycosylation-based biomarkers are currently one of the most promising areas of biomarker discovery.

ACKNOWLEDGMENTS

The authors thank Science Foundation Ireland (SFI) and Enterprise Ireland for a Technology Innovation Development Award (13/TIDA/B2687) (Michelle Kilcoyne and Marta Utratna), SFI for support of the Alimentary Glycoscience Research Cluster (08/SRC/B1393) (Lokesh Joshi), and Cancer Care West (Maura Burke fellowship) and the Hardiman Fellowship, NUI Galway, for funding a PhD scholarship (Alan Kirwan).

REFERENCES

1. L. A. Torre, F. Bray, R. L. Siegel, J. Ferlay, J. Lortet-Tieulent, and A. Jemal, "Global cancer statistics, 2012," *CA: A Cancer Journal for Clinicians*, vol. 65, no. 2, pp. 87–108, 2015.
2. K. Strimbu and J. A. Tavel, "What are biomarkers?" *Current Opinion in HIV and AIDS*, vol. 5, no. 6, pp. 463–466, 2010.
3. J. A. Ludwig and J. N. Weinstein, "Biomarkers in cancer staging, prognosis and treatment selection," *Nature Reviews Cancer*, vol. 5, no. 11, pp. 845–856, 2005.
4. S. M. Hanash, S. J. Pitteri, and V. M. Faca, "Mining the plasma proteome for cancer biomarkers," *Nature*, vol. 452, no. 7187, pp. 571–579, 2008.
5. S. Yotsukura and H. Mamitsuka, "Evaluation of serum-based cancer biomarkers: a brief review from a clinical and computational viewpoint," *Critical Reviews in Oncology/Hematology*, vol. 93, no. 2, pp. 103–115, 2014.
6. M. N. Christiansen, J. Chik, L. Lee, M. Anugraham, J. L. Abrahams, and N. H. Packer, "Cell surface protein glycosylation in cancer," *Proteomics*, vol. 14, no. 4-5, pp. 525–546, 2014.
7. S. R. Stowell, T. Z. Ju, and R. D. Cummings, "Protein glycosylation in cancer," *Annual Review of Pathology: Mechanisms of Disease*, vol. 10, no. 1, pp. 473–510, 2015.
8. R. G. Spiro, "Protein glycosylation: nature, distribution, enzymatic formation, and disease implications of glycopeptide bonds," *Glycobiology*, vol. 12, no. 4, pp. 43R–56R, 2002.
9. S. V. Glavey, D. Huynh, M. R. Reagan et al., "The cancer glycome: carbohydrates as mediators of metastasis," *Blood Reviews*, 2015.
10. N. L. Anderson and N. G. Anderson, "The human plasma proteome: history, character, and diagnostic prospects," *Molecular & Cellular Proteomics*, vol. 1, no. 11, pp. 845–867, 2002.
11. S. Saussez, H. Marchant, N. Nagy et al., "Quantitative glycohistochemistry defines new prognostic markers for cancers of the oral cavity," *Cancer*, vol. 82, no. 2, pp. 252–260, 1998.
12. K. Shiraki, K. Takase, Y. Tameda, M. Hamada, Y. Kosaka, and T. Nakano, "A clinical study of lectin-reactive alpha-fetoprotein as an early indicator of hepatocellular carcinoma in the follow-up of cirrhotic patients," *Hepatology*, vol. 22, no. 3, pp. 802–807, 1995.

13. C. Ohyama, M. Hosono, K. Nitta et al., "Carbohydrate structure and differential binding of prostate specific antigen to *Maackia amurensis* lectin between prostate cancer and benign prostate hypertrophy," *Glycobiology*, vol. 14, no. 8, pp. 671–679, 2004.
14. Y. Mechref, Y. Hu, A. Garcia, and A. Hussein, "Identifying cancer biomarkers by mass spectrometry-based glycomics," *Electrophoresis*, vol. 33, no. 12, pp. 1755–1767, 2012.
15. S. A. Svarovsky and L. Joshi, "Cancer glycan biomarkers and their detection—past, present and future," *Analytical Methods*, vol. 6, no. 12, pp. 3918–3936, 2014.
16. S. V. Glavey, S. Manier, A. Natoni et al., "The sialyltransferase ST3GAL6 influences homing and survival in multiple myeloma," *Blood*, vol. 124, no. 11, pp. 1765–1776, 2014.
17. F. Dall'Olio, N. Malagolini, M. Trinchera, and M. Chiricolo, "Mechanisms of cancer-associated glycosylation changes," *Frontiers in Bioscience*, vol. 17, no. 2, pp. 670–699, 2012.
18. H. Ghazarian, B. Idoni, and S. B. Oppenheimer, "A glycobiology review: carbohydrates, lectins and implications in cancer therapeutics," *Acta Histochemica*, vol. 113, no. 3, pp. 236–247, 2011.
19. Z. Kyselova, Y. Mechref, P. Kang et al., "Breast cancer diagnosis and prognosis through quantitative measurements of serum glycan profiles," *Clinical Chemistry*, vol. 54, no. 7, pp. 1166–1175, 2008.
20. R. K. Sterling, L. Jeffers, F. Gordon et al., "Clinical utility of AFP-L3% measurement in North American patients with HCV-related cirrhosis," *The American Journal of Gastroenterology*, vol. 102, no. 10, pp. 2196–2205, 2007.
21. S. Zhang, K. Jiang, Q. Zhang, K. Guo, and Y. Liu, "Serum fucosylated paraoxonase 1 as a potential glycobiomarker for clinical diagnosis of early hepatocellular carcinoma using ELISA Index," *Glycoconjugate Journal*, vol. 32, no. 3-4, pp. 119–125, 2015.
22. C. G. Bergstrand and B. Czar, "Demonstration of a new protein fraction in serum from the human fetus," *Scandinavian Journal of Clinical and Laboratory Investigation*, vol. 8, no. 2, p. 174, 1956.
23. G. J. Mizejewski, "Alpha-fetoprotein (AFP)-derived peptides as epitopes for hepatoma immunotherapy: a commentary," *Cancer Immunology, Immunotherapy*, vol. 58, no. 2, pp. 159–170, 2009.
24. S. Saito, H. Ojima, H. Ichikawa, S. Hirohashi, and T. Kondo, "Molecular

background of α-fetoprotein in liver cancer cells as revealed by global RNA expression analysis," *Cancer Science*, vol. 99, no. 12, pp. 2402–2409, 2008.
25. S. Gupta, S. Bent, and J. Kohlwes, "Test characteristics of α-fetoprotein for detecting hepatocellular carcinoma in patients with hepatitis C: a systematic review and critical analysis," *Annals of Internal Medicine*, vol. 139, no. 1, pp. 46–50, 2003.
26. R. Saffroy, P. Pham, M. Reffas, M. Takka, A. Lemoine, and B. Debuire, "New perspectives and strategy research biomarkers for hepatocellular carcinoma," *Clinical Chemistry and Laboratory Medicine*, vol. 45, no. 9, pp. 1169–1179, 2007.
27. J. A. Marrero, "Modern diagnosis of hepatocellular carcinoma: utilization of liver biopsy and genomic markers," *Journal of Hepatology*, vol. 50, no. 4, pp. 659–661, 2009.
28. M. Kobayashi, T. Kuroiwa, T. Suda et al., "Fucosylated fraction of alpha-fetoprotein, L3, as a useful prognostic factor in patients with hepatocellular carcinoma with special reference to low concentrations of serum alpha-fetoprotein," *Hepatology Research*, vol. 37, no. 11, pp. 914–922, 2007.
29. T. Nakagawa, E. Miyoshi, T. Yakushijin et al., "Glycomic analysis of alpha-fetoprotein L3 in hepatoma cell lines and hepatocellular carcinoma patients," *Journal of Proteome Research*, vol. 7, no. 6, pp. 2222–2233, 2008.
30. H. Toyoda, T. Kumada, T. Tada et al., "Clinical utility of highly sensitive Lens culinaris agglutinin-reactive alpha-fetoprotein in hepatocellular carcinoma patients with alpha-fetoprotein <20 ng/mL," *Cancer Science*, vol. 102, no. 5, pp. 1025–1031, 2011.
31. K. Oda, A. Ido, T. Tamai et al., "Highly sensitive lens culinaris agglutinin-reactive α-fetoprotein is useful for early detection of hepatocellular carcinoma in patients with chronic liver disease," *Oncology Reports*, vol. 26, no. 5, pp. 1227–1233, 2011.
32. T. Tada, T. Kumada, H. Toyoda et al., "Relationship between *Lens culinaris* agglutinin-reactive α-fetoprotein and pathologic features of hepatocellular carcinoma," *Liver International*, vol. 25, no. 4, pp. 848–853, 2005.
33. M. G. Giardina, M. Matarazzo, R. Morante et al., "Serum α-L-fucosidase activity and early detection of hepatocellular carcinoma: a

prospective study of patients with cirrhosis," *Cancer*, vol. 83, no. 12, pp. 2468–2474, 1998.

34. T. Isono, T. Tanaka, S. Kageyama, and T. Yoshiki, "Structural diversity of cancer-related and non-cancer-related prostate-specific antigen," *Clinical Chemistry*, vol. 48, no. 12, pp. 2187–2194, 2002.
35. M. V. Dwek, A. Jenks, and A. J. C. Leathem, "A sensitive assay to measure biomarker glycosylation demonstrates increased fucosylation of prostate specific antigen (PSA) in patients with prostate cancer compared with benign prostatic hyperplasia," *Clinica Chimica Acta*, vol. 411, no. 23-24, pp. 1935–1939, 2010.
36. W. J. Catalona, D. S. Smith, T. L. Ratliff et al., "Measurement of prostate-specific antigen in serum as a screening test for prostate cancer," *The New England Journal of Medicine*, vol. 324, no. 17, pp. 1156–1161, 1991.
37. H. B. Carter, P. C. Albertsen, M. J. Barry et al., "Early detection of prostate cancer: AUA guideline," *Journal of Urology*, vol. 190, no. 2, pp. 419–426, 2013.
38. M. J. G. Bussemakers, A. van Bokhoven, G. W. Verhaegh et al., "DD3: a new prostate-specific gene, highly overexpressed in prostate cancer," *Cancer Research*, vol. 59, no. 23, pp. 5975–5979, 1999.
39. M. Tajiri, C. Ohyama, and Y. Wada, "Oligosaccharide profiles of the prostate specific antigen in free and complexed forms from the prostate cancer patient serum and in seminal plasma: a glycopeptide approach," *Glycobiology*, vol. 18, no. 1, pp. 2–8, 2008.
40. A. Sarrats, J. Comet, G. Tabarés et al., "Differential percentage of serum prostate-specific antigen subforms suggests a new way to improve prostate cancer diagnosis," *Prostate*, vol. 70, no. 1, pp. 1–9, 2010.
41. R. Saldova, Y. Fan, J. M. Fitzpatrick, R. W. G. Watson, and P. M. Rudd, "Core fucosylation and α2-3 sialylation in serum N-glycome is significantly increased in prostate cancer comparing to benign prostate hyperplasia," *Glycobiology*, vol. 21, no. 2, pp. 195–205, 2011.
42. Z. Kyselova, Y. Mechref, M. M. Al Bataineh et al., "Alterations in the serum glycome due to metastatic prostate cancer," *Journal of Proteome Research*, vol. 6, no. 5, pp. 1822–1832, 2007.
43. R. Peracaula, S. Barrabés, A. Sarrats, P. M. Rudd, and R. De Llorens, "Altered glycosylation in tumours focused to cancer diagnosis," *Disease Markers*, vol. 25, no. 4-5, pp. 207–218, 2008.

44. R. Peracaula, G. Tabarés, L. Royle et al., "Altered glycosylation pattern allows the distinction between prostate-specific antigen (PSA) from normal and tumor origins," *Glycobiology*, vol. 13, no. 6, pp. 457–470, 2003.
45. Q. K. Li, L. Chen, M. Ao et al., "Serum Fucosylated Prostate-specific Antigen (PSA) improves the differentiation of aggressive from non-aggressive prostate cancers," *Theranostics*, vol. 5, no. 3, pp. 267–276, 2015.
46. T. Yoneyama, C. Ohyama, S. Hatakeyama et al., "Measurement of aberrant glycosylation of prostate specific antigen can improve specificity in early detection of prostate cancer," *Biochemical and Biophysical Research Communications*, vol. 448, no. 4, pp. 390–396, 2014.
47. A. Vegvari, M. Rezeli, C. Sihlbom et al., "Molecular microheterogeneity of prostate specific antigen in seminal fluid by mass spectrometry," *Clinical Biochemistry*, vol. 45, no. 4-5, pp. 331–338, 2012.
48. S. Goč and M. Janković, "Evaluation of molecular species of prostate-specific antigen complexed with immunoglobulin m in prostate cancer and benign prostatic hyperplasia," *Disease Markers*, vol. 35, no. 6, pp. 847–855, 2013.
49. R. C. Bast Jr., M. Feeney, H. Lazarus, L. M. Nadler, R. B. Colvin, and R. C. Knapp, "Reactivity of a monoclonal antibody with human ovarian carcinoma," *The Journal of Clinical Investigation*, vol. 68, no. 5, pp. 1331–1337, 1981.
50. F.-G. Hanisch, G. Uhlenbruck, J. Peter-Katalinic, and H. Egge, "Structural studies on oncofetal carbohydrate antigens (Ca 19-9, Ca 50, and Ca 125) carried by O-linked sialyl-oligosaccharides on human amniotic mucins," *Carbohydrate Research*, vol. 178, no. 1, pp. 29–47, 1988.
51. H. M. Davis, V. R. Zurawski Jr., R. C. Bast Jr., and T. L. Klug, "Characterization of the CA 125 antigen associated with human epithelial ovarian carcinomas," *Cancer Research*, vol. 46, part 1, no. 12, pp. 6143–6148, 1986.
52. R. C. Bast, T. L. Klug, E. St John et al., "A radioimmunoassay using a monoclonal antibody to monitor the course of epithelial ovarian cancer," *The New England Journal of Medicine*, vol. 309, no. 15, pp. 883–887, 1983.

53. M. Muyldermans, F. J. Cornillie, and P. R. Koninckx, "CA125 and endometriosis," *Human Reproduction Update*, vol. 1, no. 2, pp. 173–187, 1995.
54. C. S. Marcus, G. L. Maxwell, K. M. Darcy, C. A. Hamilton, and W. P. McGuire, "Current approaches and challenges in managing and monitoring treatment response in ovarian cancer," *Journal of Cancer*, vol. 5, no. 1, pp. 25–30, 2014.
55. K. Akita, S. Yoshida, Y. Ikehara et al., "Different levels of sialyl-Tn antigen expressed on MUC16 in patients with endometriosis and ovarian cancer," *International Journal of Gynecological Cancer*, vol. 22, no. 4, pp. 531–538, 2012.
56. K. Chen, A. Gentry-Maharaj, M. Burnell et al., "Microarray Glycoprofiling of CA125 improves differential diagnosis of ovarian cancer," *Journal of Proteome Research*, vol. 12, no. 3, pp. 1408–1418, 2013.
57. S. Ricardo, L. Marcos-Silva, D. Pereira et al., "Detection of glyco-mucin profiles improves specificity of MUC16 and MUC1 biomarkers in ovarian serous tumours," *Molecular Oncology*, vol. 9, no. 2, pp. 503–512, 2014.
58. C. Kirchhoff, "Molecular characterization of epididymal proteins," *Reviews of Reproduction*, vol. 3, no. 2, pp. 86–95, 1998.
59. C. Kirchhoff, I. Habben, R. Iveli, and N. Krull, "A major human epididymis-specific cDNA encodes a protein with sequence homology to extracellular proteinase inhibitors," *Biology of Reproduction*, vol. 45, no. 2, pp. 350–357, 1991.
60. M. Schummer, W. V. Ng, R. E. Bumgarner et al., "Comparative hybridization of an array of 21,500 ovarian cDNAs for the discovery of genes overexpressed in ovarian carcinomas," *Gene*, vol. 238, no. 2, pp. 375–385, 1999.
61. C. D. Hough, C. A. Sherman-Baust, E. S. Pizer et al., "Large-scale serial analysis of gene expression reveals genes differentially expressed in ovarian cancer," *Cancer Research*, vol. 60, no. 22, pp. 6281–6287, 2000.
62. R. Drapkin, H. H. von Horsten, Y. Lin et al., "Human epididymis protein 4 (HE4) is a secreted glycoprotein that is overexpressed by serous and endometrioid ovarian carcinomas," *Cancer Research*, vol. 65, no. 6, pp. 2162–2169, 2005.

63. K. D. Steffensen, M. Waldstrøm, I. Brandslund, and A. Jakobsen, "Prognostic impact of prechemotherapy serum levels of HER2, CA125, and HE4 in ovarian cancer patients," *International Journal of Gynecological Cancer*, vol. 21, no. 6, pp. 1040–1047, 2011.
64. S. Ferrarow, F. Braga, M. Lanzoni, P. Boracchi, E. M. Biganzoli, and M. Panteghini, "Serum human epididymis protein 4 vs carbohydrate antigen 125 for ovarian cancer diagnosis: a systematic review," *Journal of Clinical Pathology*, vol. 66, no. 4, pp. 273–281, 2013.
65. M. Montagnana, E. Danese, O. Ruzzenente et al., "The ROMA (Risk of Ovarian Malignancy Algorithm) for estimating the risk of epithelial ovarian cancer in women presenting with pelvic mass: is it really useful?" *Clinical Chemistry and Laboratory Medicine*, vol. 49, no. 3, pp. 521–525, 2011.
66. M. A. Karlsen, N. Sandhu, C. Høgdall et al., "Evaluation of HE4, CA125, risk of ovarian malignancy algorithm (ROMA) and risk of malignancy index (RMI) as diagnostic tools of epithelial ovarian cancer in patients with a pelvic mass," *Gynecologic Oncology*, vol. 127, no. 2, pp. 379–383, 2012.
67. M. T. Sandri, F. Bottari, D. Franchi et al., "Comparison of HE4, CA125 and ROMA algorithm in women with a pelvic mass: correlation with pathological outcome," *Gynecologic Oncology*, vol. 128, no. 2, pp. 233–238, 2013.
68. N. Chhikara, M. Saraswat, A. K. Tomar, S. Dey, S. Singh, and S. Yadav, "Human epididymis protein-4 (HE-4): a novel cross-class protease inhibitor," *PLoS ONE*, vol. 7, no. 11, Article ID e47672, 2012.
69. L. Hua, Y. Liu, S. Zhen, D. Wan, J. Cao, and X. Gao, "Expression and biochemical characterization of recombinant human epididymis protein 4," *Protein Expression and Purification*, vol. 102, pp. 52–62, 2014.
70. S. Nath and P. Mukherjee, "MUC1: a multifaceted oncoprotein with a key role in cancer progression," *Trends in Molecular Medicine*, vol. 20, no. 6, pp. 332–342, 2014.
71. M. Shimizu and K. Yamauchi, "Isolation and characterization of mucin-like glycoprotein in human milk fat globule membrane," *Journal of Biochemistry*, vol. 91, no. 2, pp. 515–524, 1982.
72. J. Hilkens, F. Buijs, J. Hilgers et al., "Monoclonal antibodies against human milk-fat globule membranes detecting differentiation antigens

of the mammary gland and its tumors," *International Journal of Cancer*, vol. 34, no. 2, pp. 197–206, 1984.

73. S. B. Ho, G. A. Niehans, C. Lyftogt et al., "Heterogeneity of mucin gene expression in normal and neoplastic tissues," *Cancer Research*, vol. 53, no. 3, pp. 641–651, 1993.
74. M. Gion, R. Mione, A. E. Leon et al., "CA27.29: a valuable marker for breast cancer management. A confirmatory multicentric study on 603 cases," *European Journal of Cancer*, vol. 37, no. 3, pp. 355–363, 2001.
75. S. Schoonooghe, I. Burvenich, L. Vervoort, F. De Vos, N. Mertens, and J. Grooten, "PH1-derived bivalent bibodies and trivalent tribodies bind differentially to shed and tumour cell-associated MUC1," *Protein Engineering, Design & Selection*, vol. 23, no. 9, pp. 721–728, 2010.
76. S. E. Baldus, K. Engelmann, and F.-G. Hanisch, "MUC1 and the MUCs: a family of human mucins with impact in cancer biology," *Critical Reviews in Clinical Laboratory Sciences*, vol. 41, no. 2, pp. 189–231, 2004.
77. D. F. Hayes, H. Sekine, T. Ohno, M. Abe, K. Keefe, and D. W. Kufe, "Use of a murine monoclonal antibody for detection of circulating plasma DF3 antigen levels in breast cancer patients," *The Journal of Clinical Investigation*, vol. 75, no. 5, pp. 1671–1678, 1985.
78. J.-H. Park, T. Nishidate, K. Kijima et al., "Critical roles of mucin 1 glycosylation by transactivated polypeptide *N*-acetylgalactosaminyltransferase 6 in mammary carcinogenesis," *Cancer Research*, vol. 70, no. 7, pp. 2759–2769, 2010.
79. S. J. Storr, L. Royle, C. J. Chapman et al., "The O-linked glycosylation of secretory/shed MUC1 from an advanced breast cancer patient's serum," *Glycobiology*, vol. 18, no. 6, pp. 456–462, 2008.
80. U. Westerlind, A. Hobel, N. Gaidzik, E. Schmitt, and H. Kunz, "Synthetic vaccines consisting of tumor-associated MUC1 glycopeptide antigens and a T-cell epitope for the induction of a highly specific humoral immune response," *Angewandte Chemie*, vol. 47, no. 39, pp. 7551–7556, 2008.
81. N. Gaidzik, U. Westerlind, and H. Kunz, "The development of synthetic antitumour vaccines from mucin glycopeptide antigens," *Chemical Society Reviews*, vol. 42, no. 10, pp. 4421–4442, 2013.
82. A. Kaiser, N. Gaidzik, U. Westerlind et al., "A synthetic vaccine consisting of a tumor-associated sialyl-T_N-MUC1 tandem-repeat

glycopeptide and tetanus toxoid: induction of a strong and highly selective immune response," *Angewandte Chemie International Edition*, vol. 48, no. 41, pp. 7551–7555, 2009.

83. V. Ludovini, S. Gori, M. Colozza et al., "Evaluation of serum HER2 extracellular domain in early breast cancer patients: correlation with clinicopathological parameters and survival," *Annals of Oncology*, vol. 19, no. 5, pp. 883–890, 2008.
84. B. A. Gusterson, "Identification and interpretation of epidermal growth factor and c-erbB-2 overexpression," *European Journal of Cancer*, vol. 28, no. 1, pp. 263–267, 1992.
85. A. C. Wolff, M. E. Hammond, J. N. Schwartz et al., "American Society of Clinical Oncology/College of American Pathologists guideline recommendations for human epidermal growth factor receptor 2 testing in breast cancer," *Archives of Pathology & Laboratory Medicine*, vol. 131, no. 1, pp. 18–43, 2007.
86. D. Di Gioia, M. Dresse, D. Mayr, D. Nagel, V. Heinemann, and P. Stieber, "Serum HER2 in combination with CA 15-3 as a parameter for prognosis in patients with early breast cancer," *Clinica Chimica Acta*, vol. 440, pp. 16–22, 2015.
87. K. Oyama, S. Fushida, T. Tsukada et al., "Evaluation of serum HER2-ECD levels in patients with gastric cancer," *Journal of Gastroenterology*, vol. 50, no. 1, pp. 41–45, 2015.
88. J. T. Garrett, S. Rawale, S. D. Allen et al., "Novel engineered trastuzumab conformational epitopes demonstrate in vitro and in vivo antitumor properties against HER-2/neu," *Journal of Immunology*, vol. 178, no. 11, pp. 7120–7131, 2007.
89. H. Gadler, K. Bremme, B. Wahren, and S. Hammarstrom, "CEA and NCA in amniotic fluid of normal and abnormal pregnancies," *Cancer*, vol. 42, no. 3, pp. 1579–1584, 1978.
90. X. G. Ni, X. F. Bai, Y. L. Mao et al., "The clinical value of serum CEA, CA19-9, and CA242 in the diagnosis and prognosis of pancreatic cancer," *European Journal of Surgical Oncology*, vol. 31, no. 2, pp. 164–169, 2005.
91. R. Goslin, G. Steele Jr., J. MacIntyre et al., "The use of preoperative plasma CEA levels for the stratification of patients after curative resection of colorectal cancers," *Annals of Surgery*, vol. 192, no. 6, pp. 747–751, 1980.

92. M. J. Duffy, "Carcinoembryonic antigen as a marker for colorectal cancer: is it clinically useful?" *Clinical Chemistry*, vol. 47, no. 4, pp. 624–630, 2001.

93. R. J. Paxton, G. Mooser, H. Pande, T. D. Lee, and J. E. Shively, "Sequence analysis of carcinoembryonic antigen: identification of glycosylation sites and homology with the immunoglobulin supergene family," *Proceedings of the National Academy of Sciences of the United States of America*, vol. 84, no. 4, pp. 920–924, 1987.

94. M. Garcia, C. Seigner, C. Bastid, R. Choux, M. J. Payan, and H. Reggio, "Carcinoembryonic antigen has a different molecular weight in normal colon and in cancer cells due to *N*-glycosylation differences," *Cancer Research*, vol. 51, no. 20, pp. 5679–5686, 1991.

95. S. N. Thomas, F. Zhu, R. L. Schnaar, C. S. Alves, and K. Konstantopoulos, "Carcinoembryonic antigen and CD44 variant isoforms cooperate to mediate colon carcinoma cell adhesion to E- and L-selectin in shear flow," *Journal of Biological Chemistry*, vol. 283, no. 23, pp. 15647–15655, 2008.

96. H. Koprowski, M. Herlyn, Z. Steplewski, and H. F. Sears, "Specific antigen in serum of patients with colon carcinoma," *Science*, vol. 212, no. 4490, pp. 53–55, 1981.

97. C. C. Hsu, A. Goyal, A. Iuga et al., "Elevated CA19-9 is associated with increased mortality in a prospective cohort of hepatocellular carcinoma patients," *Clinical and Translational Gastroenterology*, vol. 6, no. 2, p. e74, 2015.

98. E. M. Vestergaard, H. O. Hein, H. Meyer et al., "Reference values and biological variation for tumor marker CA 19-9 in serum for different Lewis and secretor genotypes and evaluation of secretor and Lewis genotyping in a Caucasian population," *Clinical Chemistry*, vol. 45, no. 1, pp. 54–61, 1999.

99. D. P. O'Brien, N. S. Sanayake, C. Jenkinson et al., "Serum CA19-9 is significantly upregulated up to 2 years before diagnosis with pancreatic cancer: implications for early disease detection," *Clinical Cancer Research*, vol. 21, no. 3, pp. 622–631, 2014.

100. F. Safi, W. Schlosser, G. Kolb, and H. G. Beger, "Diagnostic value of CA 19-9 in patients with pancreatic cancer and nonspecific gastrointestinal symptoms," *Journal of Gastrointestinal Surgery*, vol. 1, no. 2, pp. 106–112, 1997.

101. D. Marrelli, E. Pinto, A. de Stefano, M. Farnetani, L. Garosi, and F. Roviello, "Clinical utility of CEA, CA 19-9, and CA 72-4 in the follow-up of patients with resectable gastric cancer," *The American Journal of Surgery*, vol. 181, no. 1, pp. 16–19, 2001.

102. Y. Takahashi, T. Takeuchi, J. Sakamoto et al., "The usefulness of CEA and/or CA19-9 in monitoring for recurrence in gastric cancer patients: a prospective clinical study," *Gastric Cancer*, vol. 6, no. 3, pp. 142–145, 2003.

103. W.-K. Cai, J. J. Lin, G. H. He, H. Wang, J. H. Lu, and G. S. Yang, "Preoperative serum CA19-9 levels is an independent prognostic factor in patients with resected hilar cholangiocarcinoma," *International Journal of Clinical and Experimental Pathology*, vol. 7, no. 11, pp. 7890–7898, 2014.

104. S.-X. Yang, H. G. Pollock, and A. B. Rawitch, "Glycosylation in human thyroglobulin: location of the N-linked oligosaccharide units and comparison with bovine thyroglobulin," *Archives of Biochemistry and Biophysics*, vol. 327, no. 1, pp. 61–70, 1996.

105. J.-D. Lin, "Thyroglobulin and human thyroid cancer," *Clinica Chimica Acta*, vol. 388, no. 1-2, pp. 15–21, 2008.

106. R. Petric, A. Perhavec, B. Gazic, and N. Besic, "Preoperative serum thyroglobulin concentration is an independent predictive factor of malignancy in follicular neoplasms of the thyroid gland," *Journal of Surgical Oncology*, vol. 105, no. 4, pp. 351–356, 2012.

107. C. A. Spencer, "Challenges of serum thyroglobulin (Tg) measurement in the presence of Tg autoantibodies," *Journal of Clinical Endocrinology and Metabolism*, vol. 89, no. 8, pp. 3702–3704, 2004.

108. N. Panza, G. Lombardi, M. de Rosa, G. Pacilio, L. Lapenta, and M. Salvatore, "High serum thyroglobulin levels. Diagnostic indicators in patients with metastases from unknown primary sites," *Cancer*, vol. 60, no. 9, pp. 2233–2236, 1987.

109. J. Hrafnkelsson, H. Tulinius, M. Kjeld, H. Sigvaldason, and J. G. Jónasson, "Serum thyroglobulin as a risk factor for thyroid carcinoma," *Acta Oncologica*, vol. 39, no. 8, pp. 973–977, 2000.

110. N. Besic, G. Pilko, R. Petric, M. Hocevar, and J. Zgajnar, "Papillary thyroid microcarcinoma: prognostic factors and treatment," *Journal of Surgical Oncology*, vol. 97, no. 3, pp. 221–225, 2008.

111. M. Maruyama, R. Kato, S. Kobayashi, and Y. Kasuga, "A method to differentiate between thyroglobulin derived from normal thyroid tissue and from thyroid carcinoma based on analysis of reactivity to lectins," *Archives of Pathology and Laboratory Medicine*, vol. 122, no. 8, pp. 715–720, 1998.
112. K. Shimizu, K. Nakamura, S. Kobatake et al., "The clinical utility of *Lens culinaris* agglutinin-reactive thyroglobulin ratio in serum for distinguishing benign from malignant conditions of the thyroid," *Clinica Chimica Acta*, vol. 379, no. 1-2, pp. 101–104, 2007.
113. T. Kanai, M. Amakawa, R. Kato et al., "Evaluation of a new method for the diagnosis of alterations of Lens culinaris agglutinin binding of thyroglobulin molecules in thyroid carcinoma," *Clinical Chemistry and Laboratory Medicine*, vol. 47, no. 10, pp. 1285–1290, 2009.
114. U.-H. Stenman, H. Alfthan, and K. Hotakainen, "Human chorionic gonadotropin in cancer," *Clinical Biochemistry*, vol. 37, no. 7, pp. 549–561, 2004.
115. H. O. Smith, C. Wiggins, C. F. Verschraegen et al., "Changing trends in gestational trophoblastic disease," *The Journal of Reproductive Medicine*, vol. 51, no. 10, pp. 777–784, 2006.
116. L. A. Cole, D. Dai, S. A. Butler, K. K. Leslie, and E. I. Kohorn, "Gestational trophoblastic diseases: 1. Pathophysiology of hyperglycosylated hCG," *Gynecologic Oncology*, vol. 102, no. 2, pp. 145–150, 2006.
117. H. Alfthan, C. Haglund, P. Roberts, and U.-H. Stenman, "Elevation of free beta subunit of human choriogonadotropin and core beta fragment of human choriogonadotropin in the serum and urine of patients with malignant pancreatic and biliary disease," *Cancer Research*, vol. 52, no. 17, pp. 4628–4633, 1992.
118. A. Kobata and M. Takeuchi, "Structure, pathology and function of the N-linked sugar chains of human chorionic gonadotropin," *Biochimica et Biophysica Acta*, vol. 1455, no. 2-3, pp. 315–326, 1999.
119. L. Valmu, H. Alfthan, K. Hotakainen, S. Birken, and U.-H. Stenman, "Site-specific glycan analysis of human chorionic gonadotropin beta-subunit from malignancies and pregnancy by liquid chromatography—electrospray mass spectrometry," *Glycobiology*, vol. 16, no. 12, pp. 1207–1218, 2006.
120. J. R. Lurain, "Gestational trophoblastic disease II: classification and

management of gestational trophoblastic neoplasia," *American Journal of Obstetrics and Gynecology*, vol. 204, no. 1, pp. 11–18, 2011.

121. M. A. Rahman, S. Mitra, A. Sarkar, M. D. Wewers, and D. Hartl, "Alpha 1-antitrypsin does not inhibit human monocyte caspase-1," *PLOS ONE*, vol. 10, no. 2, Article ID e0117330, 2015.
122. M. Guarino, "Epithelial-mesenchymal transition and tumour invasion," *The International Journal of Biochemistry & Cell Biology*, vol. 39, no. 12, pp. 2153–2160, 2007.
123. J.-H. Rho, M. H. A. Roehrl, and J. Y. Wang, "Glycoproteomic analysis of human lung adenocarcinomas using glycoarrays and tandem mass spectrometry: differential expression and glycosylation patterns of vimentin and fetuin A isoforms," *Protein Journal*, vol. 28, no. 3-4, pp. 148–160, 2009.
124. Y. Liang, T. Ma, A. Thakur et al., "Differentially expressed glycosylated patterns of α-1-antitrypsin as serum biomarkers for the diagnosis of lung cancer," *Glycobiology*, vol. 25, no. 3, pp. 331–340, 2015.
125. E. Miyoshi and M. Nakano, "Fucosylated haptoglobin is a novel marker for pancreatic cancer: detailed analyses of oligosaccharide structures," *Proteomics*, vol. 8, no. 16, pp. 3257–3262, 2008.
126. I. U. Song, Y. D. Kim, S. W. Chung, and H. J. Cho, "Association between serum haptoglobin and the pathogenesis of Alzheimer's disease," *Internal Medicine*, vol. 54, no. 5, pp. 453–457, 2015.
127. K. Taketa, Y. Endo, C. Sekiya et al., "A collaborative study for the evaluation of lectin-reactive α- fetoproteins in early detection of hepatocellular carcinoma," *Cancer Research*, vol. 53, no. 22, pp. 5419–5423, 1993.
128. T. Fujimura, Y. Shinohara, B. Tissot et al., "Glycosylation status of haptoglobin in sera of patients with prostate cancer vs. benign prostate disease or normal subjects," *International Journal of Cancer*, vol. 122, no. 1, pp. 39–49, 2008.
129. Y. Kamada, N. Kinoshita, Y. Tsuchiya et al., "Reevaluation of a lectin antibody ELISA kit for measuring fucosylated haptoglobin in various conditions," *Clinica Chimica Acta*, vol. 417, pp. 48–53, 2013.
130. Y. Takeda, S. Shinzaki, K. Okudo, K. Moriwaki, K. Murata, and E. Miyoshi, "Fucosylated haptoglobin is a novel type of cancer biomarker linked to the prognosis after an operation in colorectal cancer," *Cancer*, vol. 118, no. 12, pp. 3036–3043, 2012.

131. K. Fujita, M. Shimomura, M. Uemura et al., "Serum fucosylated haptoglobin as a novel prognostic biomarker predicting high-Gleason prostate cancer," *Prostate*, vol. 74, no. 10, pp. 1052–1058, 2014.
132. J. S. Johansen, B. V. Jensen, A. Roslind, D. Nielsen, and P. A. Price, "Serum YKL-40, a new prognostic biomarker in cancer patients?" *Cancer Epidemiology Biomarkers and Prevention*, vol. 15, no. 2, pp. 194–202, 2006.
133. A. K. Mylin, T. Rasmussen, J. S. Johansen et al., "Serum YKL-40 concentrations in newly diagnosed multiple myeloma patients and YKL-40 expression in malignant plasma cells," *European Journal of Haematology*, vol. 77, no. 5, pp. 416–424, 2006.
134. A. K. Mylin, N. Abildgaard, J. S. Johansen et al., "High serum YKL-40 concentration is associated with severe bone disease in newly diagnosed multiple myeloma patients," *European Journal of Haematology*, vol. 80, no. 4, pp. 310–317, 2008.
135. A. K. Mylin, N. Abildgaard, J. S. Johansen et al., "Serum YKL-40: a new independent prognostic marker for skeletal complications in patients with multiple myeloma," *Leukemia & Lymphoma*, pp. 1–10, 2015.
136. S.-E. Lee, J.-H. Yoon, S.-H. Shin et al., "Bone marrow plasma cell assessment before peripheral blood stem cell mobilization in patients with multiple myeloma undergoing autologous stem cell transplantation," *BioMed Research International*, vol. 2014, Article ID 982504, 8 pages, 2014.
137. A. Mayampurath, E. Song, A. Mathur et al., "Label-free glycopeptide quantification for biomarker discovery in human sera," *Journal of Proteome Research*, vol. 13, no. 11, pp. 4821–4832, 2014.
138. K. Biskup, E. I. Braicu, J. Sehouli, R. Tauber, and V. Blanchard, "The serum glycome to discriminate between early-stage epithelial ovarian cancer and benign ovarian diseases," *Disease Markers*, vol. 2014, Article ID 238197, 10 pages, 2014.
139. S. Ozcan, D. A. Barkauskas, L. Renee Ruhaak et al., "Serum glycan signatures of gastric cancer," *Cancer Prevention Research*, vol. 7, no. 2, pp. 226–235, 2014.
140. J.-H. Kim, C. W. Park, D. Um et al., "Mass spectrometric screening of ovarian cancer with serum glycans," *Disease Markers*, vol. 2014, Article ID 634289, 9 pages, 2014.

141. D. H. Dube and C. R. Bertozzi, "Glycans in cancer and inflammation—potential for therapeutics and diagnostics," *Nature Reviews Drug Discovery*, vol. 4, no. 6, pp. 477–488, 2005.
142. J. N. Arnold, R. Saldova, U. M. Abd Hamid, and P. M. Rudd, "Evaluation of the serum *N*-linked glycome for the diagnosis of cancer and chronic inflammation," *Proteomics*, vol. 8, no. 16, pp. 3284–3293, 2008.
143. P. Niederhafner, M. Reiniš, J. Šebestík, and J. Ježek, "Glycopeptide dendrimers, Part III—a review: use of glycopeptide dendrimers in immunotherapy and diagnosis of cancer and viral diseases," *Journal of Peptide Science*, vol. 14, no. 5, pp. 556–587, 2008.
144. P.-L. Li, X. Zhang, T. Li et al., "Combined detection of sialic acid and hydroxyproline in diagnosis of ovarian cancer and its comparison with human epididymis protein 4 and carbohydrate antigen 125," *Clinica Chimica Acta*, vol. 439, pp. 148–153, 2015.
145. B. J. Marshall and J. R. Warren, "Unidentified curved bacilli in the stomach of patients with gastritis and peptic ulceration," *The Lancet*, vol. 1, no. 8390, pp. 1311–1315, 1984.
146. N. Ohnishi, H. Yuasa, S. Tanaka et al., "Transgenic expression of Helicobacter pylori CagA induces gastrointestinal and hematopoietic neoplasms in mouse," *Proceedings of the National Academy of Sciences of the United States of America*, vol. 105, no. 3, pp. 1003–1008, 2008.
147. T. Watanabe, M. Tada, H. Nagi, S. Sasaki, and M. Nakao, "*Helicobacter pylori* infection induces gastric cancer in Mongolian gerbils," *Gastroenterology*, vol. 115, no. 3, pp. 642–648, 1998.
148. L. Fuccio, R. M. Zagari, L. H. Eusebi et al., "Meta-analysis: can *Helicobacter pylori* eradication treatment reduce the risk for gastric cancer?" *Annals of Internal Medicine*, vol. 151, no. 2, pp. 121–128, 2009.
149. Helicobacter and Cancer Collaborative Group, "Gastric cancer and *Helicobacter pylori*: a combined analysis of 12 case control studies nested within prospective cohorts," *Gut*, vol. 49, no. 3, pp. 347–353, 2001.
150. R. Siegel, D. Naishadham, and A. Jemal, "Cancer statistics, 2013," *CA: A Cancer Journal for Clinicians*, vol. 63, no. 1, pp. 11–30, 2013.
151. F. Ceciliani and V. Pocacqua, "The acute phase protein α1-acid glycoprotein: a model for altered glycosylation during diseases," *Current Protein and Peptide Science*, vol. 8, no. 1, pp. 91–108, 2007.

152. E. Giménez, M. Balmaña, J. Figueras et al., "Quantitative analysis of N-glycans from human alfa-acid-glycoprotein using stable isotope labeling and zwitterionic hydrophilic interaction capillary liquid chromatography electrospray mass spectrometry as tool for pancreatic disease diagnosis," *Analytica Chimica Acta*, vol. 866, pp. 59–68, 2015.
153. R. S. Jabir, G. F. Ho, M. A. Anuar, and J. Stanslas, "Abstract 5563: docetaxel-induced mucositis in breast cancer patients: association with plasma alpha-1-acid glycoprotein level and SLCO1B3 genotype," *Cancer Research*, vol. 74, supplement 19, pp. 5563–5563, 2014.
154. A. Harazono, N. Kawasaki, S. Itoh et al., "Site-specific N-glycosylation analysis of human plasma ceruloplasmin using liquid chromatography with electrospray ionization tandem mass spectrometry," *Analytical Biochemistry*, vol. 348, no. 2, pp. 259–268, 2006.
155. M. Balmaña, A. Sarrats, E. Llop et al., "Identification of potential pancreatic cancer serum markers: Increased sialyl-Lewis X on ceruloplasmin," *Clinica Chimica Acta*, vol. 442, pp. 56–62, 2015.
156. J.-H. Rho, J. R. Mead, W. S. Wright et al., "Discovery of sialyl Lewis A and Lewis X modified protein cancer biomarkers using high density antibody arrays," *Journal of Proteomics*, vol. 96, pp. 291–299, 2014.
157. F. Hvid-Jensen, L. Pedersen, A. M. Drewes, H. T. Srøensen, and P. Funch-Jensen, "Incidence of adenocarcinoma among patients with Barrett's esophagus," *The New England Journal of Medicine*, vol. 365, no. 15, pp. 1375–1383, 2011.
158. M. Titi, A. Overhiser, O. Ulusarac et al., "Development of subsquamous high-grade dysplasia and adenocarcinoma after successful radiofrequency ablation of Barrett's esophagus," *Gastroenterology*, vol. 143, no. 3, pp. 564–566.e1, 2012.
159. A. Pierce, R. Saldova, U. M. Abd Hamid et al., "Levels of specific glycans significantly distinguish lymph node-positive from lymph node-negative breast cancer patients," *Glycobiology*, vol. 20, no. 10, pp. 1283–1288, 2010.
160. P. J. Johnson, T. C. W. Poon, N. M. Hjelm, C. S. Ho, C. Blake, and S. K. W. Ho, "Structures of disease-specific serum alpha-fetoprotein isoforms," *British Journal of Cancer*, vol. 83, no. 10, pp. 1330–1337, 2000.

5

Implication of Fructans in Health: Immunomodulatory and Antioxidant Mechanisms

Elena Franco-Robles and Mercedes G. López

Centro de Investigación y de Estudios Avanzados del IPN, Unidad Irapuato, Km 9.6 Libramiento Norte Carretera Irapuato-León, 36821 Irapuato, GTO, Mexico

ABSTRACT

Previous studies have shown that fructans, a soluble dietary fiber, are beneficial to human health and offer a promising approach for the treatment of some diseases. Fructans are nonreducing carbohydrates composed of fructosyl units and terminated by a single glucose molecule. These carbohydrates may be straight or branched with varying degrees of polymerization. Additionally, fructans are resistant to hydrolysis by human digestive enzymes but can be fermented by the colonic microbiota to produce short chain fatty acids (SCFAs), metabolic by-products that possess immunomodulatory activity. The indirect role of fructans in stimulating probiotic growth is one of the mechanisms through which fructans exert

Citation: Implication of Fructans in Health: Immunomodulatory and Antioxidant Mechanisms. Elena Franco-Robles and Mercedes G. López. Hindawi Publishing Corporation e Scientific World Journal Volume 2015, Article ID 289267, 15 pages http://dx.doi.org/10.1155/2015/289267

their prebiotic activity and improve health or ameliorate disease. However, a more direct mechanism for fructan activity has recently been suggested; fructans may interact with immune cells in the intestinal lumen to modulate immune responses in the body. Fructans are currently being studied for their potential as "ROS scavengers" that benefit intestinal epithelial cells by improving their redox environment. In this review, we discuss recent advances in our understanding of fructans interaction with the intestinal immune system, the gut microbiota, and other components of the intestinal lumen to provide an overview of the mechanisms underlying the effects of fructans on health and disease.

INTRODUCTION

Fructans are recognized as health-promoting food ingredients. They are found in a small number of mono- and dicotyledonous families of plants, such as Liliaceae, Amaryllidaceae, Gramineae, Compositae, Nolinaceae, and Agavaceae. Various fructan-containing plant species, including asparagus, garlic, leek, onion, Jerusalem artichoke, and chicory roots, are often eaten as vegetables [1–3]. Substantial variation in chemical and structural conformations makes fructans a flexible and appealing ingredient for different dietary products such as nutraceuticals.

Inulin-type fructans (ITFs) are among the most studied; ITFs are indigestible, fully soluble, fermentable food ingredients with known prebiotic properties. ITFs are linear fructose polymers with $\beta(2\rightarrow1)$ linkages found naturally in chicory roots, wheat, onion, garlic, and other foods. In the scientific literature, ITFs are frequently referenced generically but inconsistently as "inulin," "oligofructose" (OF), and "fructooligosaccharides" (FOS) [4]. *Agave* fructans have a more complex, highly branched structure, including $\beta(2\rightarrow1)$ and $\beta(2\rightarrow6)$ linkages. Thus, *Agave* fructans can contain an external glucose, characteristic of graminans, and an internal glucose, characteristic of neofructans. For this reason, this type of fructans has been called "agavins" [5].

Fructans contribute to host health through multiple mechanisms. Fructans are selective substrates for probiotic bacteria stimulating probiotic bacterial growth, which can confer health benefits to the host through the several mechanisms, including immunomodulation [6–8]. Fructans may also act as scavengers of reactive oxygen species [9], decreasing inflammation and improving redox status. Fructans are fermented to short chain fatty acids (SCFAs), which have important implications in host health. In addition, direct interaction between fructans and intestinal immune cells has recently been suggested. The aim of this review is to summarize the latest findings on studies investigating fructans as prebiotics and to provide an overall image of the mechanisms underlying the health effects of fructans.

FRUCTANS: STRUCTURE, SOURCE, AND SYNTHESIS

Approximately 15% of flowering plants store fructans as reserve carbohydrates [10]. Worldwide, the most studied and marketed fructan is inulin, which is obtained primarily from chicory roots. However, some candidate fructans, such as galactooligosaccharides (GOS) derived from lactose and lactulose, have also demonstrated potential prebiotic effects [11]. In addition to chicory root, another potential fructan source includes the more recently investigated *Agave* fructans. The *Agave tequilana* Weber *azul* variety is an economically important species of *Agave* cultivated in Mexico. Because of its high inulin concentration, this variety is the only species in the Agavaceae family that is appropriate for tequila production. The high inulin concentrations, specifically in the head (pine), provide added economic and environmental value to this species of *Agave* [12].

Fructans have been classified into 4 groups based on their structural bonds: inulin, levans, graminans, and neoseries fructans (inulin neoseries and levan neoseries mixture) [13]. Inulin is the simplest linear fructan, consisting of β(2→1)-linked fructose residues. Inulin is usually found in plants such as *Cichorium intybus* (15–20% fructans), *Jerusalem artichoke* (15–20% fructans), *Helianthus tuberosus* (15–20% fructans), and *Dahlia variabilis* (15–20% fructans) (Figure 1) [13–15]. Levan-type fructans (also called phleins in plants) can be found in grasses (Poaceae). Levan fructans contain a linear β(2→6)-linked fructose polymer and are found in big bluegrass (*Poa secunda*) [16, 17]. Graminan-type fructans consist of β(2→6)-linked fructose residues with β(2→1) branches or can consist of more complex structures in which neosugars are combined with branched fructan chains. These complex fructans are usually found in plants such as *Avena sativa*, *Lolium sp*., and *Agave sp.* (15–22% fructans) (Figure 1) [5, 18–20]. The inulin neoseries are linear (2-1)-linked β-d-fructosyl units linked to both C1 and C6 on the glucose moiety of the sucrose (Suc) molecule. This results in a fructan polymer with a fructose chain ((mF2-1F2-6G1-2F1-2Fn); F (fructose), G (glucose)) on both ends of the glucose molecule. These fructans are found in plants belonging to the Liliaceae family (e.g., onion and asparagus (10–15% fructans)) [15, 21]. The smallest inulin neoseries molecule is called neokestose. The levan neoseries consists of polymers with predominantly β(2→6)-linked fructosyl residues on either end of the glucose moiety of the sucrose molecule. These fructans are rare, although they have been found in a few plant species belonging to the Poales (e.g., oat) [18].

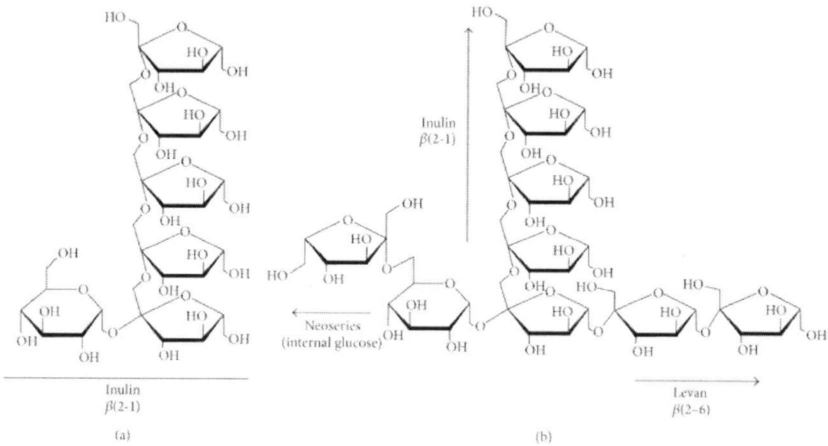

Figure 1. Structural comparison of the (a) inulin from *Cichorium intybus* and (b) agavin from *Agave* spp.

The length of fructosyl chains varies greatly in plants; plant fructosyl chains are much shorter than those of bacterial fructans. In general, the chain length or degree of polymerization (DP) is between 30 and 50 fructosyl residues in plants but can occasionally exceed 200 [13]. Fructans can also be classified according to their DP into small (2 to 4), medium (5 to 10), and relatively large chain lengths (11 to 60 fructose units). The term fructooligosaccharides (FOS) is used for short fructans with a DP of 3–5 derived from sucrose [22]; oligofructose (OF) is used for molecules with a DP of 3–10 derived from native inulin [23].

The biosynthesis of fructans begins with sucrose (Suc), to which fructose residues are added [4]. In plants, fructans are synthesized from Suc by the action of two or more enzymes known as fructosyltransferases. The first enzyme, 1-SST (sucrose:sucrose fructosyltransferase), initiates *de novo* fructan synthesis by catalyzing the transfer of a fructosyl residue from one Suc molecule to another, resulting in the formation of the trisaccharide 1-kestose. The second enzyme, 1-FFT (fructan:fructan

1-fructosyltransferase), transfers fructosyl residues from a fructan molecule with a DP of ≤3 to either another fructan molecule or a Suc. The actions of 1-SST and 1-FFT result in the formation of a mixture of fructan molecules with different chain lengths [13].

FUNCTIONAL EFFECTS OF FRUCTANS

Worldwide, over 60% of functional food products are directed toward intestinal health, and additional therapeutic benefits of these products to human health are constantly being explored. Prebiotics are defined as "selectively fermented ingredients that allow specific changes, both in the composition and/or activity in the gastrointestinal microbiota that confers benefits upon host well-being and health" [24]. Moreover, prebiotics may suppress pathogen growth to improve overall health [25]. Current evidence indicates that beneficial bacteria reduce the risk of diseases through diverse mechanisms, including modulation of gut microbiota composition or function, and regulation of host epithelial and immunological responses. These effects may be revealed through changes in bacterial populations or metabolic activity [26]. Bacterial metabolism can confer a number of advantageous effects to the host, including the production of vitamins, modulation of the immune system, enhancement of digestion and absorption, inhibition of harmful bacterial species, and removal of carcinogens and other toxins. The resident microbiota is also known to consist of pathogens that can disrupt normal gut function and predispose the host toward disease if allowed to overgrow [27].

Fructans play protective roles in plants subjected to drought, salt, or cold stress [14]. However, the therapeutic potential of fructans in human health has only recently been explored. As described above, fructans are the most widely known and used prebiotics [28]. Of the many nondigestible food ingredients studied for their prebiotic potential, human trials favor ITFs, FOS, OF, and GOS [29–32]. Fructans have been proposed as modulators of the microbial ecology and host physiology in animals and humans [33, 34] because they are not digested [9]. Although they are subjected to minor hydrolysis in the stomach, the human gut lacks the hydrolytic enzymes capable of digesting β linkages [35]. Therefore, fructans reach the colon relatively intact and eventually trigger a decrease in the pH, thereby altering the colonic environment [36]. The rate and extent of ITFs fermentation appear to be strongly influenced by the DP. FOS (low DP) are rapidly fermented in the proximal colon [37], whereas inulin (high DP) appears to have a more sustained fermentation profile that potentially enables protective effects in the distal colon [4, 38]. Acting as prebiotics, inulin, FOS, and GOS improve glucose, reduce triglycerides, modify lipid metabolism, and reduce plasma LPS. Additionally, they stimulate *Lactobacillus* and *Bifidobacterium* species to reduce the presence of pathogens in the gut and relieve constipation (Table 1). Other fructans, including soluble gut oligosaccharides, mimic the sugar chains found in the glycoproteins and glycolipids of gut epithelial cells, thereby preventing the adhesion of pathogenic microorganisms [39] and exerting direct antimicrobial effects [40] (Table 1).

Table 1. Main prebiotic effects of fructans in *in vitro* and *in vivo* studies.

Effect	Type of fructan	Dose/duration	Model	Results	Reference
Decreasing blood glucose	FOS, inulin	8 g/d for 14 days; 10% for 4 weeks	Diabetic subjects; animal models	Significant reduction of mean fasting blood glucose levels. Improving glucose tolerance	[49–51]
Reduction in blood serum triacylglycerol levels	FOS, inulin	4–34 g/d for 21–60 days; 10% for 3–5 weeks	Healthy humans; obese animal models	Significant reduction in blood serum triacylglycerol levels	[52–54]
Improved lipid metabolism	FOS, GOS, inulin, and agavins	5%–10% for 21 day to 8 weeks	Obese animal models	Decrease in body weight gain. Decrease in epididymal adipose tissue, inguinal adipose tissue, and subcutaneous adipose tissue. Reducing fat-mass development	[41, 50, 51, 55–59]
Stimulation of lactobacilli and bifidobacteria and decreasing pathogens	FOS, GOS, and inulin	2.5–34 g/d for 14–64 days	Healthy subjects and animal models	Stimulating the growth of bifidobacteria and contributing to the suppression of potential pathogenic bacteria	[46, 60, 61]
Relief of constipation	Inulin, FOS, and GOS	20–40 g/d for 19 days	Constipated humans and animal models	Inulin showing a better laxative effect than lactose and reducing functional constipation with only mild discomfort	[62, 63]

Increased production of SCFAs and decreasing colon pH	Inulin, FOS, and agavins	24 g/d for 5 weeks; 10% for 28 days	Healthy subjects; animal models	Significant increase of acetate, propionate, and butyrate. Significantly increasing activity of bacterial enzymes and decreasing the pH of digesta	[36, 64, 65]
Improving mineral uptake	Inulin, FOS, and agavins	1–40 g/d for 9 days; 50–100 g/kg diet for 4 weeks	Male healthy adolescents; animal models	FOS stimulating fractional calcium absorption in male adolescents. A combination of different carbohydrates showing synergistic effects on intestinal Ca absorption and balance in rats	[66–69]
Regulated gut peptides	Inulin, FOS, and agavins	24 g/d for 5 weeks; 10% for 5 weeks	Healthy subjects; animals models	Increasing plasma glucagon-like peptide-1 (GLP-1) concentrations and reducing ghrelin. Increasing endogenous GLP-2 production and consequently improving gut barrier functions	[36, 41, 50, 57, 59]
Reducing body weight and energy intake	Agavins	10% for 5 weeks	Male healthy animal model	*Agave* fructans showing indications of prebiotic activity, particularly in relation to satiety and GLP-1 and ghrelin secretion. In this same study, the levels of butyric acid were higher for *Agave potatorum* fructans	[43]
Growth inhibition and prevention of adhesion of pathogenic microorganisms	FOS	170 mg/kg, 2 weeks of lactation	Breast-fed infant; cocultures of *Pseudomonas aeruginosa*	Oligosaccharides in human milk interfering with microbial adhesion. Reduction of exotoxin A in cultures of *P. aeruginosa*	[39, 40]

Reduction of oxidative stress by reducing ROS levels	FOS, agavins	10% for 4–8 weeks	Male obese animal models	FOS reducing TBARS urine. Lipopolysaccharides reduction in plasma. Improving the redox status by reducing the malondialdehyde serum levels and protein oxidative damage	[9, 42, 65]
Stimulation of the immune system	FOS, GOS, and inulin				See Table 2

FOS: fructooligosaccharides; GOS: galactooligosaccharides; SCFAs: short chain fatty acids.

Interestingly, fructans from *Dasylirion* spp. (DAS) and *A. tequilana* Gto. (TEQ) increased SCFAs production and decreased colon pH in *in vitro* studies [41]. Furthermore, supplementation of the mouse diet with *Agave* fructans (TEQ and DAS) has been shown to increase secretion of GLP-1 and its precursor, proglucagon mRNA, in all colonic segments of the mouse. These results suggest that fermentable fructans of different botanical origins and with differing chemical structures are able to promote the production of satietogenic/incretin peptides in the lower part of the gut [41] (Table 1). Moreover, *Agave* fructans have been shown to have physiological effects on lipid metabolism [41, 42] and reduce oxidative stress in conjunction with phenolic compounds in *in vitro* and *in vivo* assays [42] (Table 1). For the first time, the effect of agavins from *Agave angustifolia* and *Agave potatorum* as prebiotics has been reported showing satiety effect as well as an increment on GLP-1 and a decrement on ghrelin in an animal model [43] (Table 1).

Studies have been performed to determine whether probiotics reduce cancer risk. To maximize the effect of a prebiotic compound, the prebiotic would need to ferment in the distal colon, where proteolytic fermentation predominates and toxic metabolites such as ammonia, hydrogen sulfide, and cresol are produced [44, 45]. A recent study by Gomez et al. was the first to investigate the effect of *Agave* fructan fermentation on complex fecal microbiota *in vitro* [46] (Table 1). The first clinical trial in humans with *Agave* fructans was very promising, as *Agave* treatment improved laxation [47]. Other carbohydrates, including glucooligosaccharides, isomaltooligosaccharides, lactulose, mannanoligosaccharides (MOS), nigerooligosaccharides, oat β-glucans, raffinose, soybean oligosaccharides, transgalactooligosaccharides, and xylooligosaccharides, are considered candidate prebiotics [31, 48]; however, more research is required.

IMMUNOMODULATORY EFFECTS OF FRUCTANS

The consumption of prebiotics can modulate immune parameters in gut-associated lymphoid tissue (GALT), secondary lymphoid tissues, and peripheral circulation [70]. GALT functions to distinguish between harmful and innocuous agents and protects against infections while simultaneously avoiding the generation of hypersensitivity reactions to commensal bacteria and harmless antigens [71–73]. In inductive GALT, more structured and localized sites of antigen processing and presentation are distinguished in areas such as Peyer's patches (PPs), mesenteric lymph nodes (MLNs), the appendix, and isolated lymph nodes. GALT also contains effector sites with more diffuse organization, containing previously activated and differentiated cells that performed effector functions (Figure 2). Joint activity of the inductive and effector sites generates a rich response in immunoglobulin A (IgA) and cellular immunity, with robust cytotoxic regulatory functions and memory at the level of the mucosa and serum [74]. The intestinal epithelium provides a physical barrier that separates the trillions of commensal bacteria in the intestinal lumen from the underlying lamina propria (LP) and the deeper intestinal layers. Microfold cells (M cells), B cells (especially IgA-producing plasma cells), T cells, macrophages, and dendritic cells (DCs) in the LP are located directly below the intestinal epithelium (Figure 2). M cells are part of the epithelial layer covering the PP and specialize in transporting antigens from the lumen to GALT [75].

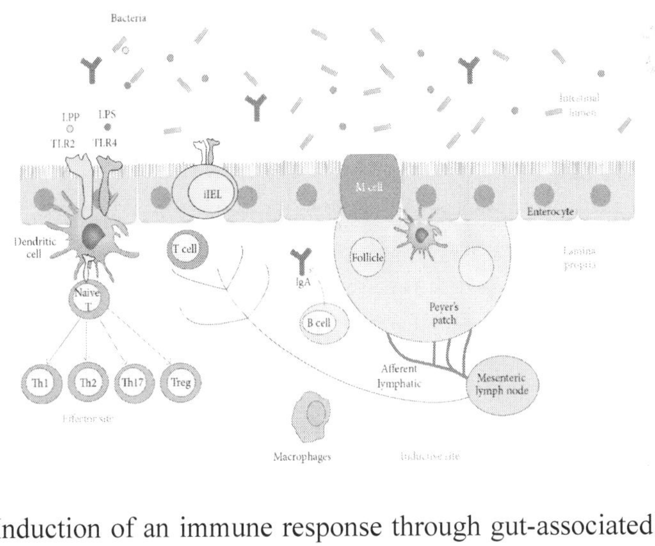

Figure 2. Induction of an immune response through gut-associated lymphoid tissue (GALT).

T and B cells are activated after initial contact with the antigen at inductive sites. These cells then proliferate, differentiate, and migrate to various effector sites, such as the LP or the intestinal epithelium, where a single population of iIELs (intestinal intraepithelial lymphocytes) and some DCs are located between the enterocytes [76–78] (Figure 2).

In fact, iIELs provide a cellular defense against any individual antigen [79]. Meanwhile, DCs are potent antigen-presenting cells critical for the induction of downstream adaptive immune responses [80]. For instance, several subsets of DCs have been identified within the PP that possess either Th1- or Th2-polarizing ability [81]. The CD103+ subset has been found within the small intestinal LP, MLN, and PP, as well as the colonic LP. CD103+ DCs have FoxP3+ Treg-polarizing ability, as well as the ability to imprint gut-homing T cells; expression of the a4b7 integrin on conventional T cells and Treg cells involved in directing gut tropism ensures their ability to be imprinted [82, 83]. CD103+ DC subsets have also been shown to induce Th17 polarization and IgA class switching [84, 85]. Moreover, all DC subsets and antigen-presenting cells, including macrophages, are equipped with a battery of pattern recognition receptors (PRRs). These receptors can detect molecular patterns of invading microorganisms or endogenous "danger" signals and stimulate the immune response. PRRs are expressed on the cell surface and intracellularly are extremely diverse and capable of detecting a wide range of molecular species, including proteins, carbohydrates, lipids, and nucleic acids [86]. The Toll-like receptor (TLRs) family is the most intensely studied family of PRRs on DCs. Triggering TLRs on DCs is thought to be critical for their functional maturation to immunogenic DCs and for their ability to prime naive T cells in response to infection. Therefore, TLR activation couples innate and adaptive immunity [87]. TLR-mediated recognition of commensal microorganisms may also play important roles in tissue homeostasis, as recent studies have shown that TLR signaling by DCs was required to maintain immune homeostasis and tolerance to gut microbiota [88]. Interestingly, Tregs are also abundant at host-microbiota interfaces. Studies have suggested that commensal microbiota can stimulate the generation of Tregs and Th17 cells [89]. These results highlight the importance of diet and the microbiota in the establishment and configuration of the immune system of the intestinal mucosa. However, whether prebiotic compounds directly affect immune components or whether they act exclusively through the modulation of the endogenous intestinal microbiota remains unclear.

Indirect Mechanisms of Fructan Health Effects

Prebiotics and probiotics may have indirect immunomodulatory functions through their actions on nonimmune cells, such as epithelial cells. However, they may also exert immune system-independent effects by selectively stimulating the growth and/or activity of beneficial intestinal bacteria, such as *Lactobacillus* and *Bifidobacterium* species, which results in the restoration of the normal composition of the intestinal microbiota [90]. Mutualism between the host and its microbiota is fundamental for maintaining homeostasis in a healthy individual [91]. Commensal bacteria provide the host with essential nutrients. They also metabolize indigestible compounds, defend against the colonization of opportunistic pathogens, and contribute to the development of intestinal architecture in addition to stimulating the immune system [92]. In fact, intestinal immune and metabolic homeostasis in mammals is largely maintained by interactions between the gut microbiota and GALT [93]. The host actively engages the gut microbiota and controls its composition by secreting antimicrobial peptides and immunoglobulins. Conversely, commensals shape the gut-associated immune system by controlling the prevalence of distinct T cell populations [94]. *Bacteroides fragilis* protects mice from infection by *Helicobacter hepaticus* through several immunological mechanisms, including suppression of IL-17 production [95]. These commensals also express capsular zwitterionic polysaccharide A, which is a cognate antigen to effector $CD4^+$ T cells [92]. Other zwitterionic polysaccharides, such as type 1 capsule of *Streptococcus pneumoniae*, can also modify inflammatory responses in animal models by stimulating IL-10-producing $CD4^+$ T cells [96]. Moreover, bacterial symbionts, such as *Bacteroides*, *Barnesiella*, and *Turicibacter*, interact with $CD8^+$ cytotoxic T cells in the mucosal compartment of the small intestine and colon [97].

Other indirect pathways by which fructans exert immunomodulatory effects include the production of SCFAs, which are the fermentation products of fructans. Inulin fermentation increases the production of SCFAs (acetate, propionate, and butyrate), lactic acid, and hydrogen (H_2), while decreasing the pH of the colonic environment [36]. *Bifidobacterium* species are able to use some monosaccharides in a unique manner to ultimately generate SCFAs [98] and acidify the colonic environment. The increase in SCFAs antagonizes the growth of some pathogenic bacterial strains [99] and favors mucin production in the colon [100]. SCFAs bind to SCFAs receptors on GALT immune cells [101–103], activating G protein-coupled receptors (GPR) [104], such as GPR41 and GPR43 [101, 102, 104]. This binding affects the recruitment of leukocytes to inflammatory sites [105, 106] and suppresses the production of proinflammatory cytokines and chemokines [106–108]. GPR43 is highly expressed in polymorphonuclear cells (PMNs, i.e., neutrophils) and is lowly expressed in peripheral blood mononuclear cells (PBMCs) and purified monocytes. Conversely, GPR41 is expressed in PBMCs but not in PMNs, monocytes, or DCs [102]. Importantly, butyrate decreases the glutamine requirement for epithelial cells and alters epithelial cell gene expression [71, 109]. The mechanism for the indirect effect of fructans on the immune system is shown in Figure 3.

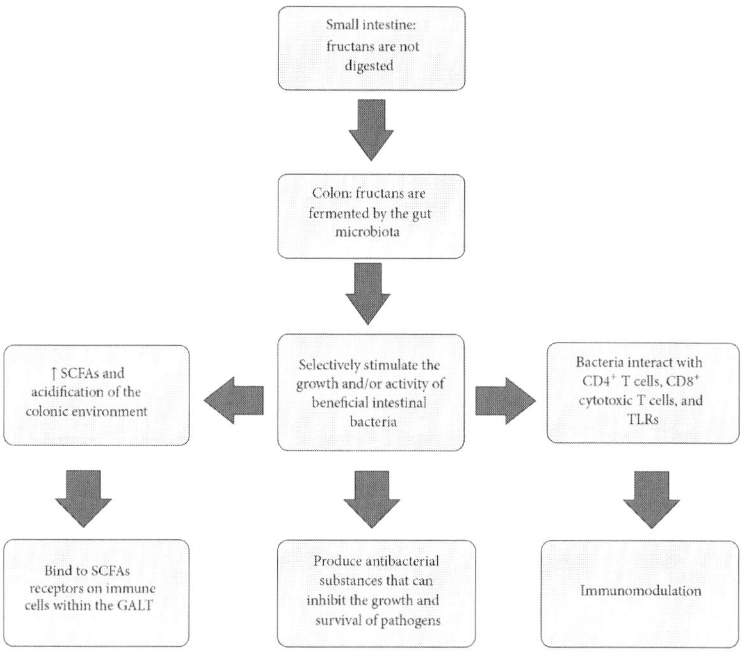

Figure 3. Mechanism for the indirect effect of fructans on the immune system.

Direct Mechanism: Pattern Recognition Receptors

In addition to the indirect effects of fructans and their fermentation products on the microbiota, the direct effects of fructans on the signaling of immune cells have gained attention as an additional pathway of immunomodulation. ITFs have been reported to interact directly with GALT components, such as gut dendritic cells (DCs) and intraepithelial lymphocytes (iIELs), through receptor ligation of PRRs [7]. Signaling through PRRs, such as TLRs (Toll-like receptors), is considered the starting point of innate immune system activation against various environmental factors, including microbes and antigens. The innate immune system enables appropriate adaptive immune responses to be generated through the activation of multiple specific immunocompetent clones [110]. TLRs play an important role in initial innate immune responses, which includes cytokine synthesis and activating acquired immunity. The β(2→1)-linked fructans can provide a direct signal to human immune cells primarily by activating TLR2 and to a lesser extent TLR4, TLR5, TLR7, TLR8, and NOD2. β(2→1)-linked fructans stimulation results in NF-κB/AP-1 activation, further suggesting that β(2→1)-fructans are specific ligands for TLR2. However, chain length is important for the induced activation pattern and IL-10/IL-12 ratios stimulated by β(2→1)-fructans [111, 112]. In fact, ITFs increase the proportion of DCs in PPs and increase the secretion of IL-2, IL-10, and interferon- from the spleen and MLNs. Additionally, ITFs reduce the number and proportion of T cell receptor (TCR-) αβ $^{+}$ CD8^{+} cells in the spleen and CD45RA^{+} cells in the MLNs [113] (Table 2). Furthermore, TLR4 appears to be involved in levan β(2→6)-fructans pattern recognition. Oral administration of levans *in vivo* significantly reduced IgE serum levels and Th2 response in mice immunized with ovalbumin [8].

Table 2. Effect of fructans on the immune function in healthy animal and human models.

Effects of fructans	Dose fructan/duration	Model	Reference
↑ DC and γδ T cells in lamina propria of the caecum and ↓ PGE2 in small intestine, colon, and caecum	3% FOS for 12 days	Mice treated with antibiotics and conventionalized with *Clostridium difficile*	[118]
In peripheral blood: ↑ CD4^{+}/CD8^{+} ratio and ↓ B cells. In GALT: ↑ proportion of CD4^{+} cells and CD8^{+} cells, PP, and lamina propria cells and ↓ CD4^{+}/CD8^{+} ratio in lamina propria	0.87% FOS for 14 days	Adult dogs	[119]
Synbiotics ↑ whole blood phagocyte activation level.	1% FOS for 28 days	Piglets infected with *S. typhimurium*	[120]
↑ counts of leucocytes, lymphocytes, neutrophils, CD2^{+} T cells, CD4^{+} T cells, CD8^{+} T cells, B cells, and macrophages in blood, ↑ % phagocytic activity of leucocytes and neutrophils in blood.	3 g/d OF for 20 days	Newborn piglets	[121]

↑ ileal IgA concentration.	2 g/d FOS and/or MOS for 14 days	Adult dogs	[122]
↓ blood neutrophils and ↑ blood lymphocytes.	2 g FOS plus/1 g MOS for 14 days	Adult dogs	[123]
↑ rotavirus-specific IgA levels in serum and ↓ duration of a strong rotavirus-specific IgA response in faeces and % IgA and IgG positive B cell in the PP. ↑ serum rotavirus-specific IgG and Rhesus rotavirus antigen concentration in stools.	1.25 g/L OF for 7 weeks	Mice (pups) infected with *Rhesus* rotavirus	[124]
No change in protein, alb, serum Ig, secreting IgA, and IL-4 and IFN- secretion, ↑ antibodies against influenza B and pneumococcus.	6 g OF/ITFs for 28 weeks	Healthy elderly (>70 years)	[125]
↑ % CD4 and CD8 lymphocytes, ↓ phagocytic activity in granulocytes and monocytes and IL-6 mRNA expression in PBMCs.	8 g/day FOS, 3 weeks	Nursing home elderly (77–97 years)	[126]
↑ total faecal IgA, size of PP, total IgA secretion by PP cells and IL-10 and IFN- production from PP $CD4^+$ T cells.	0–7.5% FOS for 6 weeks	Female mice	[127]
↓ leucocyte counts, ↑ NK activity of splenocytes and peritoneal macrophage phagocytosis of *Listeria monocytogenes*.	2.5–10% FOS or OF for 6 weeks	Female mice	[128]
↑ total number of immune cells in PP, B lymphocytes in PP and T lymphocytes and $CD4^+/CD8^+$ ratio in PP in endotoxemic mice only.	10% FOS for 16 days	Female mice healthy or endotoxemic	[129]
↓ peripheral blood lymphocyte concentration.	1% ITFs/MOS for 4 weeks	Senior dogs	[130]
↑ total intestinal IgA, ileal and colonic polymeric Ig receptor expression, ileal IgA secretion rate, IgA response of PP cells, and % of $B220^+$ IgA^+ cells.	5% FOS for 23–44 days	Newborn mice	[131]
↑ IL-10 and IFN- production in PP, secretory IgA concentration in ileum and caecum.	10% FOS-enriched ITFs for 4 weeks	Male rats	[132]
↑ NK activity. Prevention of the decrease in proportion of T cells with NK activity.	6 g/d OF and ITFs (2 : 1 ratio) for 1 year	Elderly free-living adults (age ≤ 70 years)	[133]
Improved response to some vaccine components and increased lymphocyte proliferation to influenza vaccine components.	4.95% FOS for 183 days	Healthy adults (age ≤ 65 years)	[134]

↑ T cells, MHCII on antigen-presenting cells in spleen, MLN, and thymus, IL-2 and IL-4 in blood.	10% FOS/ITFs for 4 months	Male rats	[135]
Trend towards higher fecal sIgA.	0.6 g (GOS/FOS)/100 mL formula for 32 weeks	Newborn non-breast-fed infants	[136]
Improved response to ↑ B cells, ↓ memory cytotoxic T cells, ↑ influenza-activated lymphocytes (CD69 and CD25) and IL-6 and ↓ IL10.	4.95% FOS for 4 weeks	Healthy adults (age ≤ 65 years)	[137]
In pregnant females and pups no effect on serum IgG1, IgG2, IgA, or IgM. In colostrum and milk ↑ IgM.	0.1% OF during lactation	Pregnant female dogs and pups	[138]
↓ severity of enterocyte sloughing.	1% FOS or ITFs for 14 days	Puppies	[139]
↑ % CD19 (B) cells, $CD3^+$ HLA-DR^+ (activated T cells) and ↓ % $ICAM^{-1}$ bearing lymphocytes and % $CD3^+$ NK^+ cells.	9 g/d ITFs for 5 weeks	Adults smokers and nonsmokers	[140]
↑ vaccine-specific faecal IgA and plasma IgG levels, peritoneal macrophage activity, mean fluorescence intensity of $MHCII^+$ cells in spleen, IL-12 and IFN- production by splenocytes, and survival from *Salmonella* infection when given vaccine.	5% mix (ITFs, FOS, and OF) for 1 week	Female mice	[141]
↑ fecal sIgA.	6 g/L GOS/FOS (9 : 1) for 26 weeks	Newborn healthy infants	[142]
↑ NK activity, and IL-10, ↓ IL-6, IL-1β, and TNF-α.	5.5 g GOS/d for 10 weeks	Elderly (64–79 years)	[143]
↑ DCs in PP, ↑ IL-2, IL-10, and IFN-δ from spleen and MNL cells. ↓ number and proportion of T cell receptor (TCR-) $\alpha\beta^+CD8^+$ cells in spleen and $CD45RA^+$ cells in MLN.	5% ITFs for 4 weeks	Female rats	[113]
↓ total IgE, IgG1, IgG2, and IgG3; ↓ cow's milk protein-specific IgG1.	8 g/L GOS/FOS for 6 months	Newborn infants at risk for allergy	[144]
↓ intestinal sIgA.	2.51–0.42 g/kg/d mix of GOS, XOS, OF, and ITFs (3.6 : 1 : 0.4 : 5) for 12 days	Female rats induced with diphenoxylate	[145]
↓ IL-1β in macrophage cultures and ↑ fecal IgA.	3–5% FOS for 30 days	Female mice	[146]

↓ LPS in blood and ↓ LPS-induced increases in gene expression in IL-1β and LPS-induced decreases in gene expression in IL-13 in blood.	5 g XOS, ITFs–XOS (3 : 1) for 4 weeks	Healthy volunteers	[147]
↓ serum cortisol, TNF-α and IL-6 after a LPS injection.	0.10% levan-type fructan for 42 days	Growing pigs	[63]
↑ fecal secretory IgA and ↓ fecal calprotectin and plasma C-reactive protein.	5.5 g/d B-GOS (Bi-^{2}muno) for 12 weeks	Overweight adults	[148]
↑ TGF-β secretion by splenocytes and IFN- production and ↓ IL-5.	GOS/ITFs (dose and duration data not shown)	Healthy mice	[149]
↓ CD16/56 on natural killer T cells and ↓ IL-10 secretion, XOS and Bi-07 supplementation ↓ CD19 on B cells.	8 g XOS or with 10^9 CFU Bi-07/d for 21 days	Healthy adults (25–65 years)	[150]
↑ cell-mediated immunity in terms of skin indurations and CD4$^+$ T-lymphocyte population.	20–60 g/kg FOS/ITFs for 12 weeks	Healthy rats	[151]

FOS: fructooligosaccharides; PGE2: prostaglandin E2; GALT: gut-associated lymphocyte tissue; CD: cluster of differentiation; PP: Peyer's patch; OF: oligofructose; MOS: mannanoligosaccharides; IgA: immunoglobulin A; IgG; immunoglobulin G; ITFs: inulin-type fructan; IL: interleukin; PMBCs: peripheral blood mononuclear cells; NK: natural killer cells; MHC II: major histocompatibility complex II; GOS: galactooligosaccharides; HLA: human leukocyte antigen; ICAM-1: intercellular adhesion molecule 1; IFN- : interferon gamma; DC: dendritic cell; TCR: T cell receptor; MLN: mesenteric lymph nodes; XO: xylooligosaccharides; LPS: lipopolysaccharides.

A fructose receptor may exist on immune cells, as receptors for β-glucan [114] and mannose [115] have been identified on the surface of immune cells. Oligofructose has also been shown to bind to receptors on pathogenic bacteria, preventing them from attaching to the epithelial membrane [116]. Furthermore, ITFs treatment of gut epithelial cells can modulate the innate immune barrier by modifying the integrity of epithelial tight junctions or by altering signals from the epithelial cells to the underlying immune cells [117].

Thirty-six fructan studies reporting immune outcomes have been conducted in mice, rats, pigs, dogs, and humans, and these investigations are summarized in Table 2. These reports show that fructans may have specific effects on different immune system components.

FRUCTANS ACT AS ROS SCAVENGERS

Because inulins and agavins have health benefits, improve blood metabolic parameters [41, 52], reduce colonic pH [152], increase SCFAs production [36, 43, 69], and stimulate the immune system [48], interest has developed in the antioxidant capacity of fructans. As in plants, fructans and other carbohydrates have been shown to scavenge ROS [153–157]. ROS include free radicals such as the superoxide anion ($O_2^{\bullet-}$), hydroxyl radical ($^{\bullet}OH$), and nonradical molecules such as hydrogen peroxide (H_2O_2) and singlet oxygen ($^1O_2$). These molecules attack DNA, lipids, and proteins resulting in cellular damage [158]. Fructans, galactooligosaccharides (GOS), arabinoxylans, β-glucans, and fructooligosaccharides (FOS) might act as ROS scavengers in plants [159] because they have strong antioxidant activity *in vitro*. Raffinose appears to be a moderate ROS scavenger [160].

Recently reports have suggested that fructans possess antioxidant activity in *in vivo* models. A putative role for oligofructoses in counteracting the prooxidative effects of a high fructose diet has been demonstrated in rats. The addition of fructans to the diet may provide an early defense against oxidative stress and may act before the activation of the endogenous ROS detoxification systems [65]. In an indirect mechanism, these nondigestible carbohydrates might serve as ROS scavengers, which suggests that inulin can protect the colonic mucosa by acting as a barrier against oxidative stress in addition to its positive prebiotic effect. This hypothesis is consistent with the recently proposed ROS scavenging capability of inulin [65, 161] and the reported effects of SCFAs, which induce the expression of crucial antioxidant enzymes, such as glutathione S-transferases (GSTs) [162]. Li et al. showed that, in aged mice, synthetic oligosaccharides increase the activity of antioxidant enzymes [161]. By contrast, oligofructose has been shown to reduce the expression of NADPH oxidase in the colons of obese mice [51]. Moreover, intraperitoneal administration of synthetic oligosaccharides stimulates a dose-dependent decrease in lipid peroxidation, which supports the *in vivo* ROS scavenging capability of certain sugars [161]. Furthermore, agavins from *Agave tequilana* have been shown to improve the redox status in hypercholesterolemic mice by reducing malondialdehyde serum levels and oxidative protein damage. These results could be attributed to a reduction in the generation of oxidative products during digestion and colonic fermentation [42]. Additionally, polyphenol studies have indicated that metabolism in the large intestine is positively affected by prebiotic fructooligosaccharides, which have a synergistic effect with polyphenol to counteract oxidative stress in *in vivo* models [163].

CONCLUSION

Prebiotic consumption is undoubtedly associated with several health benefits. In this review, we assessed the potential immunomodulatory and antioxidants mechanisms of the prebiotic fructans as well as the impact of fructans on immune health. Some preliminary data have convincingly suggested that fructan consumption can modulate immune parameters in GALT. Additionally, fructans may act as ROS scavengers providing an increase in antioxidant defenses partially through the activation of endogenous ROS detoxification systems. Further studies will be required to fully understand and elucidate the mechanisms of action for fructans on GALT in various disease models.

REFERENCES

1. Lopez M. G., Mancilla-Margalli N. A., Mendoza-Diaz G. Molecular structures of fructans from *Agave tequilana* Weber var. *azul* . *Journal of Agricultural and Food Chemistry*. 2003;51(27):7835–7840. doi: 10.1021/jf030383v.
2. van Loo J., Coussement P., de Leenheer L., Hoebregs H., Smits G. On the presence of inulin and oligofructose as natural ingredients in the western diet. *Critical Reviews in Food Science and Nutrition*. 1995;35(6):525–552. doi: 10.1080/10408399509527714.
3. Pollock C. J., Cairns A. J. Fructan metabolism in grasses and cereals. *Annual Review of Plant Physiology and Plant Molecular Biology*. 1991;42(1):77–101. doi: 10.1146/annurev.pp.42.060191.000453.
4. Kelly G. Inulin-type prebiotics—a review: part 1. *Alternative Medicine Review*. 2008;13(4):315–329.
5. Mancilla-Margalli N. A., Lopez M. G. Water-soluble carbohydrates and fructan structure patterns from *Agave* and *Dasylirion* species. *Journal of Agricultural and Food Chemistry*. 2006;54(20):7832–7839. doi: 10.1021/jf060354v.
6. Tsai C.-C., Lin C.-R., Tsai H.-Y., et al. The immunologically active oligosaccharides isolated from wheatgrass modulate monocytes via toll-like receptor-2 signaling. *The Journal of Biological Chemistry*. 2013;288(24):17689–17697. doi: 10.1074/jbc.m112.448381.
7. Vogt L., Meyer D., Pullens G., et al. Immunological properties of inulin-type fructans. *Critical Reviews in Food Science and Nutrition*. 2014;55(3):414–436. doi: 10.1080/10408398.2012.656772.
8. Xu Q., Yajima T., Li W., Saito K., Ohshima Y., Yoshikai Y. Levan (β-2, 6-fructan), a major fraction of fermented soybean mucilage, displays immunostimulating properties via Toll-like receptor 4 signalling: induction of interleukin-12 production and suppression of T-helper type 2 response and immunoglobulin E production. *Clinical & Experimental Allergy*. 2006;36(1):94–101. doi: 10.1111/j.1365-2222.2006.02401.x.
9. van den Ende W., Peshev D., de Gara L. Disease prevention by natural antioxidants and prebiotics acting as ROS scavengers in the gastrointestinal tract. *Trends in Food Science and Technology*. 2011;22(12):689–697. doi: 10.1016/j.tifs.2011.07.005.
10. Hendry G. A. F. Evolutionary origins and natural functions of fructans—a climatological, biogeographic and mechanistic appraisal.

New Phytologist. 1993;123(1):3–14.

11. Cardelle-Cobas A., Corzo N., Olano A., Peláez C., Requena T., Ávila M. Galactooligosaccharides derived from lactose and lactulose: influence of structure on *Lactobacillus*, *Streptococcus* and *Bifidobacterium* growth. *International Journal of Food Microbiology*. 2011;149(1):81–87. doi: 10.1016/j.ijfoodmicro.2011.05.026.
12. Iniguez-Covarrubias G., Díaz-Teres R., Sanjuan-Duenas R., Anzaldo-Hernández J., Rowell R. M. Utilization of by-products from the tequila industry. Part 2: potential value of *Agave tequilana* Weber azul leaves. *Bioresource Technology*. 2001;77(2):101–108. doi: 10.1016/s0960-8524(00)00167-x.
13. Vijn I., Smeekens S. Fructan: more than a reserve carbohydrate? *Plant Physiology*. 1999;120(2):351–359. doi: 10.1104/pp.120.2.351.
14. Ritsema T., Smeekens S. Fructans: beneficial for plants and humans. *Current Opinion in Plant Biology*. 2003;6(3):223–230. doi: 10.1016/s1369-5266(03)00034-7.
15. Kaur N., Gupta A. K. Applications of inulin and oligofructose in health and nutrition. *Journal of Biosciences*. 2002;27(7):703–714. doi: 10.1007/BF02708379.
16. Chatterton N. J., Harrison P. A. Fructan oligomers in *Poa ampla* . *New Phytologist*. 1997;136(1):3–10. doi: 10.1046/j.1469-8137.1997.00716.x.
17. Wei J.-Z., Chatterton N. J., Harrison P. A., Wang R. R.-C., Larson S. R. Characterization of fructan biosynthesis in big bluegrass (*Poa secunda*) *Journal of Plant Physiology*. 2002;159(7):705–715. doi: 10.1078/0176-1617-0781.
18. Livingston D. P., Chatterton N. J., Harrison P. A. Structure and quantity of fructan oligomers in oat (*Avena* spp.) *New Phytologist*. 1993;123(4):725–734. doi: 10.1111/j.1469-8137.1993.tb03783.x.
19. Sims I. M., Pollock C. J., Horgan R. Structural analysis of oligomeric fructans from excised leaves of *Lolium temulentum* . *Phytochemistry*. 1992;31(9):2989–2992. doi: 10.1016/0031-9422(92)83432-x.
20. Pavis N., Chatterton N. J., Harrison P. A., et al. Structure of fructans in roots and leaf tissues of *Lolium perenne* . *New Phytologist*. 2001;150(1):83–95. doi: 10.1046/j.1469-8137.2001.00069.x.
21. Shiomi N. Properties of fructosyltransferases involved in the synthesis of fructan in liliaceous plants. *Journal of Plant Physiology*.

1989;134(2):151–155. doi: 10.1016/s0176-1617(89)80048-3.

22. Carabin I. G., Gary Flamm W. Evaluation of safety of inulin and oligofructose as dietary fiber. *Regulatory Toxicology and Pharmacology*. 1999;30(3):268–282. doi: 10.1006/rtph.1999.1349.
23. Roberfroid M. B. Concepts in functional foods: the case of inulin and oligofructose. *Journal of Nutrition*. 1999;129(7):1398S–1401s.
24. Roberfroid M. B. Inulin-type fructans: functional food ingredients. *Journal of Nutrition*. 2007;137(11, supplement):2493S–2502S.
25. Roberfroid M. B. Prebiotics: preferential substrates for specific germs? *American Journal of Clinical Nutrition*. 2001;73(2, supplement):406S–409S.
26. Sanders M. E., Guarner F., Guerrant R., et al. An update on the use and investigation of probiotics in health and disease. *Gut*. 2013;62(5):787–796. doi: 10.1136/gutjnl-2012-302504.
27. Kolida S., Gibson G. R. Prebiotic capacity of inulin-type fructans. *Journal of Nutrition*. 2007;137(11, supplement):2503S–2506S.
28. Al-Sheraji S. H., Ismail A., Manap M. Y., Mustafa S., Yusof R. M., Hassan F. A. Prebiotics as functional foods: a review. *Journal of Functional Foods*. 2013;5(4):1542–1553. doi: 10.1016/j.jff.2013.08.009.
29. Gibson G. R., Probert H. M., Van Loo J., Rastall R. A., Roberfroid M. B. Dietary modulation of the human colonic microbiota: updating the concept of prebiotics. *Nutrition Research Reviews*. 2004;17(2):259–275. doi: 10.1079/nrr200479.
30. Menne E., Guggenbuhl N., Roberfroid M. Fn-type chicory inulin hydrolysate has a prebiotic effect in humans. *Journal of Nutrition*. 2000;130(5):1197–1199.
31. Bouhnik Y., Flourié B., D'Agay-Abensour L., et al. Administration of transgalacto-oligosaccharides increases fecal bifidobacteria and modifies colonic fermentation metabolism in healthy humans. *Journal of Nutrition*. 1997;127(3):444–448.
32. Ito M., Deguchi Y., Matsumoto K., Kimura M., Onodera N., Yajima T. Influence of galactooligosaccharides on the human fecal microflora. *Journal of Nutritional Science and Vitaminology*. 1993;39(6):635–640. doi: 10.3177/jnsv.39.635.
33. Cani P. D., Bibiloni R., Knauf C., et al. Changes in gut microbiota control metabolic endotoxemia-induced inflammation in high-fat diet-induced obesity and diabetes in mice. *Diabetes*. 2008;57(6):1470–

1481. doi: 10.2337/db07-1403.

34. Delzenne N. M., Neyrinck A. M., Cani P. D. Modulation of the gut microbiota by nutrients with prebiotic properties: consequences for host health in the context of obesity and metabolic syndrome. *Microbial Cell Factories*. 2011;10(supplement 1, article S10) doi: 10.1186/1475-2859-10-s1-s10.
35. Di Bartolomeo F., Startek J. B., van den Ende W. Prebiotics to fight diseases: reality or fiction? *Phytotherapy Research*. 2013;27(10):1457–1473. doi: 10.1002/ptr.4901.
36. Tarini J., Wolever T. M. S. The fermentable fibre inulin increases postprandial serum short-chain fatty acids and reduces free-fatty acids and ghrelin in healthy subjects. *Applied Physiology, Nutrition and Metabolism*. 2010;35(1):9–16. doi: 10.1139/H09-119.
37. Rumessen J. J., Bodé S., Hamberg O., Gudmand-Høyer E. Fructans of Jerusalem artichokes: intestinal transport, absorption, fermentation, and influence on blood glucose, insulin, and C-peptide responses in healthy subjects. *American Journal of Clinical Nutrition*. 1990;52(4):675–681.
38. van de Wiele T., Boon N., Possemiers S., Jacobs H., Verstraete W. Inulin-type fructans of longer degree of polymerization exert more pronounced *in vitro* prebiotic effects. *Journal of Applied Microbiology*. 2007;102(2):452–460. doi: 10.1111/j.1365-2672.2006.03084.x.
39. Dai D., Nanthkumar N. N., Newburg D. S., Walker W. A. Role of oligosaccharides and glycoconjugates in intestinal host defense. *Journal of Pediatric Gastroenterology and Nutrition*. 2000;30(supplement 2):S23–S33. doi: 10.1097/00005176-200003002-00005.
40. Ortega-González M., Sánchez De Medina F., Molina-Santiago C., et al. Fructooligosacharides reduce *Pseudomonas aeruginosa* PAO1 pathogenicity through distinct mechanisms. *PLoS ONE*. 2014;9(1) doi: 10.1371/journal.pone.0085772.e85772
41. Urías-Silvas J. E., Cani P. D., Delmée E., Neyrinck A., López M. G., Delzenne N. M. Physiological effects of dietary fructans extracted from *Agave tequilana* Gto. and *Dasylirion* spp. *British Journal of Nutrition*. 2008;99(2):254–261.
42. Sáyago-Ayerdi S. G., Mateos R., Ortiz-Basurto R. I., et al. Effects of consuming diets containing *Agave tequilana* dietary fibre and jamaica calyces on body weight gain and redox status in hypercholesterolemic rats. *Food Chemistry*. 2014;148:54–59. doi:

10.1016/j.foodchem.2013.10.004.

43. Santiago-García P. A., López M. G. Agavins from *Agave angustifolia* and *Agave potatorum* affect food intake, body weight gain and satiety-related hormones (GLP-1 and ghrelin) in mice. *Food & Function*. 2014;5(12):3311–3319. doi: 10.1039/c4fo00561a.

44. McBurney M. I., Van Soest P. J., Jeraci J. L. Colonic carcinogenesis: the microbial feast or famine mechanism. *Nutrition and Cancer*. 1987;10(1-2):23–28. doi: 10.1080/01635588709513937.

45. De Preter V., Vanhoutte T., Huys G., et al. Effects of Lactobacillus casei Shirota, Bifidobacterium breve, and oligofructose-enriched inulin on colonic nitrogen-protein metabolism in healthy humans. *The American Journal of Physiology—Gastrointestinal and Liver Physiology*. 2007;292(1):G358–G368. doi: 10.1152/ajpgi.00052.2006.

46. Gomez E., Tuohy K. M., Gibson G. R., Klinder A., Costabile A. *In vitro* evaluation of the fermentation properties and potential prebiotic activity of Agave fructans. *Journal of Applied Microbiology*. 2010;108(6):2114–2121. doi: 10.1111/j.1365-2672.2009.04617.x.

47. Holscher H. D., Doligale J. L., Bauer L. L., et al. Gastrointestinal tolerance and utilization of agave inulin by healthy adults. *Food & Function*. 2014;5(6):1142–1149. doi: 10.1039/c3fo60666j.

48. Lomax A. R., Calder P. C. Prebiotics, immune function, infection and inflammation: a review of the evidence. *British Journal of Nutrition*. 2009;101(5):633–658. doi: 10.1017/s0007114508055608.

49. Yamashita K., Kawai K., Itakura M. Effects of fructo-oligosaccharides on blood glucose and serum lipids in diabetic subjects. *Nutrition Research*. 1984;4(6):961–966. doi: 10.1016/s0271-5317(84)80075-5.

50. Cani P. D., Knauf C., Iglesias M. A., Drucker D. J., Delzenne N. M., Burcelin R. Improvement of glucose tolerance and hepatic insulin sensitivity by oligofructose requires a functional glucagon-like peptide 1 receptor. *Diabetes*. 2006;55(5):1484–1490. doi: 10.2337/db05-1360.

51. Everard A., Lazarevic V., Derrien M., et al. Responses of gut microbiota and glucose and lipid metabolism to prebiotics in genetic obese and diet-induced leptin-resistant mice. *Diabetes*. 2011;60(11):2775–2786. doi: 10.2337/db11-0227.

52. Brighenti F. Dietary fructans and serum triacylglycerols: a meta-analysis of randomized controlled trials. *Journal of Nutrition*. 2007;137(11, supplement):2552S–2556S.

53. Daubioul C., Rousseau N., Demeure R., et al. Dietary fructans, but not cellulose, decrease triglyceride accumulation in the liver of obese Zucker fa/fa rats. *Journal of Nutrition*. 2002;132(5):967–973.
54. Reimer R. A., Russell J. C. Glucose tolerance, lipids, and GLP-1 secretion in JCR:LA-cp rats fed a high protein fiber diet. *Obesity*. 2008;16(1):40–46. doi: 10.1038/oby.2007.16.
55. Sakaguchi E., Sakoda C., Toramaru Y. Caecal fermentation and energy accumulation in the rat fed on indigestible oligosaccharides. *British Journal of Nutrition*. 1998;80(5):469–476.
56. Daubioul C. A., Taper H. S., de Wispelaere L. D., Delzenne N. M. Dietary oligofructose lessens hepatic steatosis, but does not prevent hypertriglyceridemia in obese Zucker rats. *Journal of Nutrition*. 2000;130(5):1314–1319.
57. Delmée E., Cani P. D., Gual G., et al. Relation between colonic proglucagon expression and metabolic response to oligofructose in high fat diet-fed mice. *Life Sciences*. 2006;79(10):1007–1013. doi: 10.1016/j.lfs.2006.05.013.
58. Jamieson J. A., Ryz N. R., Taylor C. G., Weiler H. A. Dietary long-chain inulin reduces abdominal fat but has no effect on bone density in growing female rats. *British Journal of Nutrition*. 2008;100(2):451–459.
59. Cani P. D., Possemiers S., van de Wiele T., et al. Changes in gut microbiota control inflammation in obese mice through a mechanism involving GLP-2-driven improvement of gut permeability. *Gut*. 2009;58(8):1091–1103. doi: 10.1136/gut.2008.165886.
60. Kleessen B., Schwarz S., Boehm A., et al. Jerusalem artichoke and chicory inulin in bakery products affect faecal microbiota of healthy volunteers. *British Journal of Nutrition*. 2007;98(3):540–549. doi: 10.1017/s0007114507730751.
61. Wang S., Zhu H., Lu C., et al. Fermented milk supplemented with probiotics and prebiotics can effectively alter the intestinal microbiota and immunity of host animals. *Journal of Dairy Science*. 2012;95(9):4813–4822. doi: 10.3168/jds.2012-5426.
62. Hidaka H., Tashiro Y., Eida T. Proliferation of bifidobacteria by oligosaccharides and their useful effect on human health. *Bifidobacteria and Microflora*. 1991;10(1):65–79. doi: 10.12938/bifidus1982.10.1_65.
63. Li J., Kim I. H. Effects of levan-type fructan supplementation on

growth performance, digestibility, blood profile, fecal microbiota, and immune responses after lipopolysaccharide challenge in growing pigs. *Journal of Animal Science*. 2013;91(11):5336–5343. doi: 10.2527/jas.2013-6665.

64. Zduńczyk Z., Juśkiewicz J., Estrella I. Cecal parameters of rats fed diets containing grapefruit polyphenols and inulin as single supplements or in a combination. *Nutrition*. 2006;22(9):898–904. doi: 10.1016/j.nut.2006.05.010.

65. Busserolles J., Gueux E., Rock E., Demigné C., Mazur A., Rayssiguier Y. Oligofructose protects against the hypertriglyceridemic and pro-oxidative effects of a high fructose diet in rats. *Journal of Nutrition*. 2003;133(6):1903–1908.

66. van den Heuvel E. G. H. M., Muys T., van Dokkum W., Schaafsma G. Oligofructose stimulates calcium absorption in adolescents. *The American Journal of Clinical Nutrition*. 1999;69(3):544–548.

67. Younes H., Coudray C., Bellanger J., Demigné C., Rayssiguier Y., Rémésy C. Effects of two fermentable carbohydrates (inulin and resistant starch) and their combination on calcium and magnesium balance in rats. *British Journal of Nutrition*. 2001;86(4):479–485. doi: 10.1079/BJN2001430.

68. Scholz-Ahrens K. E., Ade P., Marten B., et al. Prebiotics, probiotics, and synbiotics affect mineral absorption, bone mineral content, and bone structure. *Journal of Nutrition*. 2007;137(3, supplement 2):838S–846S.

69. García-Vieyra M. I., del Real A., López M. G. Agave fructans: their effect on mineral absorption and bone mineral content. *Journal of Medicinal Food*. 2014;17(11):1247–1255. doi: 10.1089/jmf.2013.0137.

70. Bodera P. Influence of prebiotics on the human immune system (GALT) *Recent Patents on Inflammation and Allergy Drug Discovery*. 2008;2(2):149–153. doi: 10.2174/187221308784543656.

71. Sanderson I. R. Dietary modulation of GALT. *Journal of Nutrition*. 2007;137(supplement 11):2557S–2562S.

72. Mayer L. Mucosal immunity. *Pediatrics*. 2003;111(6, part 3):1595–1600.

73. MacDonald T. T., Monteleone G. Immunity, inflammation, and allergy in the gut. *Science*. 2005;307(5717):1920–1925. doi: 10.1126/science.1106442.

74. Mowat A. M. Anatomical basis of tolerance and immunity to intestinal antigens. *Nature Reviews Immunology*. 2003;3(4):331–341. doi: 10.1038/nri1057.

75. Ramiro-Puig E., Pérez-Cano F. J., Castellote C., Franch A., Castell M. The bowel: a key component of the immune system. *Revista Espanola de Enfermedades Digestivas*. 2008;100(1):29–34.

76. Rescigno M., Urbano M., Valzasina B., et al. Dendritic cells express tight junction proteins and penetrate gut epithelial monolayers to sample bacteria. *Nature Immunology*. 2001;2(4):361–367. doi: 10.1038/86373.

77. Jabri B., Ebert E. Human CD8$^+$ intraepithelial lymphocytes: a unique model to study the regulation of effector cytotoxic T lymphocytes in tissue. *Immunological Reviews*. 2007;215(1):202–214. doi: 10.1111/j.1600-065x.2006.00481.x.

78. Niess J. H., Brand S., Gu X., et al. CX3CR1-mediated dendritic cell access to the intestinal lumen and bacterial clearance. *Science*. 2005;307(5707):254–258. doi: 10.1126/science.1102901.

79. Cheroutre H. IELs: enforcing law and order in the court of the intestinal epithelium. *Immunological Reviews*. 2005;206(1):114–131. doi: 10.1111/j.0105-2896.2005.00284.x.

80. Steinman R. M. Decisions about dendritic cells: past, present, and future. *Annual Review of Immunology*. 2012;30(1):1–22. doi: 10.1146/annurev-immunol-100311-102839.

81. Shiu J., Blanchard T. G. Dendritic cell function in the host response to *Helicobacter pylori* infection of the gastric mucosa. *Pathogens and Disease*. 2013;67(1):46–53. doi: 10.1111/2049-632x.12014.

82. Coombes J. L., Powrie F. Dendritic cells in intestinal immune regulation. *Nature Reviews Immunology*. 2008;8(6):435–446. doi: 10.1038/nri2335.

83. Sun C. M., Hall J. A., Blank R. B., et al. Small intestine lamina propria dendritic cells promote de novo generation of Foxp3 T reg cells via retinoic acid. *Journal of Experimental Medicine*. 2007;204(8):1775–1785. doi: 10.1084/jem.20070602.

84. Uematsu S., Fujimoto K., Jang M. H., et al. Regulation of humoral and cellular gut immunity by lamina propria dendritic cells expressing Toll-like receptor 5. *Nature Immunology*. 2008;9(7):769–776. doi: 10.1038/ni.1622.

85. Denning T. L., Norris B. A., Medina-Contreras O., et al. Functional

specializations of intestinal dendritic cell and macrophage subsets that control Th17 and regulatory T cell responses are dependent on the T cell/APC ratio, source of mouse strain, and regional localization. *The Journal of Immunology*. 2011;187(2):733–747. doi: 10.4049/jimmunol.1002701.

86. Takeuchi O., Akira S. Pattern recognition receptors and inflammation. *Cell*. 2010;140(6):805–820. doi: 10.1016/j.cell.2010.01.022.
87. Dalod M., Chelbi R., Malissen B., Lawrence T. Dendritic cell maturation: functional specialization through signaling specificity and transcriptional programming. *The EMBO Journal*. 2014;33(10):1104–1116. doi: 10.1002/embj.201488027.
88. Han D., Walsh M. C., Cejas P. J., et al. Dendritic cell expression of the signaling molecule TRAF6 is critical for gut microbiota-dependent immune tolerance. *Immunity*. 2013;38(6):1211–1222. doi: 10.1016/j.immuni.2013.05.012.
89. Chen X., Oppenheim J. J. Th17 cells and Tregs: unlikely allies. *Journal of Leukocyte Biology*. 2014;95(5):723–731. doi: 10.1189/jlb.1213633.
90. Reid G., Younes J. A., van der Mei H. C., Gloor G. B., Knight R., Busscher H. J. Microbiota restoration: natural and supplemented recovery of human microbial communities. *Nature Reviews Microbiology*. 2011;9(1):27–38. doi: 10.1038/nrmicro2473.
91. Leser T. D., Mølbak L. Better living through microbial action: the benefits of the mammalian gastrointestinal microbiota on the host. *Environmental Microbiology*. 2009;11(9):2194–2206. doi: 10.1111/j.1462-2920.2009.01941.x.
92. Mazmanian S. K., Liu C. H., Tzianabos A. O., Kasper D. L. An immunomodulatory molecule of symbiotic bacteria directs maturation of the host immune system. *Cell*. 2005;122(1):107–118. doi: 10.1016/j.cell.2005.05.007.
93. Garrett W. S., Gordon J. I., Glimcher L. H. Homeostasis and inflammation in the intestine. *Cell*. 2010;140(6):859–870. doi: 10.1016/j.cell.2010.01.023.
94. Ivanov I. I., Honda K. Intestinal commensal microbes as immune modulators. *Cell Host and Microbe*. 2012;12(4):496–508. doi: 10.1016/j.chom.2012.09.009.
95. Mazmanian S. K., Round J. L., Kasper D. L. A microbial symbiosis factor prevents intestinal inflammatory disease. *Nature*. 2008;453(7195):620–

625. doi: 10.1038/nature07008.

96. Ruiz-Perez B., Chung D. R., Sharpe A. H., et al. Modulation of surgical fibrosis by microbial zwitterionic polysaccharides. *Proceedings of the National Academy of Sciences of the United States of America*. 2005;102(46):16753–16758. doi: 10.1073/pnas.0505688102.
97. Presley L. L., Wei B., Braun J., Borneman J. Bacteria associated with immunoregulatory cells in mice. *Environmental Microbiology*. 2010;76(3):936–941. doi: 10.1128/AEM.01561-09.
98. Pokusaeva K., Fitzgerald G. F., van Sinderen D. Carbohydrate metabolism in bifidobacteria. *Genes & Nutrition*. 2011;6(3):285–306. doi: 10.1007/s12263-010-0206-6.
99. Blaut M. Relationship of prebiotics and food to intestinal microflora. *European Journal of Nutrition*. 2002;41(1):11–16.
100. Barcelo A., Claustre J., Moro F., Chayvialle J.-A., Cuber J.-C., Plaisancié P. Mucin secretion is modulated by luminal factors in the isolated vascularly perfused rat colon. *Gut*. 2000;46(2):218–224. doi: 10.1136/gut.46.2.218.
101. Brown A. J., Goldsworthy S. M., Barnes A. A., et al. The orphan G protein-coupled receptors GPR41 and GPR43 are activated by propionate and other short chain carboxylic acids. *The Journal of Biological Chemistry*. 2003;278(13):11312–11319. doi: 10.1074/jbc.m211609200.
102. Le Poul E., Loison C., Struyf S., et al. Functional characterization of human receptors for short chain fatty acids and their role in polymorphonuclear cell activation. *The Journal of Biological Chemistry*. 2003;278(28):25481–25489. doi: 10.1074/jbc.m301403200.
103. Nilsson N. E., Kotarsky K., Owman C., Olde B. Identification of a free fatty acid receptor, FFA2R, expressed on leukocytes and activated by short-chain fatty acids. *Biochemical and Biophysical Research Communications*. 2003;303(4):1047–1052. doi: 10.1016/S0006-291X(03)00488-1.
104. Covington D. K., Briscoe C. A., Brown A. J., Jayawickreme C. K. The G-protein-coupled receptor 40 family (GPR40-GPR43) and its role in nutrient sensing. *Biochemical Society Transactions*. 2006;34(5):770–773. doi: 10.1042/bst0340770.
105. Maslowski K. M., Vieira A. T., Ng A., et al. Regulation of inflammatory responses by gut microbiota and chemoattractant receptor GPR43.

Nature. 2009;461(7268):1282–1286. doi: 10.1038/nature08530.

106. Vinolo M. A. R., Rodrigues H. G., Hatanaka E., Hebeda C. B., Farsky S. H. P., Curi R. Short-chain fatty acids stimulate the migration of neutrophils to inflammatory sites. *Clinical Science (Lond)* 2009;117(9):331–338. doi: 10.1042/cs20080642.
107. Park J.-S., Lee E.-J., Lee J.-C., Kim W.-K., Kim H.-S. Anti-inflammatory effects of short chain fatty acids in IFN- -stimulated RAW 264.7 murine macrophage cells: involvement of NF-κB and ERK signaling pathways. *International Immunopharmacology*. 2007;7(1):70–77. doi: 10.1016/j.intimp.2006.08.015.
108. Vinolo M. A. R., Rodrigues H. G., Hatanaka E., Sato F. T., Sampaio S. C., Curi R. Suppressive effect of short-chain fatty acids on production of proinflammatory mediators by neutrophils. *Journal of Nutritional Biochemistry*. 2011;22(9):849–855. doi: 10.1016/j.jnutbio.2010.07.009.
109. Jenkins D. J. A., Kendall C. W. C., Vuksan V. Inulin, oligofructose and intestinal function. *Journal of Nutrition*. 1999;129(7):1431S–1433S.
110. Sanz Y., Nadal I., Sánchez E. Probiotics as drugs against human gastrointestinal infections. *Recent Patents on Anti-Infective Drug Discovery*. 2007;2(2):148–156. doi: 10.2174/157489107780832596.
111. Vogt L., Ramasamy U., Meyer D., et al. Immune modulation by different types of beta2→1-fructans is toll-like receptor dependent. *PLoS ONE*. 2013;8(7) doi: 10.1371/journal.pone.0068367.e68367
112. Vogt L. M., Meyer D., Pullens G., et al. Toll-like receptor 2 activation by β2→1-fructans protects barrier function of T84 human intestinal epithelial cells in a chain length-dependent manner. *Journal of Nutrition*. 2014;144(7):1002–1008. doi: 10.3945/jn.114.191643.
113. Ryz N. R., Meddings J. B., Taylor C. G. Long-chain inulin increases dendritic cells in the Peyer's patches and increases ex vivo cytokine secretion in the spleen and mesenteric lymph nodes of growing female rats, independent of zinc status. *British Journal of Nutrition*. 2009;101(11):1653–1663. doi: 10.1017/S000711450812342X.
114. Herre J., Gordon S., Brown G. D. Dectin-1 and its role in the recognition of β-glucans by macrophages. *Molecular Immunology*. 2004;40(12):869–876. doi: 10.1016/j.molimm.2003.10.007.
115. Taylor M. E., Conary J. T., Lennartz M. R., Stahl P. D., Drickamer K. Primary structure of the mannose receptor contains multiple

motifs resembling carbohydrate-recognition domains. *The Journal of Biological Chemistry*. 1990;265(21):12156–12162.

116. Ouwehand A. C., Derrien M., de Vos W., Tiihonen K., Rautonen N. Prebiotics and other microbial substrates for gut functionality. *Current Opinion in Biotechnology*. 2005;16(2):212–217. doi: 10.1016/j.copbio.2005.01.007.
117. Cummings J. H., Macfarlane G. T., Englyst H. N. Prebiotic digestion and fermentation. *The American Journal of Clinical Nutrition*. 2001;73(2, supplement):415S–420S.
118. Gaskins H. R., Mackie R. I., May T., Garleb K. A. Dietary fructo-oligosaccharide modulates large intestinal inflammatory responses to *Clostridium difficile* in antibiotic-compromised mice. *Microbial Ecology in Health & Disease*. 1996;9(4):157–166. doi: 10.3402/mehd.v9i4.8365.
119. Field C. J., McBurney M. I., Massimino S., Hayek M. G., Sunvold G. D. The fermentable fiber content of the diet alters the function and composition of canine gut associated lymphoid tissue. *Veterinary Immunology and Immunopathology*. 1999;72(3-4):325–341. doi: 10.1016/S0165-2427(99)00148-8.
120. Letellier A., Messier S., Lessard L., Chénier S., Quessy S. Host response to various treatments to reduce salmonella infections in swine. *Canadian Journal of Veterinary Research*. 2001;65(3):168–172.
121. Herich R., Révajová V., Levkut M., et al. The effect of *Lactobacillus paracasei* and Raftilose P95 upon the non-specific immune response of piglets. *Food and Agricultural Immunology*. 2002;14:171–179.
122. Swanson K. S., Grieshop C. M., Flickinger E. A., et al. Supplemental fructooligosaccharides and mannanoligosaccharides influence immune function, ileal and total tract nutrient digestibilities, microbial populations and concentrations of protein catabolites in the large bowel of dogs. *Journal of Nutrition*. 2002;132(5):980–989.
123. Swanson K. S., Grieshop C. M., Flickinger E. A., et al. Effects of supplemental fructooligosaccharides plus mannanoligosaccharides on immune function and ileal and fecal microbial populations in adult dogs. *Archiv für Tierernährung*. 2002;56(4):309–318. doi: 10.1080/00039420214344.
124. Qiao H., Duffy L. C., Griffiths E., et al. Immune responses in rhesus rotavirus-challenged Balb/c mice treated with bifidobacteria and

prebiotic supplements. *Pediatric Research*. 2002;51(6):750–755. doi: 10.1203/01.pdr.0000017481.64723.1f.

125. Bunout D., Hirsch S., de la Maza M. P., et al. Effects of prebiotics on the immune response to vaccination in the elderly. *Journal of Parenteral and Enteral Nutrition*. 2002;26(6):372–376. doi: 10.1177/0148607102026006372.
126. Guigoz Y., Rochat F., Perruisseau-Carrier G., Rochat I., Schiffrin E. J. Effects of oligosaccharide on the faecal flora and non-specific immune system in elderly people. *Nutrition Research*. 2002;22(1-2):13–25. doi: 10.1016/S0271-5317(01)00354-2.
127. Hosono A., Ozawa A., Kato R., et al. Dietary fructooligosaccharides induce immunoregulation of intestinal IgA secretion by murine peyer's patch cells. *Bioscience, Biotechnology and Biochemistry*. 2003;67(4):758–764. doi: 10.1271/bbb.67.758.
128. Kelly-Quagliana K. A., Nelson P. D., Buddington R. K. Dietary oligofructose and inulin modulate immune functions in mice. *Nutrition Research*. 2003;23(2):257–267. doi: 10.1016/S0271-5317(02)00516-X.
129. Manhart N., Spittler A., Bergmeister H., Mittlböck M., Roth E. Influence of fructooligosaccharides on Peyer's patch lymphocyte numbers in healthy and endotoxemic mice. *Nutrition*. 2003;19(7-8):657–660. doi: 10.1016/s0899-9007(03)00059-5.
130. Grieshop C. M., Flickinger E. A., Bruce K. J., Patil A. R., Czarnecki-Maulden G. L., Fahey G. C., Jr. Gastrointestinal and immunological responses of senior dogs to chicory and mannan-oligosaccharides. *Archives of Animal Nutrition*. 2004;58(6):483–493. doi: 10.1080/00039420400019977.
131. Nakamura Y., Nosaka S., Suzuki M., et al. Dietary fructooligosaccharides up-regulate immunoglobulin A response and polymeric immunoglobulin receptor expression in intestines of infant mice. *Clinical & Experimental Immunology*. 2004;137(1):52–58. doi: 10.1111/j.1365-2249.2004.02487.x.
132. Roller M., Rechkemmer G., Watzl B. Prebiotic Inulin Enriched with Oligofructose in Combination with the Probiotics *Lactobacillus rhamnosus* and *Bifidobacterium lactis* modulates intestinal immune Functions in Rats. *Journal of Nutrition*. 2004;134(1):153–156.
133. Bunout D., Barrera G., Hirsch S., et al. Effects of a nutritional

supplement on the immune response and cytokine production in free-living Chilean elderly. *Journal of Parenteral and Enteral Nutrition*. 2004;28(5):348–354. doi: 10.1177/0148607104028005348.

134. Langkamp-Henken B., Bender B. S., Gardner E. M., et al. Nutritional formula enhanced immune function and reduced days of symptoms of upper respiratory tract infection in seniors. *Journal of the American Geriatrics Society*. 2004;52(1):3–12. doi: 10.1111/j.1532-5415.2004.52003.x.

135. Trushina E. N., Martynova E. A., Nikityk D. B., Mustafina O. K., Baygarin E. K. The influence of dietary inulin and oligofructose on the cell-mediated and humoral immunity in rats. *Voprosy Pitaniia*. 2005;74(3):22–27.

136. Bakker-Zierikzee A. M., van Tol E. A. F., Kroes H., Alles M. S., Kok F. J., Bindels J. G. Faecal SIgA secretion in infants fed on pre- or pro-biotic infant formula. *Pediatric Allergy and Immunology*. 2006;17(2):134–140. doi: 10.1111/j.1399-3038.2005.00370.x.

137. Langkamp-Henken B., Wood S. M., Herlinger-Garcia K. A., et al. Nutritional formula improved immune profiles of seniors living in nursing homes. *Journal of the American Geriatrics Society*. 2006;54(12):1861–1870. doi: 10.1111/j.1532-5415.2006.00982.x.

138. Adogony V., Respondek F., Biourge V., et al. Effects of dietary scFOS on immunoglobulins in colostrums and milk of bitches. *Journal of Animal Physiology and Animal Nutrition*. 2007;91(5-6):169–174. doi: 10.1111/j.1439-0396.2007.00688.x.

139. Apanavicius C. J., Powell K. L., Vester B. M., et al. Fructan supplementation and infection affect food intake, fever, and epithelial sloughing from salmonella challenge in weanling puppies. *Journal of Nutrition*. 2007;137(8):1923–1930.

140. Seidel C., Boehm V., Vogelsang H., et al. Influence of prebiotics and antioxidants in bread on the immune system, antioxidative status and antioxidative capacity in male smokers and non-smokers. *British Journal of Nutrition*. 2007;97(2):349–356. doi: 10.1017/s0007114507328626.

141. Benyacoub J., Rochat F., Saudan K.-Y., et al. Feeding a diet containing a fructooligosaccharide mix can enhance Salmonella vaccine efficacy in mice. *Journal of Nutrition*. 2008;138(1):123–129.

142. Scholtens P. A. M. J., Alliet P., Raes M., et al. Fecal secretory

immunoglobulin A is increased in healthy infants who receive a formula with short-chain galacto-oligosaccharides and long-chain fructo-oligosaccharides. *Journal of Nutrition*. 2008;138(6):1141–1147.

143. Vulevic J., Drakoularakou A., Yaqoob P., Tzortzis G., Gibson G. R. Modulation of the fecal microflora profile and immune function by a novel trans-galactooligosaccharide mixture (B-GOS) in healthy elderly volunteers. *The American Journal of Clinical Nutrition*. 2008;88(5):1438–1446. doi: 10.3945/ajcn.2008.26242.
144. van Hoffen E., Ruiter B., Faber J., et al. A specific mixture of short-chain galacto-oligosaccharides and long-chain fructo-oligosaccharides induces a beneficial immunoglobulin profile in infants at high risk for allergy. *Allergy*. 2009;64(3):484–487. doi: 10.1111/j.1398-9995.2008.01765.x.
145. Li Y., Zong Y., Qi J., Liu K. Prebiotics and oxidative stress in constipated rats. *Journal of Pediatric Gastroenterology and Nutrition*. 2011;53(4):447–452. doi: 10.1097/mpg.0b013e31821eed83.
146. Choque Delgado G. T., Thomé R., Gabriel D. L., Tamashiro W. M. S. C., Pastore G. M. Yacon (*Smallanthus sonchifolius*)-derived fructooligosaccharides improves the immune parameters in the mouse. *Nutrition Research*. 2012;32(11):884–892. doi: 10.1016/j.nutres.2012.09.012.
147. Lecerf J.-M., Dépeint F., Clerc E., et al. Xylo-oligosaccharide (XOS) in combination with inulin modulates both the intestinal environment and immune status in healthy subjects, while XOS alone only shows prebiotic properties. *British Journal of Nutrition*. 2012;108(10):1847–1858. doi: 10.1017/s0007114511007252.
148. Vulevic J., Juric A., Tzortzis G., Gibson G. R. A mixture of *trans*-galactooligosaccharides reduces markers of metabolic syndrome and modulates the fecal microbiota and immune function of overweight adults1-3. *The Journal of Nutrition*. 2013;143(3):324–331. doi: 10.3945/jn.112.166132.
149. Bouchaud G., Castan L., Chabauty J., Aubert P., Neunlist M., Bodinier M. 11: perinatal exposure to galactooligosaccharides/inulin prebiotics prevent food allergy by protecting intestine and promoting tolerance. *Cytokine*. 2014;70(1):30. doi: 10.1016/j.cyto.2014.07.018.
150. Childs C. E., Röytiö H., Alhoniemi E., et al. Xylo-oligosaccharides alone or in synbiotic combination with *Bifidobacterium animalis* subsp. lactis induce bifidogenesis and modulate markers of immune

function in healthy adults: a double-blind, placebo-controlled, randomised, factorial cross-over study. *British Journal of Nutrition*. 2014;111(11):1945–1956. doi: 10.1017/s0007114513004261.

151. Samal L., Chaturvedi V. B., Saikumar G., Somvanshi R., Pattanaik A. K. Prebiotic potential of Jerusalem artichoke (*Helianthus tuberosus* L.) in Wistar rats: effects of levels of supplementation on hindgut fermentation, intestinal morphology, blood metabolites and immune response. *Journal of the Science of Food and Agriculture*. 2014 doi: 10.1002/jsfa.6873.
152. Huazano-García A., López M. G. Metabolism of short chain fatty acids in colon and faeces of mice after a supplementation of diets with agave fructans. In: Valenzuela Baez R., editor. *Lipid Metabolism*. Vol. 8. Rijeka, Croatia: InTech; 2013. pp. 163–182.
153. Chen S. K., Tsai M. L., Huang J. R., Chen R. H. In vitro antioxidant activities of low-molecular-weight polysaccharides with various functional groups. *Journal of Agricultural and Food Chemistry*. 2009;57(7):2699–2704. doi: 10.1021/jf804010w.
154. Hernandez-Marin E., Martínez A. Carbohydrates and their free radical scavenging capability: a theoretical study. *Journal of Physical Chemistry B*. 2012;116(32):9668–9675. doi: 10.1021/jp304814r.
155. Keunen E., Peshev D., Vangronsveld J., van den Ende W., Cuypers A. Plant sugars are crucial players in the oxidative challenge during abiotic stress: extending the traditional concept. *Plant, Cell and Environment*. 2013;36(7):1242–1255. doi: 10.1111/pce.12061.
156. Van den Ende W., Peshev D. Sugars as antioxidants in plants. In: Tuteja N., Gill S. S., editors. *Crop Improvement Under Adverse Conditions*. New York, NY, USA: Springer; 2013. pp. 285–307.
157. Peshev D., Vergauwen R., Moglia A., Hideg É., Van den Ende W. Towards understanding vacuolar antioxidant mechanisms: a role for fructans? *Journal of Experimental Botany*. 2013;64(4):1025–1038. doi: 10.1093/jxb/ers377.
158. Foyer C. H., Lelandais M., Kunert K. J. Photooxidative stress in plants. *Physiologia Plantarum*. 1994;92(4):696–717.
159. Nishizawa A., Yabuta Y., Shigeoka S. Galactinol and raffinose constitute a novel function to protect plants from oxidative damage. *Plant Physiology*. 2008;147(3):1251–1263. doi: 10.1104/pp.108.122465.
160. Stoyanova S., Geuns J., Hideg É., Van den Ende W. The food additives

inulin and stevioside counteract oxidative stress. *International Journal of Food Sciences and Nutrition*. 2011;62(3):207–214. doi: 10.3109/09637486.2010.523416.

161. Li X. M., Shi Y. H., Wang F., Wang H. S., Le G. W. In vitro free radical scavenging activities and effect of synthetic oligosaccharides on antioxidant enzymes and lipid peroxidation in aged mice. *Journal of Pharmaceutical and Biomedical Analysis*. 2007;43(1):364–370. doi: 10.1016/j.jpba.2006.06.041.
162. Glei M., Hofmann T., Küster K., Hollmann J., Lindhauer M. G., Pool-Zobel B. L. Both wheat (*Triticum aestivum*) bran arabinoxylans and gut flora-mediated fermentation products protect human colon cells from genotoxic activities of 4-hydroxynonenal and hydrogen peroxide. *Journal of Agricultural and Food Chemistry*. 2006;54(6):2088–2095. doi: 10.1021/jf052768e.
163. Juśkiewicz J., Zduńczyk Z., Zary-Sikorska E., Król B., Milala J., Jurgoński A. Effect of the dietary polyphenolic fraction of chicory root, peel, seed and leaf extracts on caecal fermentation and blood parameters in rats fed diets containing prebiotic fructans. *British Journal of Nutrition*. 2011;105(5):710–720. doi: 10.1017/s0007114510004344.

6

Using Carbohydrate Interaction Assays to Reveal Novel Binding Sites in Carbohydrate Active Enzymes

Darrell Cockburn, Casper Wilkens, Adiphol Dilokpimol, Hiroyuki Nakai, Anna Lewińska, Maher Abou Hachem, and Birte Svensson

Enzyme and Protein Chemistry, Department of Systems Biology, Technical University of Denmark, Lyngby, Denmark

ABSTRACT

Carbohydrate active enzymes often contain auxiliary binding sites located either on independent domains termed carbohydrate binding modules (CBMs) or as so-called surface binding sites (SBSs) on the catalytic module at a certain distance from the active site. The SBSs are usually critical for the activity of their cognate enzyme, though they are not readily detected in the sequence of a protein, but normally require a crystal structure of a complex for their identification. A variety of methods, including affinity electrophoresis (AE), insoluble polysaccharide pulldown (IPP) and surface plasmon resonance (SPR) have been used to study auxiliary binding sites.

Citation: Using Carbohydrate Interaction Assays to Reveal Novel Binding Sites in Carbohydrate Active Enzymes. Cockburn D, Wilkens C, Dilokpimol A, Nakai H, Lewińska A, Abou Hachem M, et al. (2016) PLoS ONE 11(8): e0160112. doi:10.1371/journal. pone.0160112

These techniques are complementary as AE allows monitoring of binding to soluble polysaccharides, IPP to insoluble polysaccharides and SPR to oligosaccharides. Here we show that these methods are useful not only for analyzing known binding sites, but also for identifying new ones, even without structural data available. We further verify the chosen assays discriminate between known SBS/CBM containing enzymes and negative controls. Altogether 35 enzymes are screened for the presence of SBSs or CBMs and several novel binding sites are identified, including the first SBS ever reported in a cellulase. This work demonstrates that combinations of these methods can be used as a part of routine enzyme characterization to identify new binding sites and advance the study of SBSs and CBMs, allowing them to be detected in the absence of structural data.

INTRODUCTION

Carbohydrates exist in many forms in nature, from the *O*- and *N*-linked glycans of proteins to the cell walls of plants, fungi and bacteria as well as storage polymers such as starch and glycogen. Enzymes active towards these carbohydrates must have the means to identify and act on their particular substrate within an often complex milieu. A common strategy for accomplishing this recognition is to utilize binding regions outside of the active site that are less constrained as they need not bind the substrate in an optimal mode for catalysis. These non-catalytic binding sites can operate in conjunction with the active site in catalysis, or may simply serve to localize the enzymes to the substrates. Such external sites interacting with carbohydrates exist on a continuum from independent domains, termed carbohydrate binding modules (CBMs) [1], to regions or spots termed surface binding sites (SBSs) situated on the catalytic domain at a certain distance from the active site [2–4]. Enzymes can contain multiple CBMs, multiple SBSs or combinations of both and these carbohydrate binding sites can perform diverse functions related to enzyme activity [2]. The importance of CBMs has long been established, particularly for enzymes acting on insoluble polysaccharides, and they are formally organized into families in the CAZy database [5,6] (www.cazy.org). SBSs, however, are much less studied and also much less frequently occurring, with the majority of the investigated enzymes participating in starch or xylan degradation. Thus of the approximately 60 enzymes from 20 CAZy families to date identified to possess one or more SBSs, almost half belong to GH13, the α-amylase family. It is likely, however, that since SBSs have commonly been detected only by observation in crystal structures of ligand complexes, they are in fact more widespread among the CAZy families and have to now gone unappreciated.

A variety of methods have been used to study the characteristics of SBSs and CBMs, including isothermal titration calorimetry (ITC) [7,8], surface plasmon resonance (SPR) [9–11], insoluble polysaccharide pulldown (IPP) [9,10,12], affinity electrophoresis (AE) [9,13], confocal laser scanning microscopy (CLSM) [9,14] and enzyme activity progress curves [15]. This set of assays enables measurement of binding properties of known sites, AE and CLSM having been applied to qualitatively demonstrate binding; although AE can also be used to quantify the affinity [13]. Mutational analysis has been used to confirm the presence and location of the binding site and disclose its impact on enzyme activity [10,16–19]. Many sequences suspected to perform binding functions within carbohydrate active enzymes remain uncharacterized and there are probably many more that are unsuspected. Importantly, just as IPP, AE and SPR are useful to describe the known binding sites, in this work we show these techniques are suitable for revealing the presence of and to further characterize hitherto unknown SBSs as well as CBMs, even in the absence of a solved crystal structure.

MATERIALS AND METHODS

Enzymes

The investigated carbohydrate active enzymes are designated by their CAZy family (www.cazy.org), followed by a number to separate these cases where more than one enzyme from a particular family were studied. The identification, properties and origins of the enzymes, which either stemmed from our research group, were donated by other researchers or were commercially available are listed in Table 1. GH5-1 [20], GH13-1 [21], GH13-2 [22], GH15-1/GH15-2 [23], GH36-1 [24], GH62-1 [18], GH65-1 [25] and GH94-1/GH94-2 [26] were produced in-house as previously described. AA09-1, GH10-1 and GH11-1 were cloned from *Aspergillus nidulans* FGSC A4 cDNA [27], and GH32-1 and GH32-2 from *Lactobacillus acidophilus* NCFM genomic DNA (primer pairs are listed in S1 Table). See the Supplementary Methods section (S1 File) for the full details of the expression and purification protocols.

Table 1. Origins and binding properties of enzymes in this study.

ID	Activity	Organism	CBM	AE Binding[h]	IPP[h]
AA9-1[a]	Lytic polysaccharide monoxygenase	*Aspergillus nidulans*	No CBM	None	None
CE2-1[b]	Acetyl xylan esterase	*Cellvibrio japonicus*	No CBM	None	None
GH1-1[b]	β-glucosidase	*Rhizobium etli*	No CBM	None	None
GH3-1[c]	Exo-glucanase	*Hordeum vulgare*	No CBM	None	None
GH5-1[a]	Endo-β-1,4-mannanase	*Aspergillus nidulans*	No CBM	KGM	None
GH6-1[d]	Endoglucanase	*Thermobifida fusca*	No CBM	HEC, KGM, LAM, XYG	None
GH8-1[e]	Endoglucanase	*Clostridium thermo-cellum*	No CBM	HEC, KGM, WAX	None
GH10-1[a]	Xylanase	*Aspergillus nidulans*	No CBM	XYG	None
GH11-1[a]	Xylanase	*Aspergillus nidulans*	No CBM	BWX, LAM, PUL, OSX, WAX	None
GH13-1[a]	α-amylase	*Hordeum vulgare*	No CBM	PUL, AP, GLY	STA
GH13-2[a]	Limit dextrinase	*Hordeum vulgare*	CBM48	AM	None
GH13-3[b]	α-amylase	*Escherichia coli*	No CBM	NT[i]	CHI
GH13-4[b]	Isoamylase	*Escherichia coli*	No CBM	None	NT[i]
GH13-5[b]	Pullulanase	*Bacillus subtilis*	CBM48	NT[i]	None
GH13-6[f]	α-amylase	*Sus scrofa*	No CBM	HEC, HA, AM, GAL, ARA, PUL, GLY	None
GH14-1[f]	β-amylase	*Hordeum vulgare*	No CBM	None	OSX
GH15-1[a]	Glucoamylase	*Aspergillus niger*	CBM20	AM, GAL, PUL, AP, GLY	STA
GH15-2[a]	Glucoamylase	*Aspergillus niger*	No CBM	None	None
GH26-1[b]	β-mannanase	*Cellvibrio japonicus*	No CBM	None	None
GH27-1[b]	α-galactosidase	*Clostridium cellulo-lyticum*	CBM6	NT[i]	None
GH31-1[f]	α-glucosidase	*Oryza sativa*	No CBM	RHG, KGM, BWX, GAL, LAM, XYG	CHI, OSX
GH32-1[a]	β-fructosidase	*Lactobacillus aci-dophilus*	No CBM	None	None
GH32-2[a]	Sucrose-6-phosphate hydrolase	*Lactobacillus aci-dophilus*	No CBM	None	None
GH36-1[a]	α-galactosidase	*Aspergillus nidulans*	No CBM	None	None
GH43-1[b]	β-xylosidase	*Bacillus subtilus*	No CBM	None	None
GH48-1[e]	Processive endoglu-canase	*Clostridium cellulo-lyticum*	No CBM	RHG, KGM, BBG, WAX	CEL

GH53-1[b]	β-galactanase	*Cellvibrio japonicus*	No CBM	None	STA
GH62-1[a]	α-L-arabinofuranosidase	*Aspergillus nidulans*	No CBM	HEC, KGM, BWX, BBG, OSX, WAX	None
GH65-1[a]	Maltose phosphorylase	*Lactobacillus acidophilus*	No CBM	None	None
GH77-1[g]	Amylomaltase	*Escherichia coli*	No CBM	None	CEL, STA
GH84-1[b]	O-GlcNAcase	*Streptococcus pyogenes*	No CBM	KGM	None
GH85-1[b]	Endo-β-*N*-acetylglucosaminidase	*Clostridium perfringens*	No CBM	None	None
GH94-1[a]	Cellobiose phosphorylase	*Clostridium thermocellum*	No CBM	None	None
GH94-2[a]	Cellodextrin phosphorylase	*Clostridium thermocellum*	No CBM	None	None
PL10-1[b]	Pectate lyase	*Caulobacter crescentus*	No CBM	None	STA

[a] produced as previously described (see materials and methods)

[b] from Prozomix (Haltwhistle, United Kingdom)

[c] from Maria Hrmova (University of Adelaide)

[d] from David Wilson (Cornell University)

[e] from NZYtech (Lisbon, Portugal)

[f] from Sigma

[g] from Alison Smith (John Innes Center)

[h] see Table 2 for polysaccharide identities

[i] Not tested

Affinity Electrophoresis

AE involves two native polyacrylamide gel electrophoresis (PAGE) experiments run together. One conventional native PAGE gauges the regular position of the protein of interest, while in a second the polysaccharide under investigation is incorporated when casting the polyacrylamide gel. Proteins migrating slower in the second gel as compared to the control are presumed to interact with the polysaccharide. To account for variation in gel pore size due to polysaccharide incorporation, reference proteins are included in both gels. The stained gels were scanned and then analyzed using the program ImageJ (http://imagej.nih.gov/ij/). For the target enzyme, the ratio of the migration distances in the presence of polysaccharide to that in the control gel was divided by the corresponding ratio for a reference protein chosen from the molecular weight marker to migrate closest in the control gel to the target enzyme. Thus proteins not binding the polysaccharide have ratios of approx. 1, while those binding to polysaccharides have ratios < 1. In the present study, ratios < 0.85 were considered positive for binding to account for average band thickness and average migration distance to allow stringent hypothesis testing, minimizing false positives. Most polysaccharides were used at a final concentration of 0.1%; although some were used at 0.05% due to limited solubility and/or changes in gel properties at higher concentrations (see Table 2). Proteins were analyzed in one of four regimes of conditions depending on size and isoelectric point (Table 3), all gels being buffered by 50 mM Tris-borate pH 8.8 and run at 4°C. For the enzymes active on cellulose: GH6-1, GH8-1 and GH48-1, the effect of 100 mM cellobiose was tested by including it in the gels, loading and running buffers. The migration in control, hydroxyethyl cellulose (HEC) or barley β-glucan (BBG) containing gels were compared to runs in corresponding gels without cellobiose. All of these gels were run according to the AE3 conditions (Table 3).

Table 2. Properties of polysaccharides used in this study.

Polysaccharide[a]	Monosaccharide Units	Linkage
Amylopectin (AP, 0.1%)	Glucose	α1–4 (backbone), α-1,6 (branch points)
Amylose (AM, 0.05%)	Glucose	α1–4
Avicel (crystalline cellulose, CEL)	Glucose	β1–4
Barley β-glucan (BBG, 0.1%)	Glucose	Mixture of β1–3 and β1–4
Birchwood xylan (BWX, 0.1%)	Xylose	β1–4
Chitin (CHI)	*N*-acetylglucosamine	β1–4
Curdlan (CLN, 0.05%)	Glucose	β1–3
Galactomannan (GLM, 0.05%)	Mannose, Galactose	β1–4 mannose backbone, α1–6 galactose sidechains
Glycogen (GLY, 0.1%)	Glucose	α1–4 (backbone), α1–6 (branch points)
Hyaluronic acid (HA, 0.1%)	Glucuronic acid, *N*-acetylglucosamine	β1–4 and β1–3
Hydroxyethyl cellulose (HEC, 0.1%)	Glucose	β1–4 with CH_2CH_2OH at O2, O3 or O6
Konjac glucomannan (KGM, 0.1%)	Mannose, Glucose	Mixture of β1–4 linked mannose and glucose
Laminarin (LAM, 0.1%)	Glucose	β1–3 interspersed with some β1–6
Oat spelt xylan (OSX, 0.1%)	Xylose, Arabinose, Glucose	β1–4 xylose backbone, α1–2 or α1–3 arabinose or glucose branches

Potato galactan (GAL, 0.05%)	Galactose	β1–4
Pullulan (PUL, 0.05%)	Glucose	$(\alpha1–4)_3$ repeat linked by α1–6
Rhamnogalacturonan I (RHG, 0.1%)	Galacturonic acid, Rhamnose, others	α1–2 and α1–4, others for sidechains
Starch (STA)	Glucose	α1–4 (backbone), α1–6 (branch points)
Sugar beet arabinan (ARA, 0.1%)	Arabinose	α1–5
Wheat arabinoxylan (WAX, 0.1%)	Xylose, Arabinose	β1–4 xylose backbone, α1–2 or α1–3 arabinose branches
Xyloglucan (XYG, 0.1%)	Xylose, Glucose	β1–4 glucose, α1–6 xylose sidechains

[a] in brackets are the abbreviations used in this study and the concentration used in AE

Table 3. Affinity electrophoresis regimes.

Regime	Acryl-amide conc. (%)	Poten-tial (V)	Time (h)	Enzymes
AE1	6.5	80	4.5	GH3-1, GH11-1, GH13-2, GH14-1, GH31-1, GH32-2, GH36-1, GH65-1, GH77-1, GH94-1, GH94-2
AE2	12.0	45	16.0	AA9-1, GH5-1, GH6-1, GH10-1, GH13-1, GH13-6, GH15-1, GH15-2, GH32-1, GH62-1
AE3	12.0	50	18.0	GH1-1, GH8-1, GH13-4, GH26-1, GH43-1, GH48-1, GH85-1
AE4	6.5	80	6.0	CE2-1, GH3-1, GH53-1, GH84-1, PL10-1

Insoluble Polysaccharide Pulldown

IPP was performed for avicel (approx. 50% crystalline cellulose), corn starch, chitin and oatspelt xylan. Polysaccharide (5 mg) in a 1.5 mL Eppendorf tube, was washed three times with 10 mM HEPES pH 7.0, 150 mM NaCl followed by addition of enzymes (200 μL, 0.2 mg/mL) and incubated (1 h, 4°C) with agitation. After pelleting (20 000 x g, 10 min, 4°C) insoluble polysaccharides and bound enzymes, the supernatants were decanted and centrifuged again to remove trace polysaccharides, followed by protein concentration determination (BioRad, Hercules, California; based on the Bradford method [28]). The fraction of protein bound was calculated by dividing the final protein concentration by the initial protein concentration.

For the GH6-1, GH8-1 and GH48-1 enzymes active on cellulose, the effect of 100 mM cellobiose or 4 mM cellopentaose on their binding to avicel was also measured (in duplicate).

Surface Plasmon Resonance

The affinity and binding stoichiometry for oligosaccharides of the cellulases GH6-1, GH8-1 and GH48-1 and the GH13 enzymes GH13-3, GH13-5 and GH13-6 were analyzed using SPR (Biacore T100; GE Healthcare). The cellulases were immobilized on CM5 chips (GE Healthcare) *via* amine coupling to carboxyl groups on the chip according to the manufacturer's protocol, while the GH13 enzymes were biotinylated and immobilized on streptavidin chips as previously described [17]. Some GH13 enzymes are not able to withstand the low pH used for the immobilization on CM5 chips and thus the alternative immobilization on streptavidin was used, though this requires somewhat more protein. Enzymes were immobilized at 1500 response units (RU) in the presence of 5 mM of an anticipated ligand, i.e. cellopentaose for cellulose active enzymes and β-cyclodextrin in the case of starch active enzymes to protect binding sites. The immobilized enzymes were assayed for binding to 10 — 5000 μM cellopentaose, xylotetraose and maltotetraose for the cellulases or β-cyclodextrin, maltohexaose and a DP10 limit dextrin (in house prepared as in Roberts and Whelan [29]) for the GH13 amylolytic enzymes. Stoichiometry was calculated using the formula: S=MWLMWE*RERmax where S is the stoichiometry, MW_L is the molecular weight of the ligand, MW_E is the molecular weight of the enzyme, R_E is the number of response units of enzyme bound to the chip and R_{max} is the maximum response calculated from a one site binding model fitted to the data using the Biacore software.

Surface Binding Site Conservation Analysis

For enzymes that have a related family member known to possess an SBS, conservation of the residues comprising the SBS was examined by sequence alignment. The enzymes from this study and the PDB ID of the corresponding family member are: GH1-1 (1UYQ, 1GNX), GH5-1 (2PC8), GH8-1 (2B4F), GH10-1 (1GOQ, 1B3V), GH11-1 (2QZ2, 2QZ3), GH13-3 (2D3N), GH14-1 (1B9Z), GH15-1 (2F6D), GH27-1 (2HG2) and GH31-1 (3POC). Residues involved in forming the SBSs were identified in consultation with the literature and by manual inspection of the crystal structures using PyMol. Alignments were performed in CLC Sequence viewer v7 (CLC Bio) using the default algorithm.

RESULTS

Properties of Enzymes Studied

Thirty five enzymes from 27 different families in the CAZy database [6] were examined for the possible presence of an SBS (see Table 1). The vast majority are glycoside hydrolases, but glycosyl transferases, lytic polysaccharide monoxygenases, carbohydrate esterases, polysaccharide lyases and carbohydrate phosphorylases are also represented. Most of the enzymes are from bacteria or fungi, some are from plants and one is of mammalian origin. They were chosen to embody large diversity and to have a few characteristics facilitating their study by the chosen techniques. First, most enzymes contained no CBM, which simplifies interpretation of the binding data. In case of the glucoamylase from *Aspergillus niger* forms with and without its native CBM20 were compared. Secondly, three enzymes known to have SBS(s),barley α-amylase AMY1 (GH13-1) [17,30,31], pig pancreatic α-amylase (GH13-6) [32,33] and pullulanase from *Bacillus subtilis* (GH13-5; PDB ID: 2E9B, unpublished), were selected as positive controls in the screening process. Additionally, three thoroughly structurally characterized enzymes in complex with carbohydrate ligands without a detected SBS served as pseudo-negative controls, although we cannot definitively exclude the possibility that they possess an SBS. These were barley β-amylase (GH14-1) [34,35], *Cellvibrio japonicus* β-mannanase (GH26-1) [36–39] and *Clostridium cellulolyticus* processive endoglucanase (GH48-1) [40–42]. Finally, all but one (GH94-2) of the tested enzymes have molecular mass < 100 kDa and all had an isoelectric point < pH 7, compatible with AE in the four applied regimes. It should be emphasized that these are not hard limits for the technique. Thus for higher molecular weights longer run times can be used, while for higher pI, a lower buffer pH can be used and the electrodes switched to allow migration of positively charged proteins.

Affinity Electrophoresis

Thirty five enzymes were tested for interactions through AE with 18 polysaccharides (Table 1), providing a variety of monosaccharide units and linkages (summarized in Table 2) representing carbohydrate structures relevant to the enzymes under study. Three of the 35 enzymes appeared unsuitable for AE as they did not give a distinct band in the native PAGE. These were not screened against the full set of polysaccharides, but still by IPP. Of the 18 polysaccharides, only galactomannan and curdlan do not detectably interact with any of the 32 enzymes. By contrast konjac glucomannan very interestingly retards seven different enzymes in AE, only three of which (GH5-1, GH6-1, GH8-1) are expected to have activity against glucomannan. Noticeably, glucomannan does not retard GH26-1 even though it is a substrate [39]. Nineteen of the enzymes show no significant binding with any of the polysaccharides in AE (Table 1). Of the 13 enzymes that interact with a polysaccharide in AE, nine bind to several (from three to seven) polysaccharides, often of diverse linkage type and/or monosaccharide composition. For instance GH13-6 (pig pancreas α-amylase) interacts with the starch related α-glucans glycogen, amylose and pullulan as expected, (whereas amylopectin was just outside the 0.85 polysaccharide: control ratio chosen as the cutoff indicative of binding), plus with other polysaccharides as diverse as cellulose, hyaluronic acid, arabinan and galactan. Also the α-glucosidase GH31-1 interacts with several different polysaccharides, none of these, however, are starch related. Some of the screened enzymes thus appear surprisingly promiscuous in their affinity for polysaccharides. To further investigate the utility of AE to characterize SBSs, three cellulases were tested with BBG and HEC in the presence and absence of the active site inhibitor cellobiose (Fig 1). The added cellobiose clearly competes with HEC for binding to GH6-1 and GH8-1, whereas it only slightly weakens the binding of GH48-01 to BBG.

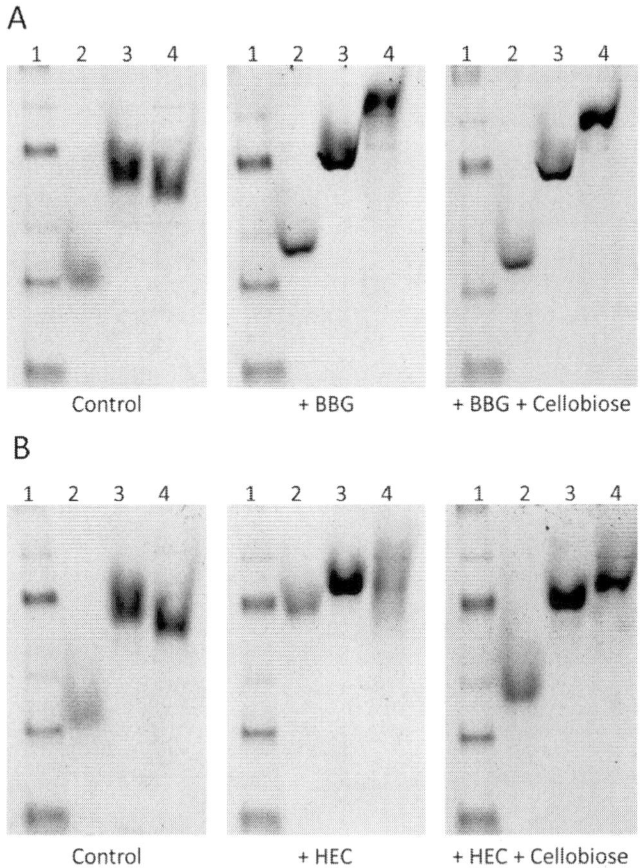

Figure 1. Affinity electrophoresis of cellulases with and without cellobiose.

AE was performed with the polysaccharides barley β-glucan (A; BBG) and hydroxyethyl cellulose (B; HEC) in the presence and absence of the active site inhibitor cellobiose. Lane 1 is a protein ladder (NativeMark from Invitrogen), lane 2 is GH6-1, lane 3 GH8-1 and lane 4 GH48-1.

Insoluble Polysaccharide Pulldown

IPP was performed as a complement to AE that monitors interactions with soluble polysaccharides. A total of 34 enzymes (1 produced inconsistent results between replicates and is not included) were tested with four carbohydrates; cellulose, insoluble starch, chitin and xylan (Table 1). The positive controls barley α-amylase (GH13-01) and glucoamylase (GH15-1; containing a starch specific CBM20) as expected bind to insoluble starch. Overall 7 out of 34 enzymes tested bind to one and 2 out of 34 enzymbes bind to two insoluble polysaccharides, respectively. As with AE, promiscuous binding occurs. Most remarkably the α-glucosidase GH31-1 binds to two non-starch compounds, but not to starch itself. The three cellulases GH6-1, GH8-1 and GH48-1 were again examined in more depth for interaction with cellulose in the presence and absence of either cellobiose or cellopentaose (Fig 2), which both bind to their active sites. Remarkably, even in the absence of these competitive ligands neither GH6-1 nor GH8-1 interact strongly with avicel, but exhibit weak binding that for GH6-1 is inhibited by cellobiose and cellopentaose, while GH8-1 binding is unaffected by these oligosaccharides. Remarkably, GH48-1, binds tightly to avicel, which is not changed by cellobiose, but efficiently reduced by cellopentaose.

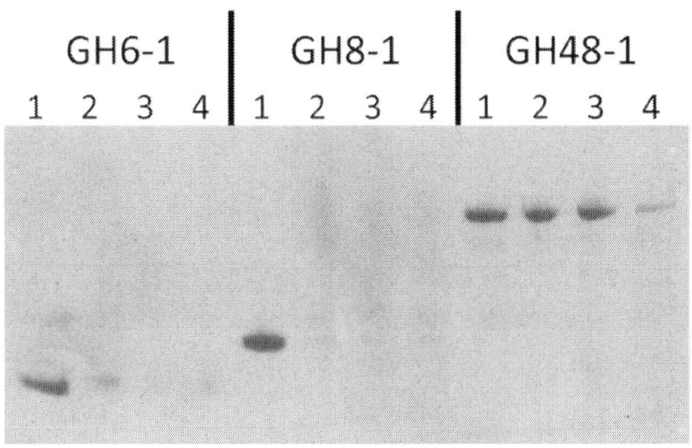

Figure 2. SDS-PAGE of fractions from insoluble polysaccharide pulldown assays of the cellulases GH6-1, GH8-1 and GH48-1 with avicel.

For each protein lane 1 is the initial protein sample, lane 2 the fraction bound to avicel in the absence of inhibitors, lane 3 the fraction bound to avicel in the presence of 100 mM cellobiose and lane 4 the fraction bound to avicel in the presence of 4 mM cellopentaose.

Surface Plasmon Resonance

AE and IPP are useful for detecting binding to soluble and insoluble polysaccharides, respectively, however, they cannot provide information about the number of binding sites. SPR conversely both monitors the affinity and quantifies the bound carbohydrate, thus revealing an apparent binding stoichiometry. To examine the utility of SPR for uncovering multiple binding sites, three GH13 family enzymes (GH13-3, GH13-5 and GH13-6) and three cellulases (GH6-1, GH8-1 and GH48-1) are analyzed for affinity and stoichiometry using different oligosaccharides (Table 4). With the GH13 enzymes are used: β-cyclodextrin (a cyclic form of maltoheptaose), maltohexaose and an α-limit dextrin (a DP10 maltooligosaccharide containing an α-1,6 branch point). Cellopentaose is tested with the cellulases and compared to the distinctly different maltotetraose and xylotetraose to probe not only the affinity, but also the specificity of recognition, given the promiscuous interactions seen via the other assays. GH13-5 (PDB ID: 2E9B, unpublished) and GH13-6 [32,33] possess SBSs as seen in crystal structures. All three GH13 enzymes give stoichiometries approaching 2 for β-cyclodextrin (Table 4), indicating the presence of at least one and possibly more SBSs, since in some GH13 enzymes β-cyclodextrin is not bound at the active site [10]. The measured fractional stoichiometry can arise because the enzymes are immobilized on the chip in different orientations, not having all binding sites well exposed [9]. Thus binding stoichiometry estimates are typically rounded up to give meaningful numbers. The *E. coli* α-amylase (GH13-3) and *B. subtilis* pullulanase (GH13-5) interact weakly with maltohexaose and α-limit dextrin, although maltohexaose is a substrate for the former [43] and α-limit dextrin for the latter enzyme [44]. In neither case were kinetics analyzed for these enzyme/substrate combinations and they are probably poor substrates. For porcine pancreatic α-amylase (GH13-6), the stoichiometry with maltohexaose is > 3 (Table 4) implying the presence of four different binding sites. This fits well with the crystal structure in complex with maltopentaose accommodated at the active site and three SBSs [33]. Notably, as a one-site binding model (not shown) fits well to the SPR data as the affinities of the individual sites are not differentiated. The binding stoichiometry of α-limit dextrin with GH13-6 is close to 1, suggesting the α1—6 linked branch in the maltooligosaccharide suppresses binding at the several SBSs known to bind linear maltooligosaccharides [33]. The cellulases GH6-1 and GH48-1 give a stoichiometry with cellopentaose below 1, probably reflecting the absence of multiple binding sites and that this interaction takes place at the active site. However, the stoichiometry of 2.3 for cellopentaose binding to the GH8-1 enzyme identifies three sites conceivably encompassing two SBSs. As expected the three cellulases show no significant affinity for maltotetraose and xylotetraose, although GH8-1 interacts weakly with these oligosaccharides.

Table 4. Binding affinities and stoichiometries for family GH13 enzymes and cellulases as determined by surface plasmon resonance.

Enzyme	β-cyclodextrin		Maltohexaose		α-Limit Dextrin	
	K_d (mM)	Stoichiometry	K_d (mM)	Stoichiometry	K_d (mM)	Stoichiometry
GH13-3	1.36 ± 0.05	1.63 ± 0.13	NB[a]		NS[b]	
GH13-5	0.97 ± 0.13	1.51 ± 0.20	NB[a]		NB[a]	
GH13-6	0.87 ± 0.40	1.69 ± 0.32	1.33 ± 0.12	3.23 ± 0.13	1.91 ± 0.22	0.88 ± 0.09
Enzyme	**Cellopentaose**		**Maltotetraose**		**Xylotetraose**	
	K_d (mM)	Stoichiometry	K_d (mM)	Stoichiometry	K_d (mM)	Stoichiometry
GH6-1	0.33 ± 0.11	0.20 ± 0.03	NB[a]		NB[a]	
GH8-1	0.86 ± 0.09	2.33 ± 0.55	NS[b]		NS[b]	
GH48-1	2.07 ± 0.78	0.85 ± 0.42	NB[a]		NB[a]	

[a] No binding detected.

[b] Binding was not saturable under the conditions tested.

DISCUSSION

Screening Strategy

In this study we have started out with a reasonably large collection of 35 enzymes and screened these for binding to 18 soluble and 4 insoluble polysaccharides using AE and IPP, respectively. These techniques have several advantages. First, they can be performed in a relatively high throughput manner, so numerous enzyme-polysaccharide interactions can be examined in a relatively short time. Second, they are inexpensive, not requiring specialized equipment and also using generally inexpensive polysaccharides. Finally, they look for interactions with polysaccharides, which in many cases are the relevant substrate for the enzyme under study. An important consideration, however, is discriminating between binding at the active site and a potential SBS or previously unrecognized CBM. Using active enzymes, as was done in this study, may help in this matter as catalysis might limit the contribution of the active site to the observed binding with these techniques. Competition with small molecules known to bind at the active site, may also help identify if binding is taking place elsewhere, however, caution is needed as these may also bind at a potential

SBS or CBM. Comparison with other related enzymes known to contain an SBS may also provide important clues. Ultimately however, these methods are used to generate the hypothesis that an SBS is present, which then must be tested by other means. In this study we have taken that next step examining a group of six enzymes by SPR. The main drawback of SPR is that oligosaccharides must be used, which depending on the enzyme, may be quite expensive or unavailable, making it more appropriate for targeted studies, rather than screening. However, SPR has several advantages in the identification of SBSs. It uses relatively little protein and can examine a wide range of binding affinities compared to an alternative technique such as ITC. It is also able to be used with active enzymes as it measures equilibrium binding with a constant flow of ligand. Its most useful attribute, however, is that binding stoichiometry can be detected. This provides direct evidence of binding taking place outside of the active site, a critical consideration as will be discussed below.

Positive Controls

Among the tested enzymes, three from GH13: GH13-1 [17,30,31], GH13-5 (PDB ID: 2E9B, unpublished) and GH13-6 [32,33] presented one or more SBSs in crystal structures and serve as positive controls for the purposes of this study. The properties of the two SBSs in GH13-1, barley α-amylase 1 (AMY1), have been extensively studied by using IPP [9,10], AE [9] and SPR assays [9,10]. However, exploitation of these techniques to detect SBSs in the absence of structural data has not been reported. In the previous studies of GH13-5 and GH13-6, the SBSs were not analyzed in depth, but SBSs studied by mutational analysis in human salivary α-amylase having a high degree of identity with porcine pancreatic α-amylase (GH13-6) [45–48], were found to be important for binding to and activity against starch [48]. SPR analysis indicates GH13-5 and 13-6 have multiple binding sites for β-cyclodextrin (Table 4), possibly representing the active site where both crystal structures show bound α-cyclodextrin [49], PDB ID: 2E8Z (unpublished), and one, perhaps more, SBSs. Noticeably, the affinity at these multiple sites are not distinguished from one another, suggesting that they have similar K_d values. This has been seen for other SBS containing enzymes, e.g. *Bacillus subtilis* [7] and *B. circulans* xylanases [50] and barley AMY1 (GH13-1) [9]. In xylanases, concurrence of active site and SBS binding affinity has been measured by NMR, SPR and ITC, using mutational analysis and active site inhibitors, indicating that the phenomenon was not an artefact of any particular technique. It may be that the affinity at SBSs and their cognate active sites have evolved in tandem to efficiently act together in substrate binding. As expected AMY1 binds to pullulan, amylopectin and glycogen in AE (its binding to amylose is just outside the cutoff chosen to define binding) and to starch granules by IPP. Neither GH13-5 that possesses a CBM48 nor GH13-6 binds to starch granules, but GH13-6 interacts with all α-1,4 linked glucan substrates and several other polysaccharides in AE (GH13-5 was not tested as it did not produce easily distinguishable bands). Binding of GH13-6 to non-substrates may result from the surface exposed nature of SBSs making them less selective in their interactions than active sites, which tend to bury a greater proportion of the substrates in a groove or tunnel in the protein.

The GH15-1 containing CBM20 interacts with all of the starch-like compounds in AE and additionally with galactan, whereas the version lacking CBM20 (GH15-2) does not. Thus binding to non-substrate polysaccharides may be a particular characteristic of SBSs as well as CBMs. While this phenomenon has not been studied before in SBS containing enzymes, it has been noted for CBMs. Recently Hernandez-Gomez et al. [51] examined the ability of crystalline cellulose specific CBMs to bind xyloglucan, a more soluble polymer thought to apply an entirely different mode of binding. They found that binding was weaker to xyloglucan than to crystalline cellulose, but still occurred on the same binding site. In another study, Strobel et al. [52] demonstrated that the cellulose binding face of a CBM1 was also binding to lignin. Furthermore, they attempted to increase the specificity for cellulose using a mutational approach, but as most mutants had reduced affinity for both cellulose and lignin, perhaps substantial overlap occurred among the binding sites for both ligands. In complex polysaccharide environments it can be advantageous for CBMs to attach onto other polysaccharides that are present in addition to the substrate of their appended enzyme. Mannanases are found with cellulose specific CBMs [53], thus targeting the far more abundant cellulose in the plant cell wall, while the active site maintains exquisite selectivity [39]. There may be evolutionary reasons for CBMs and SBSs to have lower specificity, which may go hand in hand with the highly exposed nature of these binding sites unavoidably resulting in this reduced specificity. In the CBM2 family there are both cellulose and xylan recognizing members, however it is possible to interconvert the specificity by mutation of a single residue [54]. This propensity for binding to non-substrate polysaccharides could prove useful in identifying enzymes possessing previously unrecognized CBMs or SBSs.

Negative Controls

In complement to the positive controls analyzed, several proteins that have been extensively structurally characterized without detection of an SBS were included to serve as negative controls. The barley limit dextrinase (GH13-2) has been co-crystallized with several different ligands, recently including its natural substrate, an α-limit dextrin [2]. The only structure, however, which has shown binding outside of the active site was soaked with 400 mM maltotetraose. Soaking at lower concentrations did not result in detectable binding at this SBS, suspiciously located at a symmetry molecule interface, suggesting that it was an artefact. GH13-2, moreover, demonstrates no affinity in IPP and in AE only interacts with the non-substrate amylose. Similarly, crystal structures of barley β-amylase (GH14-1) in complex with different oligosaccharides did not reveal the presence of an SBS [34,35], moreover it does not bind any of the polysaccharides in AE and only to OSX by IPP. This enzyme does not exhibit binding to its substrate, starch, and since this is binding to a single polysaccharide, which is not biologically relevant, we conclude that this is most likely irrelevant non-specific binding. In addition, a β-mannanase (GH26-1) has crystal structures solved in complex with several oligosaccharides without showing an SBS and does not bind to any polysaccharides in the screening experiments.

GH48-1, a processive endoglucanase from *Clostridium cellulolyticum*, is a cellulosomal enzyme and hence part of a multi-enzyme assembly where it can take advantage of binding sites on other enzymes in the scaffoldin-enzyme complex [55]. Although crystal structures did not reveal any additional binding sites, this enzyme indeed is retarded by several polysaccharides in AE, including the non-substrates RHG and WAX; as expected GH48-1 binds to cellulose in IPP (Table 1). Further investigation shows that cellobiose, a potential inhibitor of binding, does not significantly suppress the interaction of GH48-1 with BBG in AE or in IPP with avicel. Cellopentaose, however, virtually eliminates GH48-1's binding to avicel in IPP. Follow up by SPR reveals just one binding site for cellopentaose, which almost certainly is the active site as crystal structures show cellotetraose and cellohexaose accommodated in the active site [40]. Thus all GH48-1 interactions with cellulosic substrates seem mediated through the active site, while the possibility remains that binding to non-substrate polysaccharides takes place at an alternate location. This illustrates the point that while AE and IPP are very good for detecting the presence of binding sites, further experiments where the active site is blocked and/or the number of binding sites can be assessed are needed to rule out false positives of SBSs due to active site binding. There are a few reports on xylanases where the active site was blocked by covalent reaction with a mechanism-based inhibitor. Thus *B. circulans* xylanase treated with 2′,4′-dinitrophenyl 2-deoxy-2-fluoro-β-xylobioside ensured independent analysis of SBS binding in AE, while distinct active site data were achieved by mutation of the SBS [50]. Similarly, Cuyvers et al. [7] reacted the *B. subtilis* xylanase with the mechanism-based inhibitor 2,3-epoxypropyl β-D-xylopyranoside to investigate the SBS. Such inhibitors are absolutely useful tools in the study of non-catalytic site binding, but they are far from available for enzymes in general. Hence alternative, more universal strategies for detecting multiple binding sites, such as SPR are extremely valuable.

Screening for Novel Binding Sites

Of the 28 enzymes not discussed already as positive or negative controls, 12 display binding to at least one ligand. These are GH5-1, GH6-1, GH8-1, GH10-1, GH11-1, GH13-3, GH31-1, GH53-1, GH62-1, GH77-1, GH84-1 and PL10-1. GH62-1 [18] and GH77-1 are currently undergoing further investigations and will not be discussed below. Additionally GH53-1, GH84-1 and PL10-1 each bind only to a single non-substrate polysaccharide that do not seem to be of biological relevance. Therefore we consider these to be spurious false positives and have not examined them further.

GH6-1, an endoglucanase from *Thermobifida fusca*, normally has a cellulose binding CBM2, which however, is omitted from the recombinant form used in this study to allow detection of one or more SBSs. It interacts with several polysaccharides, but most would be expected to serve as substrates or at least to bind in the active site (both KGM and XYG contain some unsubstituted β-1,4 linked glucose). Furthermore, SPR does not give evidence of multiple binding sites, in agreement with GH6-1 crystal structures of oligosaccharide complexes that showed binding only in the active site [56].

GH5-1 [57], GH8-1 [58], GH10-1 [59,60], GH11-1 [50,61], GH13-3 [62,63] and GH31-1 (PDB ID: 3NQQ, unpublished) all have a counterpart of the enzyme within the same family (or subfamily in the case of GH13-3) that has an SBS demonstrated by X-ray crystallography (see Fig 3). For GH5-1 and GH10-1 binding is observed to only a single ligand (Table 1) expected to be capable of binding at the active site. However, in both enzymes the residues comprising the SBS in the related family member are conserved. For GH5-1 this putative SBS would be located at W290, while for GH10-1, the alignment indicates a potential SBS primarily consisting of K38. This conservation does not necessarily indicate the presence of a functional SBS as changes in the surrounding residues could influence or even eliminate binding altogether as found for SBS2 of AMY1 (GH13-1). There is a second isoform AMY2 of this barley enzyme with high sequence identity to AMY1, including key residues in the two SBSs detected in AMY1. SBS2, however, seems non-functional in AMY2 [64], likely due the altered residues surrounding the conserved SBS2 binding residues [65]. Thus further testing would be needed to confirm the presence of an SBS in GH5-1 and GH10-1.

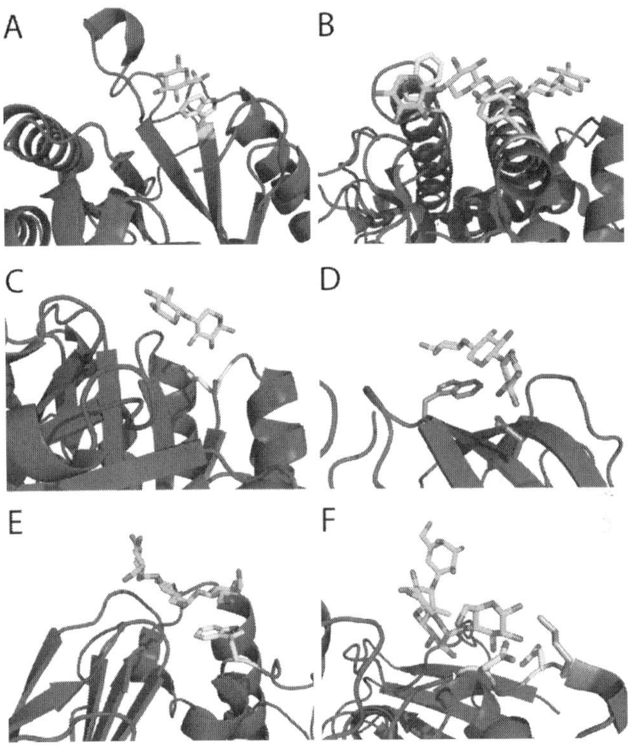

Figure 3. Conserved surface binding sites in related family members.

The crystal structure is represented as a cartoon in red, with the bound carbohydratesshown in green (carbon) and red (oxygen). Conserved residues of the SBS also found in the enzymes in this studyare shown in yellow (carbon) and red (oxygen), while non-conserved residues are shown in grey. A shows the GH5 exo-β1—3 glucanase from *Candida albicans* (2PC8), B shows the GH8 xylanase from *Pseudoalteromonas haloplanktis* (2B4F),

C shows the GH10 xylanase from *Thermoascus aurantiacus* (1GOQ), D shows the GH11 xylanase from Bacillus subtilis (2QZ3), E shows the GH13 (subfamily 5) α-amylase from *Bacillus* sp. 707, F shows the GH31 α-glucosidase from *Ruminococcus obeum* (3POC).

The xylanase GH11-1 interacts with five ligands (Table 1), including LAM and PUL which are not expected to be substrates. Three GH11 enzymes have been reported to possess an SBS, which occur, however, at two distinct structural locations [66]. None of these SBS residues are conserved within GH11-1. This does not necessarily indicate that GH11-1 has no SBS as there could be a third structural location for an SBS within GH11. Further investigation will be required to determine if GH11-1 indeed contains an SBS. In family GH13 there are several distinct structural locations for SBSs, which are often present within the same enzyme. For instance, the recently published structure of the *E. coli* branching enzyme demonstrated six different SBS positions occupied by linear maltooligosaccharides [67].

The *E. coli* α-amylase (GH13-3) is homologous to maltohexaose forming α-amylase from *Bacillus* sp. 707 known to have four SBSs [63]. Only one of these SBSs is conserved in GH13-3, which fits with the SPR β-cyclodextrin binding data yielding an average stoichiometry of 1.63, compatible with the presence of a second binding site beyond the active site. Based on alignment with the *Bacillus* enzyme, the location of the SBS is most likely centered at W346 in GH13-3.

Rice α-glucosidase (GH31-1) is the enzyme found to interact with the greatest variety of polysaccharides in both AE and IPP experiments. This may possibly be explained by the relatively wide variety of activities present within the GH31 family, including α-glucosidase, α-galactosidase, α-mannosidase and α-xylosidase, suggesting this family of enzymes is highly adaptable. Remarkably, residues comprising the SBS in *Blautia obeum* (formerly *Ruminococcus obeum*) GH31 α-glucosidase known to have an SBS are in part conserved in GH31-1 [68]. Somewhat unusually this SBS has no aromatic, but solely hydrogen bonding residues. Two of these, D651 and D653, are conserved in GH31-1 while a third, K654, is not. This in conjunction with the large variety of polysaccharides binding to GH31-1, suggests an SBS exists and most likely at a site equivalent to that in the *B. obeum* enzyme.

Family GH8 is multi-specific, containing xylanases, cellulases and chitinases among others. One enzyme, the xylanase from *Pseudoalteromonas haloplanktis* has been demonstrated by X-ray crystallography to possess an SBS [58] and its binding properties were explored in more detail [69]. GH8-1 is an endoglucanase from *Clostridium thermocellum* with the SBS from *P. haloplanktis* xylanase conserved (at W261 and Y315 in the *C. thermocellum* enzyme, see Fig 3). GH8-1 binds to several polysaccharides, including cellulose and xylan, and cellobiose suppresses the interaction with cellulose. SPR results in an average stoichiometry for cellopentaose of 2.33 suggesting the presence of two SBSs. While weak, there is also evident binding of xylotetraose and maltotetraose that probably reflects lower specificity of the SBS(s). While this is the first direct evidence of an SBS in a cellulase, both molecular dynamics [70] and crystalline cellulose binding experiments [71] previously indicated that binding outside the active site and the C-terminal CBM1 may occur in the *Trichoderma reesei* cellobiohydrolase Cel7A.

Future Perspectives

AE, IPP and SPR have all been used in the past to study SBSs previously identified by structural studies. Here we demonstrate that these techniques are useful in detecting novel enzyme binding sites, also in the absence of a solved structure. Assessment of additional binding sites is not routine in characterization of new enzymes, however, thanks to these relatively simple tools, the presence of potential SBSs can be recognized fairly quickly. Indeed, the first known SBS was detected not in a crystal structure, but through techniques monitoring binding [72]. Localizing the binding site to particular residues is more challenging, however, besides structural techniques, chemical modification in the presence and absence ligands, coupled to proteolytic digestion and mass spectrometry analysis offers an alternative path. Indeed such differential labelling was used to identify an SBSs in barley α-amylase AMY2 [73].

SBSs appear to be fairly widespread in nature, but still relatively few studies examined their properties and importance to enzymes. Binding site screening techniques such as presented here can greatly expand knowledge about these important sites if employed more routinely in enzyme characterization. Additionally, while CBMs are easier to predict than SBSs based on sequences, there are still new CBM families to be discovered and characterized and these same techniques are equally suited for detecting binding by CBMs. However, interaction at the active site must always be kept in mind, although cases of binding to non-substrates or to insoluble substrates may indicate non-active site binding. Measuring the impact of non-substrate polysaccharides on enzyme activity against natural substrates provides one way of determining binding location. Follow up with a technique such as SPR that can determine the number of binding sites is an important verification step. Methods for selectively blocking the active site are of great utility, but are only available for relatively few enzymes. In this study we have demonstrated that AE, IPP and SPR are useful for identifying binding outside of the active site. Several of the examined enzymes very likely contain SBSs, including the first cellulase ever identified to possess an SBS. These techniques provide a simple toolbox for the routine identification of binding outside of the active site at SBSs and CBMs.

REFERENCES

1. Gilbert HJ, Knox JP, Boraston AB. Advances in understanding the molecular basis of plant cell wall polysaccharide recognition by carbohydrate-binding modules. Curr Opin Struct Biol. 2013; 23:669–77. 10.1016/j.sbi.2013.05.005
2. <j/>Cockburn D, Svensson B. Surface binding sites in carbohydrate active enzymes: an emerging picture of structural and functional diversity In: Lindhorst TK, Rauter AP, editors. Carbohydrate Chemistry: Chemical and Biological Approaches Cambridge: Royal Society of Chemistry; 2013. p. 204–221.
3. Cuyvers S, Dornez E, Delcour JA, Courtin CM. Occurrence and functional significance of secondary carbohydrate binding sites in glycoside hydrolases. Crit Rev Biotechnol. 2012; 32(2):93–107. 10.3109/07388551.2011.561537
4. Cockburn D, Wilkens C, Ruzanski C, Andersen S, Nielsen JW, Smith AM, et al. Analysis of surface binding sites (SBSs) in carbohydrate active enzymes with focus on glycoside hydrolase families 13 and 77 — a mini-review. Biologia. 2014; 69(6):705–12.
5. Boraston AB, Bolam DN, Gilbert HJ, Davies GJ. Carbohydrate-binding modules: fine-tuning polysaccharide recognition. Biochem J. 2004; 382:769–81.
6. Lombard V, Golaconda Ramulu H, Drula E, Coutinho PM, Henrissat B. The carbohydrate-active enzymes database (CAZy) in 2013. Nucleic Acids Res. 2014; 42(1):D490–5.
7. Cuyvers S, Dornez E, Abou Hachem M, Svensson B, Hothorn M, Chory J, et al. Isothermal titration calorimetry and surface plasmon resonance allow quantifying substrate binding to different binding sites of *Bacillus subtilis* xylanase. Anal Biochem. 2012; 420(1):90–2. 10.1016/j.ab.2011.09.005
8. Koropatkin NM, Smith TJ. SusG: a unique cell-membrane-associated α-amylase from a prominent human gut symbiont targets complex starch molecules. Structure. 2010; 18(2):200–15. 10.1016/j.str.2009.12.010
9. Cockburn D, Nielsen MM, Christiansen C, Andersen JM, Rannes JB, Blennow A, et al. Surface binding sites in amylase have distinct roles in recognition of starch structure motifs and degradation. Int J Biol Macromol. 2015; 75:338–45. 10.1016/j.ijbiomac.2015.01.054
10. Nielsen MM, Bozonnet S, Seo E, Mótyán JA, Andersen JM, Dilokpimol

A, et al. Two secondary carbohydrate binding sites on the surface of barley α-amylase 1 have distinct functions and display synergy in hydrolysis of starch granules. Biochemistry. 2009; 48(32):7686–97. 10.1021/bi900795a

11. Ruzanski C, Smirnova J, Rejzek M, Cockburn D, Pedersen HL, Pike M, et al. A bacterial glucanotransferase can replace the complex maltose metabolism required for starch to sucrose conversion in leaves at night. J Biol Chem. 2013; 288(40):28581–98. 10.1074/jbc.M113.497867
12. Yaniv O, Jindou S, Frolow F, Lamed R, Bayer EA. A simple method for determining specificity of carbohydrate-binding modules for purified and crude insoluble polysaccharide substrates. Methods Mol Biol. 2012; 908:101–7. 10.1007/978-1-61779-956-3_10
13. Tomme P, Boraston A, Kormos JM, Warren RA, Kilburn DG. Affinity electrophoresis for the identification and characterization of soluble sugar binding by carbohydrate-binding modules. Enzyme Microb Technol. 2000; 27(7):453–8.
14. Christiansen C, Abou Hachem M, Glaring MA, Viksø-Nielsen A, Sigurskjold BW, Svensson B, et al. A CBM20 low-affinity starch-binding domain from glucan, water dikinase. FEBS Lett. 2009; 583(7):1159–63. 10.1016/j.febslet.2009.02.045
15. Nielsen JW, Kramhøft B, Bozonnet S, Abou Hachem M, Stipp SLS, Svensson B, et al. Degradation of the starch components amylopectin and amylose by barley α-amylase 1: role of surface binding site 2. Arch Biochem Biophys. 2012; 528(1):1–6. 10.1016/j.abb.2012.08.005
16. Søgaard M, Kadziola A, Haser R, Svensson B. Site-directed mutagenesis of histidine 93, aspartic acid 180, glutamic acid 205, histidine 290, and aspartic acid 291 at the active site and tryptophan 279 at the raw starch binding site in barley α-amylase 1. J Biol Chem. 1993; 268(30):22480–4.
17. Bozonnet S, Jensen MT, Nielsen MM, Aghajari N, Jensen MH, Kramhøft B, et al. The 'pair of sugar tongs' site on the non-catalytic domain C of barley α-amylase participates in substrate binding and activity. FEBS J. 2007; 274(19):5055–67.
18. Wilkens C, Andersen S, Petersen BO, Li A, Busse-Wicher M, Birch J, et al. An efficient arabinoxylan-debranching α-l-arabinofuranosidase of family GH62 from *Aspergillus nidulans*contains a secondary carbohydrate binding site. Appl Microbiol Biotechnol. 2016;

doi: 10.1007/ s00253-016-7417-8

19. Wilkens C, Auger KD, Anderson NT, Meekins DA, Raththagala M, Abou Hachem M, et al. Plant alpha-glucan phosphatases SEX4 and LSF2 display different affinity for amylopectin and amylose. FEBS Lett. 2016; 590(1):118–28. 10.1002/1873-3468.12027
20. Dilokpimol A, Nakai H, Gotfredsen CH, Baumann MJ, Nakai N, Abou Hachem M, et al. Recombinant production and characterisation of two related GH5 endo-β-1,4-mannanases from *Aspergillus nidulans* FGSC A4 showing distinctly different transglycosylation capacity. Biochim Biophys Acta. 2011; 1814(12):1720–9. 10.1016/j.bbapap.2011.08.003
21. Mori H, Bak-Jensen K, Gottschalk TE, Motawia MS, Damager I, Møller BL, et al. Modulation of activity and substrate binding modes by mutation of single and double subsites +1/+2 and -5/-6 of barley α-amylase 1. FEBS J. 2001; 268(24):6545–58.
22. Vester-Christensen MB, Abou Hachem M, Svensson B, Henriksen A. Crystal structure of an essential enzyme in seed starch degradation: barley limit dextrinase in complex with cyclodextrins. J Mol Biol. 2010; 403(5):739–50. 10.1016/j.jmb.2010.09.031
23. Stoffer B, Frandsen TP, Busk PK, Schneider P, Svendsen I, Svensson B. Production, purification and characterization of the catalytic domain of glucoamylase from *Aspergillus niger*. Biochem J. 1993; 292:197–202.
24. Nakai H, Baumann MJ, Petersen BO, Westphal Y, Abou Hachem M, Dilokpimol A, et al. *Aspergillus nidulans* α-galactosidase of glycoside hydrolase family 36 catalyses the formation of α-galacto-oligosaccharides by transglycosylation. FEBS J. 2010; 277(17):3538–51. 10.1111/j.1742-4658.2010.07763.x
25. Nakai H, Baumann MJ, Petersen BO, Westphal Y, Schols H, Dilokpimol A, et al. The maltodextrin transport system and metabolism in *Lactobacillus acidophilus* NCFM and production of novel α-glucosides through reverse phosphorolysis by maltose phosphorylase. FEBS J. 2009; 276(24):7353–65. 10.1111/j.1742-4658.2009.07445.x
26. Nakai H, Abou Hachem M, Petersen BO, Westphal Y, Mannerstedt K, Baumann MJ, et al. Efficient chemoenzymatic oligosaccharide synthesis by reverse phosphorolysis using cellobiose phosphorylase and cellodextrin phosphorylase from *Clostridium thermocellum*. Biochimie. 2010; 94:1–9.

27. Bauer S, Vasu P, Persson S, Mort AJ, Somerville CR. Development and application of a suite of polysaccharide-degrading enzymes for analyzing plant cell walls. Proc Natl Acad Sci U S A. 2006; 103(30):11417–22.
28. Bradford MM. A rapid and sensitive method for the quantitation of microgram quantities of protein utilizing the principle of protein-dye binding. Anal Biochem. 1976; 72:248–54.
29. Roberts PJ, Whelan WJ. The mechanism of carbohydrase action. 5. Action of human salivary α-amylase on amylopectin and glycogen. Biochem J. 1960; 76:246–53.
30. Robert X, Haser R, Gottschalk T, Ratajczek F, Driguez H, Svensson B, et al. The structure of barley α-amylase isozyme 1 reveals a novel role of domain C in substrate recognition and binding: a pair of sugar tongs. Structure. 2003; 11(8):973–84.
31. Robert X, Haser R, Mori H, Svensson B, Aghajari N. Oligosaccharide binding to barley α-amylase 1. J Biol Chem. 2005; 280(38):32968–78.
32. Qian M, Haser R, Payan F. Carbohydrate binding sites in a pancreatic α-amylase-substrate complex, derived from X-ray structure analysis at 2.1 A resolution. Protein Sci. 1995; 4(4):747–55.
33. Payan F, Qian M. Crystal structure of the pig pancreatic α-amylase complexed with malto-oligosaccharides. J Protein Chem. 2003; 22(3):275–84.
34. Mikami B, Yoon H, Yoshigi N. The crystal structure of the sevenfold mutant of barley β-amylase with increased thermostability at 2.5 Å resolution. J Mol Biol. 1999; 285(3):1235–43.
35. Rejzek M, Stevenson CE, Southard AM, Stanley D, Denyer K, Smith AM, et al. Chemical genetics and cereal starch metabolism: structural basis of the non-covalent and covalent inhibition of barley β-amylase. Mol BioSyst. 2011; 7(3):718–30. 10.1039/c0mb00204f
36. Ducros VMA, Zechel DL, Murshudov GN, Gilbert HJ, Szabó L, Stoll D, et al. Substrate distortion by a β-mannanase: snapshots of the Michaelis and covalent-intermediate complexes suggest a B2,5 conformation for the transition state. Angew Chem Int Ed Engl. 2002; 41(15):2824–7.
37. Hogg D, Woo E, Bolam DN, McKie VA, Gilbert HJ, Pickersgill RW. Crystal structure of mannanase 26A from *Pseudomonas cellulosa* and analysis of residues involved in substrate binding. J Biol Chem. 2001; 276(33):31186–92.
38. Jahn M, Stoll D, Warren RA, Szabo L, Singh P, Gilbert HJ, et

al. Expansion of the glycosynthase repertoire to produce defined manno-oligosaccharides. Chem Commun. 2003; (12):1327–9.

39. Tailford LE, Ducros VM, Flint JE, Roberts SM, Morland C, Zechel DL, et al. Understanding how diverse β-mannanases recognize heterogeneous substrates. Biochemistry. 2009; 48(29):7009–18. 10.1021/bi900515d

40. Parsiegla G, Reverbel-Leroy C, Tardif C, Belaich JP, Driguez H, Haser R. Crystal structures of the cellulase Cel48F in complex with inhibitors and substrates give insights into its processive action. Biochemistry. 2000; 39(37):11238–46.

41. Parsiegla G, Juy M, Reverbel-Leroy C, Tardif C, Belaich JP, Driguez H, et al. The crystal structure of the processive endocellulase CelF of *Clostridium cellulolyticum* in complex with a thiooligosaccharide inhibitor at 2.0 A resolution. EMBO J. 1998; 17(19):5551–62.

42. Parsiegla G, Reverbel C, Tardif C, Driguez H, Haser R. Structures of mutants of cellulase Cel48F of *Clostridium cellulolyticum* in complex with long hemithiocellooligosaccharides give rise to a new view of the substrate pathway during processive action. J Mol Biol. 2008; 375(2):499–510.

43. Raha M, Kawagishi I, Müller V, Kihara M, Macnab RM. *Escherichia coli* produces a cytoplasmic α-amylase, AmyA. J Bacteriol. 1992; 174(20):6644–52.

44. Shim JH, Park JT, Hong JS, Kim KW, Kim MJ, Auh JH, et al. Role of maltogenic amylase and pullulanase in maltodextrin and glycogen metabolism of *Bacillus subtilis* 168. J Bacteriol. 2009; 191(15):4835–44. 10.1128/JB.00176-09

45. Scannapieco FA, Bhandary K, Ramasubbu N, Levine MJ. Structural relationship between the enzymatic and streptococcal binding sites of human salivary α-amylase. Biochem Biophys Res Commun. 1990; 173(3):1109–15.

46. Ramasubbu N, Ragunath C, Mishra PJ. Probing the role of a mobile loop in substrate binding and enzyme activity of human salivary amylase. J Mol Biol. 2003; 325(5):1061–76.

47. Ramasubbu N, Ragunath C, Mishra PJ, Thomas LM, Gyémánt G, Kandra L. Human salivary α-amylase Trp58 situated at subsite -2 is critical for enzyme activity. FEBS J. 2004; 271(12):2517–29.

48. Ragunath C, Manuel SGA, Venkataraman V, Sait HBR, Kasinathan C,

Ramasubbu N. Probing the role of aromatic residues at the secondary saccharide-binding sites of human salivary α-amylase in substrate hydrolysis and bacterial binding. J Mol Biol. 2008; 384(5):1232–48. 10.1016/j.jmb.2008.09.089

49. Larson SB, Greenwood A, Casico D, Day J, McPherson A. Refined molecular structure of pig pancreatic α-amylase at 2.1 Å resolution. J Mol Biol. 1994; 235:1560–84.
50. Ludwiczek ML, Heller M, Kantner T, McIntosh LP. A secondary xylan-binding site enhances the catalytic activity of a single-domain family 11 glycoside hydrolase. J Mol Biol. 2007; 373(2):337–54.
51. Hernandez-Gomez M, Rydahl MG, Rogowski A, Morland C, Cartmell A, Crouch L, et al. Recognition of xyloglucan by the crystalline cellulose-binding site of a family 3a carbohydrate-binding module. FEBS Lett. 2015; 589:2297–303. 10.1016/j.febslet.2015.07.009
52. Strobel KL, Pfeiffer KA, Blanch HW, Clark DS. Structural insights into the affinity of Cel7A carbohydrate-binding module for lignin. J Biol Chem. 2015; 290:22818–26. 10.1074/jbc.M115.673467
53. Hägglund P, Eriksson T, Collén A, Nerinckx W, Claeyssens M, Stålbrand H. A cellulose-binding module of the *Trichoderma reesei* β-mannanase Man5A increases the mannan-hydrolysis of complex substrates. J Biotechnol. 2003; 101(1):37–48.
54. Simpson PJ, Xie H, Bolam DN, Gilbert HJ, Williamson MP. The structural basis for the ligand specificity of family 2 carbohydrate-binding modules. J Biol Chem. 2000; 275(52):41137–42.
55. Fan L, Zhang Z, Yu X, Xue Y, Tan T. Self-surface assembly of cellulosomes with two miniscaffoldins on *Saccharomyces cerevisiae* for cellulosic ethanol production. Proc Natl Acad Sci U S A. 2012; 109(33):13260–5. 10.1073/pnas.1209856109
56. Larsson AM, Bergfors T, Dultz E, Irwin DC, Roos A, Driguez H, et al. Crystal structure of *Thermobifida fusca* endoglucanase Cel6A in complex with substrate and inhibitor: the role of tyrosine Y73 in substrate ring distortion. Biochemistry. 2005; 44(39):12915–22.
57. Patrick WM, Nakatani Y, Cutfield SM, Sharpe ML, Ramsay RJ, Cutfield JF. Carbohydrate binding sites in *Candida albicans* exo-β-1,3-glucanase and the role of the Phe-Phe 'clamp' at the active site entrance. The FEBS journal. 2010; 277(21):4549–61. 10.1111/j.1742-4658.2010.07869.x

58. De Vos D, Collins T, Nerinckx W, Savvides SN, Claeyssens M, Gerday C, et al. Oligosaccharide binding in family 8 glycosidases: crystal structures of active-site mutants of the β-1,4-xylanase pXyl from *Pseudoaltermonas haloplanktis* TAH3a in complex with substrate and product. Biochemistry. 2006; 45(15):4797–807.
59. Lo Leggio L, Kalogiannis S, Eckert K, Teixeira SC, Bhat MK, Andrei C, et al. Substrate specificity and subsite mobility in *T. aurantiacus* xylanase 10A. FEBS Lett. 2001; 509(2):303–8.
60. Schmidt A, Gubitz GM, Kratky C. Xylan binding subsite mapping in the xylanase from *Penicillium simplicissimum* using xylooligosaccharides as cryo-protectant. Biochemistry. 1999; 38(8):2403–12.
61. Vandermarliere E, Bourgois TM, Rombouts S, Van Campenhout S, Volckaert G, Strelkov SV, et al. Crystallographic analysis shows substrate binding at the -3 to +1 active-site subsites and at the surface of glycoside hydrolase family 11 endo-1,4-β-xylanases. Biochem J. 2008; 410(1):71–9.
62. Lyhne-Iversen L, Hobley TJ, Kaasgaard SG, Harris P. Structure of *Bacillus halmapalus* α-amylase crystallized with and without the substrate analogue acarbose and maltose. Acta Crystallogr Sect F Struct Biol Cryst Commun. 2006; 62:849–54.
63. Kanai R, Haga K, Akiba T, Yamane K, Harata K. Biochemical and crystallographic analyses of maltohexaose-producing amylase from alkalophilic *Bacillus* sp. 707. Biochemistry. 2004; 43(44):14047–56.
64. Tranier S, Deville K, Robert X, Bozonnet S, Haser R, Svensson B, et al. Insights into the "pair of sugar tongs" surface binding site in barley α-amylase isozymes and crystallization of appropriate sugar tongs mutants. Biologia. 2005; 60 (Suppl. 16)(16):37–46.
65. Seo E, Andersen JM, Nielsen MM, Vester-Christensen MB, Christiansen C, Jensen JM, et al. New insight into structure/function relationships in plant α-amylase family GH13 members. J Appl Glycosci. 2010; 57(2):157–62.
66. Cuyvers S, Dornez E, Rezaei MN, Pollet A, Delcour JA, Courtin CM. Secondary substrate binding strongly affects activity and binding affinity of *Bacillus subtilis* and *Aspergillus niger* GH11 xylanases. FEBS J. 2011; 278(7):1098–111. 10.1111/j.1742-4658.2011.08023.x
67. Feng L, Fawaz R, Hovde S, Gilbert L, Chiou J, Geiger JH. Crystal structures of *Escherichia coli* branching enzyme in complex with

linear oligosaccharides. Biochemistry. 2015; 54:6207–18. 10.1021/acs.biochem.5b00228

68. Tan K, Tesar C, Wilton R, Keigher L, Babnigg G, Joachimiak A. Novel α-glucosidase from human gut microbiome: substrate specificities and their switch. FASEB J. 2010; 24(10):3939–49. 10.1096/fj.10-156257
69. Cuyvers S, Dornez E, Delcour JA, Courtin CM. The secondary substrate binding site of the *Pseudoalteromonas haloplanktis* GH8 xylanase is relevant for activity on insoluble but not soluble substrates. Appl Microbiol Biotechnol. 2011; 92(3):539–49. 10.1007/s00253-011-3343-y
70. Lin Y, Silvestre-Ryan J, Himmel ME, Crowley MF, Beckham GT, Chu J. Protein allostery at the solid-liquid interface: endoglucanase attachment to cellulose affects glucan clenching in the binding cleft. J Am Chem Soc. 2011; 133(41):16617–24. 10.1021/ja206692g
71. Jalak J, Väljamäe P. Multi-mode binding of cellobiohydrolase Cel7A from *Trichoderma reesei* to cellulose. PLoS One. 2014; 9(9):e108181 10.1371/journal.pone.0108181
72. Loyter A, Schramm M. Multimolecular complexes of alpha-amylase with glycogen limit dextrin. Number of binding sites of the enzyme and size of the complexes. J Biol Chem. 1966; 241(11):2611–7.
73. Gibson RM, Svensson B. Identification of tryptophanyl residues involved in binding of carbohydrate ligands to barley α-amylase 2. Carlsberg Res Commun. 1987; 52(6):373–9.

7

Sugar Alcohols, Caries Incidence, and Remineralization of Caries Lesions: A Literature Review

Kauko K. Makinen

Institute of Dentistry, University of Turku, Lemminkaisenkatu 2, 20520 Turku, Finland

ABSTRACT

Remineralization of minor enamel defects is a normal physiological process that is well known to clinicians and researchers in dentistry and oral biology. This process can be facilitated by various dietary and oral hygiene procedures and may also concern dentin caries lesions. Dental caries is reversible if detected and treated sufficiently early. Habitual use of xylitol, a sugar alcohol of the pentitol type, can be associated with significant reduction in caries incidence and with tooth remineralization. Other dietary polyols that can remarkably lower the incidence of caries include erythritol which is a tetritol-type alditol. Based on known molecular parameters of simple dietary alditols, it is conceivable to predict that their efficacy in caries prevention will follow the homologous series, that is, that the number of OH-groups

Citation: Sugar Alcohols, Caries Incidence, and Remineralization of Caries Lesions: A Literature Review. Hindawi Publishing Corporation International Journal of Dentistry Volume 2010, Article ID 981072, 23 pages doi:10.1155/2010/981072

present in the alditol molecule will determine the efficacy as follows: erythritol ≥ xylitol > sorbitol. The possible difference between erythritol and xylitol must be confirmed in future clinical trials.

INTRODUCTION

The ability of human saliva to reharden acid-softened enamel was definitively first reported by Head in 1912 [1]. His report was not appreciated owing to the thenexiguous understanding of the chemical mechanisms that underlie innate host defence reactions, including the natural rehardening of caries lesions. Later, Koulourides, Pigman, and their contemporaries started to pay attention to remineralization of dental enamel by saliva [2–4]. Other earlier works are referred to in the latter references. More recent reviews and treatises were provided, for example, by the Demineralization/Remineralization Working Group [5] and by Kashket [6]. Several researchers, including Leach et al. [7] and Edgar [8] have emphasized the role that diet can play in tooth remineralization.

Rehardening of deep dentin caries lesions is not commonly encountered in most industrialized countries because dentists normally treat the teeth long before an initial lesion reaches the dentin caries level. It is well known, however, that also untreated, open dentin caries lesions can undergo a rehardening process, provided that the microbiological and chemical environment of the carious tooth changes remarkably. Such changes can be occasioned by a significant improvement of oral hygiene and remarkable changes in dietary habits. The common clinical rule is, of course, that dentin caries should always require operative treatment in order to save the tooth. The everyday reality can, however, be very different in many juvenile populations, especially in developing countries that lack the necessary dental resources. On the other hand, comment has also been made about remineralization of advanced dentinal lesions in nonindustrialized societies where restorations are not commonly completed by a dentist [9]. Furthermore, hardening, darkening, and continued service for many years by remineralized teeth have been noted [10].

Clinical studies and laboratory experiments have shown that the usage of certain dietary sugar alcohols (polyols), notably xylitol and sorbitol, can be associated with rehardening of artificial and genuine caries lesions. It is surprising that previous remineralization reviews have not mentioned even a single scientific study in this field, that is, sugar alcoholassociated remineralization. Earlier reviews have ignored the potential that sugar alcohols, especially xylitol, can provide in the planning of caries prevention strategies that may lead to remineralization of caries lesions. Consequently, the aim of this treatise is (1) to summarize the clinical caries trials on xylitol, (2) to examine those chemical and biologic features of sugar alcohols that are assumed to play a role in caries prevention and tooth remineralization, (3) to summarize the roles that innate salivary factors can play in physiologic remineralization, and (4) to review results obtained in clinical and laboratory remineralization studies with sugar alcohols. The majority of remineralization-associated investigations have been carried out with xylitol and sorbitol. Several reviews have dealt with caries limitation and sugar alcohols (vide infra). Consequently, the present treatise focuses mainly on tooth remineralization. The cariostatic potential of erythritol, a sugar alcohol of the tetritol type, will also be discussed.

CLINICAL CARIES TRIALS ON SUGAR ALCOHOLS

The first dental caries studies on xylitol began at the Institute of Dentistry at the University of Turku in Finland in late 1969. The results of these studies showed that consumption of xylitol reduced the growth of dental plaque in participating dental students by up to 50% compared with use of sucrose, D-glucose, or D-fructose [11, 12]. Based on these observations, a two-year clinical caries study and a one-year chewing-gum trial (collectively called the Turku Sugar Studies) were set up in 1972-1973 [13]. The results showed that xylitol consumption was associated with an impressive caries reduction and prompted other researchers to repeat the Turku studies. Accordingly, a newly formed "xylitol concept" was presented in 1975 to the world scientific dental health community for purposes of expansion and verification. The confirmatory rounds of testing carried out during the next 30 years showed that the most important original claims of the dental efficacy of xylitol were verified by independent researchers in long-term clinical trials which were carried out under greatly varying and challenging conditions. Nineteen clinical trials and most of the nearly 300 short-term oral biologic laboratory studies on xylitol have been reviewed and commented [14–29].

The presently available information on clinical trials on the caries-limiting effects of xylitol (reviewed in the above publications) is summarized in Table 1. Several trials also investigated sorbitol. The accumulated data of these trials have in part constituted the rationale behind current public endorsements of xylitol worldwide [30, 31]. The clinical efficacy of xylitol in caries prevention was further discussed and reviewed at the U.S. National Institutes of Health Consensus Conference in 2001 [26, 32].

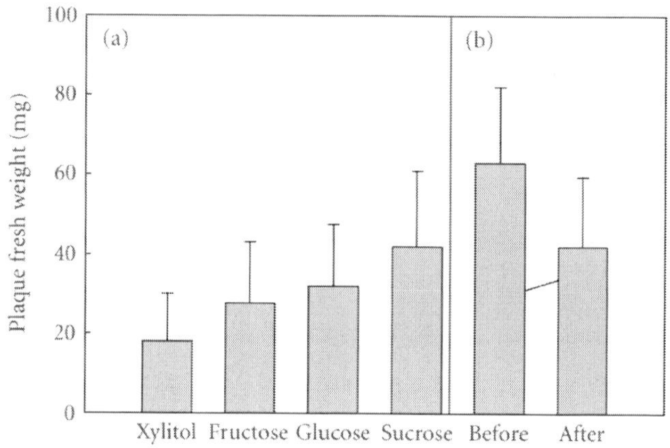

Figure 1: "How it all began": a pioneering plaque assessment study carried out in 1970 (a). Effect of dietary carbohydrates and xylitol on the growth of dental plaque after consumption of the shown sweeteners for four days (while the subjects refrained from oral hygiene), mainly in coffee or tea, and in the form of hard candies [11]. The consumption level of each sweetener was about 20 g per day and per subject. The values shown are means ± S.D. of fresh weight of plaque collected from all available tooth surfaces. (b) Inverse relationship between plaque fresh weight and its protein content. Twelve test subjects used xylitol chewing gum five times a day over a period of one month. Plaque from all available surfaces was collected following a 2-day no-oral-hygiene period. Consumption level of xylitol per day and per subject was 6.7 g. Xylitol consumption was associated with reduced plaque mass while the protein content of plaque simultaneously rose from 1.1 ± 0.2 mg to 1.4 ± 0.2 mg per mL of plaque suspension (straight line). Protein and nitrogen analyses should not be claimed to accurately determine the amount of dental plaque in clinical studies involving sugar alcohols.

The two plaque studies [11, 12], mentioned above, deserve attention since they can justifiably be regarded as pioneers of the subsequent wealth of dental investigations on xylitol. Figure 1(a) shows a rendering of the "how-itall-began" plaque study published in 1971. Four-day use of xylitol as a sweetener in subjects' diet resulted in about 50% reduction in plaque compared with the use of sucrose. The strong plaque mass-reducing effect of xylitol along with supportive biochemical effects observed in plaque in 1970- 1971 [11, 12] generated first the Turku Sugar Studies [13] and next all other xylitol caries trials shown in Table 1. Figure 1(b) additionally reveals a particular oral biologic feature of dental plaque exposed to xylitol: the increase of nitrogen- and protein-associated metabolism in plaque—although the plaque volume, mass, adhesiveness, and cariogenicity simultaneously decrease, and its alkalinity increases. The reason for emphasizing this plaque feature stems from the exploitation of protein or nitrogen determination as a single plaque assessment procedure by some authors who have been unaware of the biochemical effects of xylitol on plaque. This situation has led to erroneous conclusions on the true changes in plaque virulence as affected by xylitol [20]. In other words, plaque protein analysis should not be used as the sole (and decisive) analytical method to measure the effect of sugar alcohols on plaque.

"UTILITY VALUE" OF POLYOL STRATEGIES IN CARIES PREVENTION

Although the clinical efficacy of sugar alcohols in caries prevention is incontestable (Table 1), it is important to evaluate their true utility value assessed by independent meta-analyses and cost/benefit evaluations. A study by Deshpande and Jadad [33] on the impact of polyol-containing chewing gums on dental caries concluded that "research evidence supports using polyol-containing chewing gum as part of normal oral hygiene to prevent dental caries". The mean "preventive fraction" (PR; with 95% confidence interval) for the use of xylitol, xylitol-sorbitol blend, and sorbitol was 58.66% (35.42–81.90), 52.82% (39.64–66.00), and 20.01% (12.74–27.27), respectively. For the sorbitol-mannitol blend, the value of PR was 10.71% (−20.50–41.93), which is statistically insignificant. The appraisal of Mickenautsch et al. [34] arrived at an almost identical conclusion.

One of the studies evaluated was that of Machiulskiene et al. [35], carried out in Lithuania. It was not possible to deduce from the data of Deshpande and Jadad [33] and Mickenautsch et al. [34] whether the subsequent "rectifying" paper of Hayes [36] had been considered; Hayes determined that xylitol gum (among the several gums tested in the Lithuanian study) was the only gum that lowered the DMFS increment compared with the no-gum group after three years. The original authors [35] of the Lithuanian study had failed to recognize the potential of xylitol gum in their own clinical trial. Other public health evaluations [29, 37] have supported the above contentions on the potential that dental-protective chewing gums can provide in oral health interventions.

Further support for the utility value of xylitol interventions was provided by Milgrom et al. [38–40], who were among the first to determine the dose and frequency response of xylitol usage on mutans streptococci growth. A commendable effort was made by Milgrom et al. [39], who examined all xylitol products available on the United States market. Another practical achievement was the use of a xylitol-containing gummy bear snack as a means to reduce the growth of S. mutans/sobrinus levels [41]. The efficacy of xylitol in a dentifrice has been demonstrated in several studies [42–45], and in a pacifier as a slow-release mechanism in infants [46]. Further support for the practicality of xylitol usage has been obtained from its suggested synergistic effects with fluoride [47, 48], its use in combination with chlorhexidine [49, 50], as an alternative to fissure sealing [51], and as a sweetener in hard candies [52]. The practicality of the use of sorbitol may be somewhat affected by its low sweetness, the adaptation of oral streptococci to sorbitol [53, 54], and its customary support of the growth of dental plaque and mutans streptococci. Use of xylitol in caries limitation has been investigated from the point of view of public health [55–57]. These surveys provided information on factual usage of xylitol as a caries-limiting agent. They reviewed public endorsements and use of xylitol for caries prevention programs worldwide [30]. It is significant that researchers have emphasized the role that xylitol-containing chewing gums could potentially have as a preventive measure in public health [58]. Basic research began to provide meaningful explanations for the observed xylitol effects about twenty years ago [15, 59, 60].

LITERATURE SEARCHES ON TOOTH REMINERALIZATION

Online literature searches provide interesting historical views into tooth remineralization research. Table 2 shows a typical PubMed search using the program's own MeSH vocabulary, whose term for the search was "tooth remineralization". The number of references per year presumed during the 1980s the current reference rate of more than 40 per year. These figures tend to change over time, however, since more detailed searches retroactively add new references to the database. The Science Citation Index gave slightly different figures. Selecting "enamel rehardening" as the key word resulted in 16 references through the ages, the oldest one from 1964, while simple "remineralization" yielded a total of 2268 references (as of December 17, 2008), the oldest one from 1910, but this figure includes a large number of nondental references. The increase in the number of dental remineralization papers reflects general acceptance within the relevant clinical and scientific circles that the remineralization of caries lesions is a normal physiologic repair process that can also be aided by various dietary measures.

NONSPECIFIC XYLITOL EFFECTS VERSUS SPECIFIC ONES

The use of xylitol in food and oral hygiene products has been found to result in significant caries reduction (Table 1 and the above references). Remineralization and rehardening are terms that have most often been employed to describe the type of caries arrest demonstrated in clinical xylitol programs and laboratory studies. Several studies suggest that xylitol can exert specific effects on dental caries not shown by hexitols. Authors who have opposed the existence of specific xylitolassociated effects in caries reduction have contended that caries reduction observed after xylitol use can simply be explained in terms of the following passive xylitol effects.

- Involvement of mere salivary effects, that is, the increase in salivation regularly associated with the consumption of sweet items, constitutes the only reason that explains the clinical observations made with xylitol in caries prevention studies.
- Mere partial removal of a caries-inducive agent (sugar, notably sucrose) from the diet and substituting it with an essentially non-fermentable sweetener (xylitol) explains the observed caries reduction. In other words, in the presence of xylitol the cariogenic organisms are merely deprived of their normal growth substrate. The growth of dental plaque and the progression of caries will reduce only as a result of partial removal of the cariogenic challenge.

The above passive xylitol effects naturally constitute an important cornerstone in xylitol-associated caries limitation and would even as such fully justify the promotion of xylitol as a caries-reducing agent. The scientific review papers and professional evaluations thus far published have not denied this fact. Scientific literature is, however, replete with findings that also support the involvement of active, specific xylitol effects that operate even in the presence of fermentable hexose-based carbohydrates, that is, in situations where a strong cariogenic challenge is present. In the Turku Sugar

Studies [13] sucrose and xylitol gums differed significantly from each other in their caries-limiting ability in a situation where the salivary involvement (i.e., the chewing effect) was regarded as similar in both study cohorts. The Belize studies [61] and the preceding animal experiments [62–64] also support the idea of specific xylitol effects; xylitol was found to limit dental caries even in the presence of a strong cariogenic challenge and was more effective than sorbitol.

Shyu and Hsu [63] showed in rats that alditols differed significantly in their caries-reducing potential: xylitol caused the lowest caries scores (86% reduction). D-Mannitol reduced caries by 70%, sorbitol by 48%, and plain diet by 39% (compared with sucrose added to the plain diet; all sweeteners were tested at a 10% level). This study also reflects a problem that plagues most comparative dental trials on dietary sweeteners: the use of percentage levels. When 10% of either xylitol (mol. wt. 152.1), sorbitol (182.2), mannitol (182.2), or sucrose (342.3) is expressed in molarities (connoting the number of molecules present in solution), the true chemical concentrations differ significantly, that is, 0.657 M for xylitol, 0.549 M for the hexitols, and 0.292 M for sucrose. Comparing 10% erythritol (122.1) with 10% sucrose reveals even more remarkable differences in true chemical concentrations: 0.819 M versus 0.292 M. Critical appraisal of clinical studies should thus be conducted by considering the above discrepancy in expressing concentrations. However, the conclusions of Shyu and Hsu [63] were not remarkably affected by the above criticism.

Some of the physicochemical properties of xylitol discussed below will further elucidate the complex scientific background that is assumed to lie behind the clinical effects reported in literature. All dental xylitol studies have not, however, reached positive clinical and oral biologic findings. Long-term field experience has shown that in most cases failures in demonstrating such effects can be explained in terms of the following features of the studies in question.

- Use of caries-resistant study cohorts or cohorts with extremely low caries experience.
- (Use of too-small study cohorts.
- Use of too-low concentrations of xylitol.
- Use of too-short intervention.
- Use of too-short or too-infrequent exposure to xylitol.

- Simultaneous use of other caries-limiting agents and strategies (such as fluorides).
- Use of too-insensitive analytical procedures.
- Use of a single analytical procedure to assess oral biologic parameters (such as plaque growth).

A recommended practice is to use 6 to 7 g of xylitol daily, preferably in 3 to 5 separate episodes. Regarding oral biologic measurements (such as plaque growth, microbiology, and the chemical composition of saliva and plaque), experiments lasting from a few days to several months or even years, have been implemented. Regarding dental caries outcomes, trials that last several years are recommended. A particular dilemma has indeed been occasioned by studies that have leaned toward a single plaque assessment procedure (such as protein determination, which can lead to erroneous conclusions). In plaque studies, it is advantageous to rely on simultaneous gravimetric, planimetric (before and after, using disclosing dyes and colour photography), clinical (plaque index), bacteriologic, chemical, and enzyme measurements.

Among enzyme determinations, an analysis of the combined invertase (EC 3.2.1.26) and sucrase (EC 3.2.1.48) activities have turned out promising. (The numbers shown refer to the Enzyme Commission's classification.) Consumption of a xylitol diet is normally associated with significantly decreased whole saliva and plaque invertase-sucrase activity levels [13], suggesting diminished sucrose-splitting capacity with concomitantly reduced acid production. Reduced dextranase (EC 3.2.1.11) activity of dental plaque may also result from xylitol consumption [13]; this enzyme causes endohydrolysis of 1,6-α-D-glucosidic linkages in dextran, whose levels in plaque are normally high after sugar consumption. Similarly, strongly reduced salivary α-amylase (EC 3.2.1.1) activity has been found in xylitol-consuming subjects [13]. The activity levels of α-L-fucosidase (EC 3.2.1.51) may in turn increase during xylitol consumption (possibly resulting from increased hydrolysis of the named fucoside linkages present in salivary glycoproteins; this process may be important in the formation of the acquired tooth pellicle). All of the above glycosidases can be regarded as markers of plaque metabolism. Similarly, certain proteinase and aminopeptidase activities (which are normally increased in plaque after xylitol use) can be regarded as suitable markers of plaque biochemistry. The above changes in enzyme activity can normally be encountered when analyzing plaque and plaque extracellular fluid (with the exception of α-amylase, which is derived from glandular saliva), indicating that the origin of those activities is predominantly in the oral microbiota. The proteinase, aminopeptidase, and α-fucosidase activities increased most likely because plaque microorganisms were deprived of their preferred growth substrates (hexose-based sugars) and converted their energy-yielding enzyme activities towards the proteins and glycoproteins present in saliva. This resulted in an overall increase in plaque and saliva nitrogen metabolism. The reduction in α-amylase, dextranase, and invertase/sucrase activities can in turn be interpreted as resulting from the lowering of the sucrose levels in the subjects' diet during xylitol regimen.

CHEMICAL FEATURES OF SUGAR ALCOHOLS THAT CAN BE ASSOCIATED WITH REMINERALIZATION

The term "sugar alcohol" refers in chemical colloquialism to the reduction products of "sugars", indicating that all oxygen atoms present in a simple sugar alcohol molecule are in the form of hydroxyl groups. The terms "polyol" and "polyhydric alcohol" in turn refer to chemical compounds that contain three or more hydroxyl groups. All sugar alcohols are polyols. The polyols can be divided into acyclic compounds (alditols or glycitols, which can be regarded as true sugar alcohols) and cyclic polyols. Examples of the former are erythritol, xylitol, and D-glucitol (sorbitol), while myo-inositol serves as an example of cyclic polyols. The most important dietary sugar alcohols that will be discussed in this treatise include the four-, five-, and six-carbon members of the homologous alditol series, that is, erythritol, xylitol, sorbitol, and related polyols. These molecules are based on a single monosaccharide skeleton. Among disaccharide sugar alcohols, maltitol (derived from maltose), lactitol (derived from lactose), and palatinit (equimolar mixture of α-D-glucopyranosyl-1,6-sorbitol and α-D-glucopyranosyl1,6-D-mannitol) have received attention in nutritional and special medical uses. Complex, long-chain polyols have been manufactured for various food uses by hydrogenation of starch hydrolysates. Such products (hydrogenated starch syrups) often contain varying amounts of simpler polyols (such as sorbitol, maltitol, and trimeric and even higher homologues) as by-products.

Table 1: Summary of human caries studies on xylitol that in part have constituted the justifications for public endorsements of xylitol. The percent-reductions are in comparison with a control group that received a normal diet, fluoride treatment, or sucrose products. Nondietary (dentifrice) studies and programs on multiple preventive measures that included the use of xylitol are also shown. CH = Chlorhexidine.

Study location	Product(s) tested	Duration (years)	Dose (g/day)	Caries reduction (%). Comments. (References).
Finland	Full diet	2	67	>85. Compared with sucrose diet [13, 65]. Mostly adults.
Finland	Chewing gum	1	6.7	>82. Compared with sugar gum. 1/10 of the above dosage [13, 65]. Young adults.
Soviet Union	Candies	2	30	Up to 73. Compared with sucrose candies [66].
French Polynesia	Chewing gum	3	About 20	58–62. Compared with normal diet [67].
Hungary	Gum, candies, dentifrice	2-3	14–20	37–45. Compared with fluoride [68, 69].
Canada	Chewing gum	1-2	1.0–3.9	52. [70].
Finland	Chewing gum	2	7–10	30–57. All subjects (no-gum as control)[(a)]. [71].
Finland	Chewing gum	3	7–10	59–84. High-risk subjects[(a)]. [71].
Costa Rica	Dentifrice + NaF	3	Twice/day	Up to 12.3. 10% xylitol in the product. [44].
Costa Rica	Dentifrice + Na_2FPO_3	3	Twice/day	Up to 10. 10% xylitol in the product. [45].
Belize	Chewing gum	3.3	<10.7	Up to 73. Permanent teeth[(b)]. [61].
Belize	Chewing gum	2	<10.7	Up to 63. Deciduous teeth[(c)]. [72].
USA	Gum, pastilles	1.8	8.5	80. Supragingival root surface caries. Elderly subjects [73].
Estonia	Gum, pastilles	2-3	5	50–60. Used on school days[(d)]. [51]. Pastille as effective as gum.
Finland	Chewing gum (by mothers)	ca. 1.75	6	70 (in children). NaF and CH as control [74].
Lithuania	Chewing gum	3	2.95	21–36. [35]. Rectification of initial results [36][(e)].
Sweden	Chewing gum (by mothers)	1	2	"Significant" or 40% (in children). [75, 76][(f)].
Kuwait	Hard caramels	1.5	2.3	50. Läkerol-type hard candies were used [57][(g)].
Finland	"Slow-release pacifier"	1	159 mg	No new dentinal lesions in infants [46]. A mixture of xylitol, sorbitol, and NaF was tested. The pacifier features a pocket for the sweetened tablet.
Finland	Multiple measures	About 3.4	4.6	Counselling and the use of fluoride- and xylitol products reduced caries ($P < .001$) compared with basic prevention [77][(i)].

[(a)] Long-term effects (after up to 5-year use) have been reported [78–80]. An independent analysis showed that the total number of new restored surfaces was 4.0 per child in the xylitol group and 9.3 in the controls during the decade after the onset of the trial. Participation in the xylitol gum trial thus resulted in significant reduction in the number of first restorations and hence in costs during the subsequent decade [81].

[(b)] 16-month use of xylitol gum following the 3.3-year use of sucrose gum reduced caries significantly [82]. "<10.7" indicates the maximum calculated, supervised use (at school) per day and subject.

[(c)] Two-year use of xylitol gum remarkably protected erupting permanent teeth against caries, that is, long-term effects were involved [83].

[d] Saliva stimulants were given only on school days (about 200 per school year). Gums were as effective as pastilles (hard candies of the "Lakerol-type"). ¨
[e] The original authors failed to recognize that, in their study, xylitol gum was the only gum that lowered the DMFS increment compared with the no-gum group after 3 years. "To still observe a significant caries-lowering effect of xylitol with such a small dosage is quite remarkable". The faulty conclusions were rectified by Hayes [36].
[f] In one literature source, the authors reported an 80% reduction between "test and control". Also, when the children were 18 months old, the authors reported that "maternal consumption of xylitol- and CH/xylitol-containing chewing gums significantly reduced the mother-child transmission of salivary mutans streptococci". This study actually compared a gum with high xylitol content with gums with lower xylitol content, supplemented with either CH or NaF.
[g] Xylitol hard candies were given only on school days (one piece of candy at a time, three times a day).
[h] The pacifier features a pocket from which the saliva stimulants dissolve.
[i] The Lakerol Dents brand (Leaf). The products were given to the subjects with instructions "to be used according to directions" (i.e., two pieces of candy ¨ three times a day). The calculated maximum consumption level of xylitol was about 4.6 g/day.

According to carbohydrate nomenclature, it is permissible to use the name adonitol for ribitol, arabitol for arabinitol, sorbitol for D-glucitol, and dulcitol for galactitol. The name mannitol requires D or L. Names without D and L include erythritol, xylitol, ribitol, sorbitol, dulcitol, lactitol, and maltitol (only those polyhydric alcohols are mentioned that will be discussed in the present text). Erythritol (1,2,3,4- tetrahydroxybutane) has appeared in texts as meso-erythritol and i-erythritol; meso in this case stands for optical inactivity owing to internal compensation.

Table 2: PubMed literature search for "Tooth Remineralization".

Years	Number of references	References per year
1966–1975	1	0.1
1976–1984	10	1.3
1985–1990	180	36.0
1991–1995	156	39.0
1996–1999	136	45.3
2000–2004	222	44.4
2005–2008	228	57.0

The simple alditols are crystalline substances varying in taste from faintly sweet (galactitol) to very sweet (erythritol and xylitol, which are almost isosweet with sucrose). Several alditols, notably erythritol and xylitol, exert relatively strong negative heat of solution, a physiochemical property that in practice is reflected in the perception of a "cooling effect" in the mouth as crystalline alditol dissolves in saliva; the energy which is required in the dissolution process is taken from the environment, producing a cooling effect. Of the hexitols, sorbitol and D-mannitol show specific optical rotation for sodium D line, $[\alpha_D]$, while galactitol is optically inactive. Xylitol, ribitol, and erythritol are also optically inactive. Historic and various evolutionary and chemical aspects of sugar alcohols have been discussed in other contexts [17, 18, 84–87].

For the purpose of the present discussion, it is necessary to review some of the common sugar alcohol properties as follows

Absence of a Reducing Group

The absence of a reducing carbonyl group in the alditol molecules makes them chemically somewhat less reactive than the corresponding aldoses and ketoses. Some sugar alcohols can thus avoid those chemical reactions that normally make many dietary hexose-based sugars acidogenic and cariogenic in dental plaque. Xylitol, for example, is not normally recognized by cariogenic organisms' transport mechanisms. In cases where transport occurs, as via a constitutive mechanism normally serving pentitols other than xylitol (cf. Scangos and Reiner [88] as a historic case with E. coli), the xylitol molecule is not directly involved in lactic acid production, nor is it directly involved in cariogenesis (vide infra). Some hexitols, such as sorbitol and D-mannitol can, however, be readily recognized by several strains of cariogenic mutans streptococci (Bergey's Manual of Determinative Bacteriology regards mutans streptococci as organisms that can be identified based on their positive hexitol fermentation). The absence of the reducing carbonyl group does make these hexitols less acidogenic than the corresponding aldose and ketose forms. However, sorbitol and D-mannitol normally support the growth of mutans streptococci and dental plaque.

The Reducing Power

The sugar alcohol molecules contain an "extra" number of hydrogen atoms that can be deposited on other metabolites such as coenzymes (e.g., NADP or NAD) and other acceptors to generate chemically reduced products and intermediates of metabolism. The alditol molecules' two "extra" hydrogen atoms must thus be present in products that are formed from the alditols. The general alditol structure $(CH_2O)_n \cdot 2H$ may eventually give rise to some organic acids also in dental plaque. The regular sugars (such as glucose and fructose) have an elementary composition equalling $(CH_2O)_n$. The hexitols sorbitol and D-mannitol should normally yield one mole each of lactic acid, formic acid, and ethanol, while the pentitol xylitol should normally form one mole each of acetic acid, formic acid, and ethanol. For all practical purposes, the acidogenicity of dietary xylitol and erythritol, present in dental plaque, is normally insignificant.

Complexation

Owing to the polyoxy structure of the sugar alcohol molecules they can form complex compounds with various metal cations and oxyacids. From the point of view of tooth remineralization, the complexes with Ca(II) are important. (In chemical literature, Ca(II) refers to the divalent calcium ion. In the text below, II is omitted.)

Hydrophilicity

The presence of a large number of hydroxyl groups makes most sugar alcohols readily soluble in saliva. The most hydrophilic alditols can compete with water molecules for the hydration layer of bio-molecules (such as proteins and metal cations). Some consequences of the pronounced hydrophilicity can be seen in the strengthening of hydrophobic interactions between protein molecules (and within a protein molecule). This is in practice reflected in the protection of proteins against thermal and other denaturation or damaging processes. The protected protein configurations can include α-helix and β-structures. Related to hydrophilicity is the action of sugar alcohols as chaotropic agents under certain chemical conditions. Chaotropic agents break up organized water structures (such as the primary hydration layer of proteins and metal cations) and affect reactions that obtain their energy from the release of structured water. Another concept that will be used below, that is, the stabilization of salivary Ca phosphate systems, is a prerequisite in tooth remineralization, and results from the combined effects of complexation of Ca and displacement of water molecules.

Osmoregulator's role

Owing to the relatively low molecular weight and the hydrophilic nature of alditols, they can function as osmoregulators in various biological systems. Perhaps the best known and most commonly exploited case concerns the use of intravenous D-mannitol in lowering intracranial pressure in brain surgery, in renal function studies, as a diuretic, and so forth. Sorbitol has been used as an active principal in several cathartics preparations. Erythritol (at 40 mM) exerted a significant osmoprotector effect against stress activation of corneal epithelial cells [89].

Free Radical Scavenging

Because of their polyol nature, some sugar alcohols, such as D-mannitol, xylitol, and erythritol can act as free radical scavengers in biological and experimental systems [90].

Nucleophilicity of Sugar Alcohols in some Hydrolytic Reactions

Several studies have shown that polyhydric compounds accelerate the hydrolysis of β-lactam antibiotics, pnitrophenyl esters, cephalosporins, and so forth. In aqueous solutions at neutral and alkaline pH values [91, 92]. Consequently, polyhydric alcohols have shown to be catalytically active. This effect is attributed to a nucleophilic reaction mechanism affecting the molecule under attack by an alkoxide ion derived from proton ionization of one the hydroxyl groups. It is necessary to recall that several common alditols are alcohols with pKa values in the range of 12 to 13. The nucleophilicity described is also displayed by other polyhydric compounds such as D-glucose and sucrose [92].

General Comparison between "Glucose-Polyols" and "Non-Glucose Polyols"

The successful use of xylitol in parenteral nutrition is directly associated with the molecule's "non-glucose-polyol" nature, that is, its pentitol structure [18]. Such effects cannot be observed when sorbitol or glucose is used as a source of energy in infusion therapy [18]. This decisive difference between the xylitol and sorbitol molecules is graphically presented in Figure 2. The "glucosepolyol" nature of common dietary hexitols actually constitutes an advantage in particular nutritional uses of those hexitols such as sorbitol, D-mannitol, and galactitol, whereas it naturally elicits questions about the latter's stimulating effect on the growth of plaque and certain strains of mutans streptococci.

Competition between water and Alditol Molecules for Ca

When alditols such as xylitol and sorbitol are introduced into the oral cavity, they compete with water molecules for the primary hydration layer of Ca. The latter may comprise 4 to 12 water molecules that surround the metal ion. The partial displacement of water molecules in the hydration layer of Ca results in the formation of a new layer consisting of alditol and water, as shown schematically in Figure 3. This interaction between alditols and Ca contribute to the stabilizing effect of polyols in salivary Ca phosphate systems [14, 18].

SPECIFIC ALDITOL PROPERTIES

In addition to all alditols displaying the above-mentioned common polyol properties, laboratory observations and computer-based examination of various structural features of the alditol series have shown that tetritol, pentitol, and hexitol is each also characterized by properties specific to individual alditols or alditol groups only. These properties have been adequately described in the relevant organic and physicochemical literature and are thus old news to chemists and biochemists. Such structural and physicochemical differences between alditols are inevitably also reflected in human and microbial metabolism. Since most tooth remineralization studies on polyols have dealt with xylitol and sorbitol, it is necessary to examine these alditols more closely. The differences between xylitol and sorbitol become evident when examining the alditol literature shown below.

Some properties believed to be important in the permeability of alditols and in their relationship with water molecules are shown in Table 3.

The scientific literature is indeed replete with descriptions on interesting differences between alditols. The very sweetening capacity of dietary alditols varies remarkably and presumes important differences between the detailed chemical structures of these homologues. Some alditols are optically inactive whereas for others a clear [αD] value has been determined (vide supra). The pronounced differences between the gastrointestinal tolerances of alditols in humans speak for the existence of physicochemical specificities. Extreme examples in this sense are erythritol (well tolerated) and D-mannitol (poorly tolerated). The very chromatographic separation of alditols (in their analytics and manufacturing) indicates the existence of important differences between individual alditols. The early studies of Mills [87] showed how the conformation of alditol exerts pronounced effects on the physical properties of the molecule, such as the retention times in gas chromatography. Further evidence is shown below:

Historic Remarks

The epoch-making examination of Mills [87] divided alditols into several distinct series based on their stereoregular differences; the alditol series discussed show stereoregularities with short repeating units. The alditol series differ from each other, for example, with regard to the energy content of the individual homologues. Space does not allow a detailed account of the revelations that Mills and others have found between individual alditols. For example, the hydroxymethyl end-groups of galactitol can adopt their particular conformation (the so-called A conformation, designating the conformation round the C–C bond) more readily than the end-groups in D-mannitol can. This results in an important difference between galactitol and D-mannitol in terms of their chromatographic behaviour [87]. Analogous differences can exist also in the metabolic behaviour and clinical effects of those hexitols.

Complexation

Although complexation can be regarded as a "common" polyol property, differences between individual alditols do exist. The hexitols, the pentitols, and the tetritols can form four, three, and two oxygen atom triangles, respectively. The exact nature of the Otriangles differs from alditol to alditol. Since these triangles can participate in complexation with Ca, complexes with different stabilities will form. The possible conformations by the interaction of alditols with partially hydrated Ca were presented by Schollner et al. [¨ 94]. The difference between pentitols and hexitols was striking. Complex formation between polyols and divalent metal cations is a well-studied bioinorganic research area. Some remineralization-related aspects of polyol-Ca complexes have been studied [7, 14, 18, 102–104]. The proposed structure of a xylitol-Ca complex is shown in Figure 4.

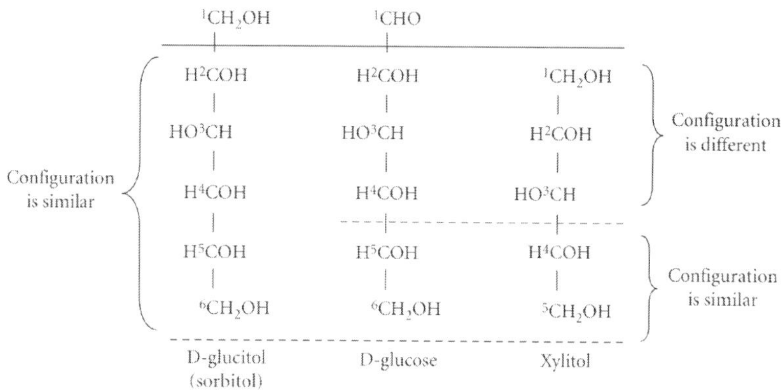

Figure 2: Relationship between the structural configurations of sorbitol (D-glucitol), D-glucose, and xylitol. The molecular configurations of sorbitol and glucose are relatively similar. Hence, sorbitol can be called a "glucose-polyol". The configuration of xylitol (a "non-glucose polyol") markedly differs from the two other configurations. The close similarity of sorbitol with glucose partly explains its plaque-promoting and mutans streptococci-stimulating effects.

Table 3: Physicochemical properties of alditols at 25◦ C.

Alditol	Molecular weight	Maxium van der Waals radius (Å)[(a)]	Partial molar volume (cm^3 mol^{-1})[(b)]	Permeability ($m\ s^{-1}$)[(c)]	"Water activity" constant K[(d)]
Glycerol	92.1	2.8	70.84	$1.49 \pm 0.40 \times 10^{-10}$	1.16
Erythritol	122.1	3.1–3.2	86.83	$4.92 \pm 0.27 \times 10^{-10}$	1.34
Xylitol	152.1	3.2–3.3	102.12	$9.9 \pm 3.4 \times 10^{-11}$	1.66
D-Arabitol	152.1	3.2			1.41
L-Arabitol	152.1	3.2			1.21
Ribitol	152.1	3.2	100.6		1.49
D-Glucitol	182.2	3.4	118.8		1.65
D-Mannitol	182.2	3.4	119.22	$7.6 \pm 4.8 \times 10^{-11}$	0.906

(a)The values for glycerol, erythritol, xylitol and D-mannitol are from Kiyosawa [93]. Other values represent estimates of the present author.
(b)At infinite dilution at 25◦C [94]. Values for ribitol and sorbitol are from Back et al. [95].
(c)Using the giant alga Chara cell membrane [96].

(d)The values of K are those of a correlating constant from the equation $a_w = x_1 \exp(-Kx_2^2)$, where x_1 and x_2 are molar fractions of water and solute, respectively, and a_w is water activity [97].

Table 4: Concentration of calcium (determined by means of atomic absorption spectrophotometry) in dental plaque of subjects who used products containing xylitol.

Study	Xylitol	Control or sucrose	Remarks
Chewing of xylitol gums (paraffin as control)	1.22 ± 0.45	0.78 ± 0.30	In μg/mg fresh weight (n = 10–12; $P < .01$). Sorbitol gave similar results [98].
Chewing of xylitol gum (compared with sucrose gum and gum base)	3.7 ± 0.5	2.4 ± 0.2	In μg/mg dry weight ($n = 83$). Gum base: 3.4 ± 0.7. Significance of differences was not given [99].
Rinsing with 0.4 M xylitol or sucrose solutions	0.90	0.67	In μg/mg protein. Plaque pools from 11 subjects in both groups. 0.01 M Na cyclamate: 0.60 [100].
Xylitol or sorbitol chewing gum compared with no gum	1.77 ± 0.99	1.70 ± 1.33	In % dry weight in plaque. No gum: 1.24 ± 0.82%. For both polyols: $P < .03$ when compared with no gum. $n = 25$ [101].

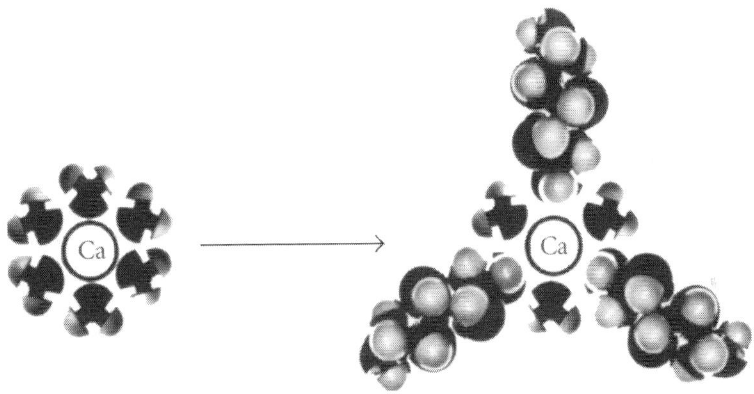

Figure 3: A simplified presentation of the competition between water and xylitol molecules for Ca, assumed to play a role in environments involving whole-mouth saliva and plaque fluid. Here, Ca has interacted with six water molecules which constitute the primary hydration layer of the metal ion (the actual number of water molecules surrounding the spherical Ca ion may vary from 4 to 12). The resulting new hydration layer consists of water molecules and xylitol molecules. This leads to stabilization of the salivary Ca phosphate systems [14, 18]. Reproduced with permission [14].

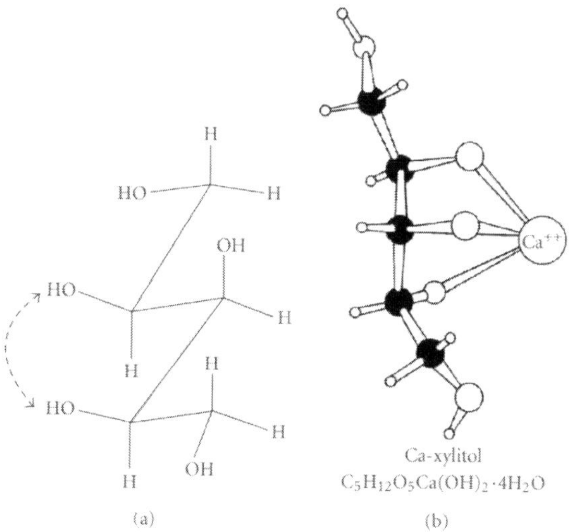

Figure 4: The zigzag structure of xylitol (a) and that of a xylitolCa complex (b) assumed to exist also in salivary environments and generally under physiologic conditions in the human body. The double-headed arrow in (a) reflects the special interaction between the oxygen atoms shown. The complex formation can facilitate the transport of Ca through membrane pores and also against weak Ca gradients. This structure may aid in the transport of Ca through the gut wall.

Molecular Parameters

The molecular volumes, the axial lengths, and the maximum molecular radii of alditols differ significantly. The apparent molar volumes and other molecular dimensions of several alditols were measured by Kiyosawa [93], who found marked differences. The differences between the molecular weights of alditols are further reflected in numerous practical situations, such as in the permeability of alditols through membranes [93, 96]. Alditols have turned out to be valuable tools in assessing the "radius" of water-filled pores in biological systems (Table 3).

Hydration and Effect on Water Structure

Even pentitols differ significantly from one another, for example, regarding hydrogen bonding within the molecule. Such differences are often reflected in pronounced differences in the hygroscopicity and other properties of the pentitols [104]. Remarkable differences between the hydration properties of alditols were observed by Carlevaro et al. [105]. In line with these studies, there is a relatively strong hydration of xylitol which should be classified as positively hydrated with an extended effect on the immediate environment. The xylitol molecule exhibits a perturbational effect on water structure. Other common alditols such as erythritol, sorbitol, and D-mannitol exhibit negative hydration. Accordingly, xylitol can be expected to protect, for example, foodstuffs against non-enzymatic browning and ascorbic acid destruction more effectively than the above hexitols do. Related to these effects are the observed differences between alditols in their "water activity" levels [97]. This property strongly depends on the chain length and the detailed molecular configuration of the alditol. The alditol-specific effects on water structure have been known since early 1970s [95]. Polyols in general interact with water to an extent which depends on their molecular structure. For example, D-mannitol behaves differently from sorbitol, and glucose behaves differently from sucrose. Polyol molecules induce structure in the water molecules, surrounding them if the orientation of OH groups is such that some of the OO spacings correspond with the O-O distance of 0.486 nm of the water lattice [95]. This effect is not a colligative property of the polyol but a property related to the OH groups.

Selective Enzyme Inhibition; Enzyme Affinity Characteristics

Xylitol inhibits more potently various microbial Dglucose isomerase-catalyzed reactions than the hexitols do; even the pentitols differ in this sense decisively from each other. Previous reports have tabulated representative inhibition constants (or percentage degrees of inhibition) that have demonstrated significant differences between alditols; the affinity constants of polyol dehydrogenases for their alditol substrates also differ significantly [15, 18]. Some authors have emphasized the selectivity of bacterial growth inhibition and the degree of pentose isomerase inhibition by pentitols already in the 1970s [106]. The simple alditols show distinctly different affinities for dehydrogenases involved in oxidation reactions. For example, the apparent Km values for D-glucitol and galactitol were 6.2 mM and 1.5 mM, respectively, in the Rhodobacter sphaeroides sorbitol dehydrogenasecatalyzed reactions [107]. The ribitol dehydrogenase of the same organism showed the following significantly different affinities: ribitol, 6.3 mM; xylitol, 77 mM [108]. Differences of the above magnitude, and even higher, are customarily found between alditols in enzyme kinetic studies of dehydrogenases. The pronounced differences, above, in enzyme affinities suggest that decisive differences between alditol effects can be encountered also in other biological reactions.

Kinetics of Oxidation of Alditols

In addition to the above dehydrogenase-associated instances, kinetic oxidation studies of alditols offer a further point of comparison between alditols. This is a well-researched area and research findings have shown significant differences between pentitols and hexitols. For example, kinetic measurements of the rate of oxidation of xylitol and galactitol by alkaline hexacyanoferrate(III) ion showed that both oxidation reactions followed first-order kinetics with respect to hydroxide ion, but with xylitol this was true only for lower hydroxyl ion concentrations, whereas with galactitol the first-order kinetics rule applied even up to manyfold variation [109]. At lower NaOH concentrations the rate constants of oxidation of xylitol and galactitol were 0.0128 and 0.0170 $mol^{-2}\ l^2\ s^{-1}$, respectively. The reaction progresses via an alkoxide ion, the end products being dicarboxylic acids. The difference between xylitol and galactitol was remarkable.

ALDITOLS ARE NOT TOOTH-DEMINERALIZING AND CALCULUS-PROMOTING AGENTS

The complexation between alditols and polyvalent cations may raise concern about the possible Ca-chelating (demineralizing) effect in the oral cavity. Readers unfamiliar with the true nature of complexation prevailing in the oral cavity may be misled by the terminology that is customarily used in chemistry. For example, $CaSO_4$ is a relatively water-insoluble compound (at 18.75°C about 0.2 parts dissolves in 100 parts water). Yet the solubility is enhanced in the presence of, for example, sorbitol. It is important to observe, however, that the stabilizing effect of polyols on the Ca phosphate systems of the oral cavity is predominantly directed to the solubility of salivary Ca and phosphate, rendering their prolonged, dissolved, supersaturated state possible, compared with the presence of, say, sucrose, which tends to initiate instantaneous precipitation of Ca and phosphate in saliva (thus eliminating a part of those substances from remineralization). The polyols' role in saliva and plaque fluid is one of stabilization; Ca and phosphate salts are stabilized in the presence of polyols and will remain in solution even at supersaturated concentrations [14, 15, 102–104].

The polyol-Ca complexes are weak compared with those formed with common food acidulants such as adipic acid, glutaric acid, ascorbic acid, succinic acid, malic acid, tartaric acid, fumaric acid, oxalic acid, and other related carboxylic acids. Research has shown that the ability of the above food acids to chelate Ca (demineralize enamel) is directly proportional to their acidity. When present in a sorbitol candy, the amount of enamel dissolution was correlated with the potential of the acids to chelate Ca. In other words, sorbitol did not chelate enamel calcium, whereas the above acids did [110]. The same notion concerns other dietary alditols.

Various techniques such as ultrasonic absorption, conductometry, solubility measurements, electrophoresis, and chromatography, have been used to determine that the Caalditol complexes are indeed relatively weak [98]. Relatively low Ca-polyol stability constants were measured for xylitol [97]. Therefore, the role normally given to alditols in the salivary environment is that of stabilization of the salivary Ca phosphate system [7, 14, 104]. This role is supposed to mimic that displayed by natural salivary peptides, such as statherin. Although the alditol-Ca complexes have thus been found to be relatively weak, they may still contribute, as Ca-carriers, to tooth remineralization and enhanced Ca absorption [18]. The consumption of xylitol has been suggested to be associated with increased Ca levels of dental plaque (Table 4). Instead of plaque hardening, (calculus formation), the extra plaque calcium is believed to enhance tooth remineralization. It is also necessary to point out that the Caalditol stability constants depend on temperature (stronger complexes are normally formed at lower temperatures) and that it is possible to calculate equilibrium constants for each of alditol's carbon atoms.

Although chemical analyses of dental plaque have thus shown increased Ca levels after xylitol consumption (Table 4), clinical studies on habitual xylitol users have not shown any increase in plaque mineralization; none of the clinical studies shown in Table 1 have reported on periodontal or calculus-forming problems. On the contrary, there are reports on the inflammation-dampening effects of xylitol in clinical and laboratory studies. In addition to xylitol generally reducing the growth and adhesiveness of dental plaque, xylitol formed plaque that was less inflammatory in a hamster cheek pouch microcirculation test than plaque grown in the presence of sucrose or fructose [111, 112]. Similar results were obtained in a bone culture study [113]. 5-Day-old "xylitol plaque" was less irritating to macrophages and bones than plaque grown in the presence of sucrose [114]. A study involving experimental gingivitis suggested that xylitol mouth rinses were periodontally less harmful than sucrose rinses (and equal to sodium cyclamate rinses) [100, 115]. Two clinical experiments on children indicated that the use of xylitol-containing chewable tablets and candies was associated with reduced plaque growth and gingival bleeding [116, 117]. More recently, xylitol was shown to inhibit cytokine expression by a lipopolysaccharide from Porphyromonas gingivalis, one of the suspected periodontopathic bacteria [118].

The above observations indicate that xylitol can be regarded as a periodontally safe dietary sweetener It is possible that xylitol's use can be augmented to comprise prevention of periodontal disease and gingival inflammation.

SALIVARY FACTORS ASSOCIATED WITH TOOTH REMINERALIZATION

The chemical conditions for tooth remineralization to occur can be summarized in the following five points.

- Sufficiently high salivary (and plaque extracellular phase) pH value.
- Sufficiently high salivary (and plaque extracellular phase) Ca level.
- Sufficiently high salivary (and plaque extracellular phase) phosphate level.
- Presence of natural salivary peptides that govern nucleation of hydroxyapatite crystals.
- Presence of the required organic and inorganic matrix (i.e., the mineral-deficient enamel or dentin sites are automatically present).

Under normal conditions human saliva meets all of the above chemical prerequisites of remineralization. Consequently, saliva is normally supersaturated with regard to Ca and phosphate, and the pH value of secreted saliva normally rises spontaneously owing to the release of carbon dioxide. These conditions will automatically facilitate the precipitation of calcium phosphate. The availability of the fluoride ion within the normal, physiologic, salivary concentration range will facilitate remineralization. The phosphate component may also be partly derived from a dietary organic source, such as casein and other protein phosphates. Typical dietary phosphate sources include milk, other dairy products (cheese), and various nuts. Dietary regimens that do not elicit too-frequent acid attacks on enamel normally guarantee that the plaque extracellular phase (plaque fluid) will not become too acidic. The nucleating and crystal growth-governing salivary factors are normally present in the saliva of all normal individuals. The concentration of Ca and inorganic phosphate (Pi) in mixed saliva, that is, 1-2 mmole/L and 2–10 mmole/L, respectively, are normally sufficient to create the required supersaturated state (of Ca phosphate) in human saliva. The overall effect of the salivary buffers gives a range of 6.2–7.4 in the saliva of most adults. These chemical conditions can be regarded as normal for tooth remineralization. Most of the buffering capacity of saliva at neutral pH values is attributed to the bicarbonate and the phosphate systems.

Saliva forms spontaneously an organic integument (the so-called acquired pellicle) on tooth surfaces. In this process extracellular enzymes secreted from plaque bacteria liberate the carbohydrate components from salivary glycoproteins (mucins), causing the residual protein structures to precipitate out of solution. The protein structures may eventually precipitate as part of the acquired pellicle film on tooth enamel. Certain salivary peptides are involved in the maintenance of the supersaturated state of Ca in saliva. Evidence has been presented on the importance of diet on the composition of the acquired pellicle; a distinct model of protein deposition on artificial hydroxyapatite discs was observed after rinses with sucrose, sorbitol, xylitol, and phosphate-buffered saline [119]. It is interesting that in the above study xylitol and sorbitol differed very significantly in Western blot tests of proteins extracted from the discs carried in the mouth for various periods of time (from 30 seconds to 20 minutes). Consequently, the alditol molecules behaved differently in the in situ studies of pellicle formation. The discs became saturated with protein very rapidly after each rinse, although clearly less salivary protein was adsorbed on to the discs after the sorbitol rinse than was adsorbed after xylitol rinse.

The important role salivary mucins and proteins may play in remineralization receives support from the studies of Kielbassa's group [120–122]. Mucin-based salivary substitutes ("artificial saliva") were considered effective remineralization-inducing adjuvants that could especially benefit hyposalivation patients. Some manufacturers have marketed mucin-based saliva substitutes that also contain xylitol.

CA-BINDING IN DENTAL PLAQUE

As shown above, xylitol consumption has been found to be associated with an increase of plaque Ca levels (Table 4). The extra Ca (compared with control situations) present in dental plaque in its entirety and in plaque extracellular fluid in particular may participate in the remineralization of mineral-deficient enamel sites. Approximately half of the concentration of plaque Ca may be ionized. It is not yet known how much of the Ca present in "xylitol plaque" is in an ionized form. It is known, however, that polyols can facilitate the solubilization of the insoluble portion of plaque Ca. It is possible that the increased Ca inhibits demineralization through a common ion effect, and may additionally facilitate remineralization during periods of high pH values.

It is necessary to recall that Ca-binding by various Grampositive plaque bacterial strains is a common phenomenon [123, 124]. Ca-binding by various surface components of oral bacteria may indeed exert significant effects on remineralization/demineralization processes. Various oralcare products such as mouth rinses and chewing gum have been used to increase the plaque levels of a combination of casein phosphopeptides and amorphous Ca phosphate (CPP-ACP), and to facilitate remineralization of enamel [125, 126]. Incorporation of CCP-ACP into plaque naturally also increases the plaque levels of Ca and Pi. In one mouth rinse study, the plaque Ca and Pi levels increased by 118% and 57%, respectively [126]. These authors strongly believe that the extra plaque Ca can significantly contribute to tooth remineralization.

The xylitol-associated increase of plaque Ca levels (Table 4) is most likely a general polyol-associated reaction; xylitol and sorbitol may not differ remarkably in this sense. The extra Ca is believed to contribute to tooth mineralization during a polyol regimen. Sorbitol, however, normally supports the growth of dental plaque and mutans streptococci.

Enamel permeability naturally plays a role in rendering Ca available for remineralization. The permeability of common salivary and dietary inorganic ions (such as Ca and Pi) has been discussed in several contexts, but less attention may have been paid to the permeability of common dietary carbohydrates. In this sense it is necessary recall that according to some permeability studies twice as much xylitol goes through the enamel as sucrose [127].

Polyols generally stabilize the calcium phosphate solutions in saliva [97]. This phenomenon results from the formation of complexes between

Ca and the polyol. Although ketoses, aldoses, disaccharides, and other carbohydrates also form complexes, those formed between xylitol and Ca are of particular interest owing to the general nonacidogenicity of xylitol in the human oral cavity. The stabilizing effect of xylitol on the salivary Ca-phosphate system can easily be demonstrated by letting acellular (Millipore-treated) wholemouth saliva stand at room temperature in the presence or either sucrose or xylitol. Sucrose allows almost instantaneous precipitation of Ca-proteinates, whereas the turbidity formation is significantly delayed in the presence of xylitol [102, 103]. In other words, xylitol mimics some of the salivary peptides (such as statherin) which control crystal formation. It has been assumed that sucrose effectively eliminates from the solution the "extra" calcium and phosphate ions that in the presence of xylitol can maintain their supersaturated, natural concentration level, a prerequisite of remineralization. In one study, saliva precipitates contained a crystalline phase that had the structure of apatite [104]. Xylitol maintains a higher pH value in saliva and plaque fluid and simultaneously maintains a supersaturated Ca-level in saliva. The combined effect can manifest as remineralization that is governed in the same way as salivary peptides.

FORMATION OF BASIC EQUIVALENTS

Because the rate of tooth remineralization generally increases already in slightly alkaline conditions, it is pertinent to recall the increase in dental plaque and whole saliva of various basic substances during xylitol regimen. The first clue of this type biochemical effects was obtained in the Turku Sugar Studies [13] which showed that the consumption of a xylitol diet was associated with a general increase of nitrogen- and proteincontaining substances in saliva. Among such substances were amino acids (which in turn can generate ammonia), and the ammonium ion itself [13, 14]. The pie chart in Figure 5 demonstrates the general increase in ninhydrin-positive substances (mostly amino acids) present in the whole-mouth saliva of subjects who habitually consumed a xylitol diet for 12 to 16.5 months. Other studies have also shown that the stimulation of saliva with xylitol products increases the ammonia [116] and the bicarbonate [13] content of wholemouth saliva.

DESCRIPTION OF INDIVIDUAL REMINERALIZATION STUDIES

The list below summarizes in chronological order 27 separate remineralization-associated clinical, basic science, and animal studies on xylitol and other sugar alcohols (two studies, #19 and #26, investigated enamel erosion). The list begins with the Turku Sugar Studies [13] whose authors used the phrase "remineralizing and therapeutic effect of xylitol" in their first clinical reports on xylitol-associated limitation of dental caries [128].

Remineralization of Caries Lesions

During a Two-Year Xylitol Feeding Trial and in a One-Year Chewing Gum Study. The final report of the two-year Turku feeding trial showed that the consumption of a xylitol diet was associated with the reversal of the caries process [65]. Because essentially similar observations were also made in a simultaneously conducted one-year chewing gum study [129], the authors of the trials were led to conclude that the usage of xylitol was indeed associated with remineralization of caries lesions [13]. Habitual use of xylitol gum resulted in a negative development of the DMFS index scores, an indication of remineralization. Subsequent planimetric evaluations of the lesion size reductions after xylitol usage in both studies supported the original conclusions made [130, 131]. An interesting, partly accidental outcome of these trials was that while the daily consumption level of xylitol in the twoyear feeding trial was estimated to be 67 g per subject, the intake of xylitol in the one-year gum study amounted to one tenth of the above value, that is, 6.7 g. Partly owing to these consumption levels, the Finnish health authorities have recommended the use of 6 to 7 g xylitol daily in the prevention of caries. This recommendation has been widely followed in the instructions issued by national dental associations in several Asian countries [30].

Effect of Xylitol-Supplemented Diet on the Regression of Fissure Caries in Rats

Dentinal molar fissure caries in rats, produced by initial exposure to dietary sucrose, were significantly reversed by subsequent exposure to a xylitolsupplemented (3% and 6%; about 0.2 M and 0.4 M, resp.) starch diet. Continued exposure to a xylitol-starch diet produced a successive regression of caries rates, which the authors interpreted as remineralization and as a therapeutic effect [7, 62].

Remineralization of Artificial Caries-Like Lesions by a Xylitol-Containing Mouth Rinse

One of the mouth rinses tested by Featherstone et al. [132] contained 2.5% (0.164 M) xylitol "to mask the mineral taste and produce a palatable rinse". Other ingredients included 0.6 mM NaF, and K, Ca, Sr, and Zn salts. Human enamel slabs with demineralized lesions were embedded in intra-oral appliances. Complete rehardening of the inner 40–50 μm and twofold rehardening of the remaining body of the lesion occurred in 160-μm deep lesions with a 1-min mouth rinse on each of 14 consecutive days. Saliva re-hardened the inner 20 μm only. The authors did not pay attention to the possible role of xylitol as a remineralization-contributing factor.

Remineralizing Properties of Xylitol in Combination with Sucrose in Rats

Rats, inoculated with Streptococcus mutans, were fed a diet containing 20% sucrose, 5% glucose, and 5% xylitol (SX diet), or solely a sucrose-glucose diet (S). (In molar terms, the above concentrations are about 0.58 M, 0.28 M, and 0.33 M, resp.) SX induced significantly fewer fissure lesions than S. Initial lesions induced by S were significantly reduced (remineralized) by subsequent exposure to SX. Change from SX to S resulted in substantial caries progression [64].

Rehardening Properties of Xylitol-CMC-Containing Saliva Substitute

The rehardening of artificially softened human enamel by nine different saliva substitutes was investigated by micro-hardness measurements [133]. All saliva substitutes exhibited a rehardening potential. The largest reduction of the rehardening potential was observed after addition of high concentrations of carboxymethylcellulose (CMC) or mucins and the addition of sorbitol (3% in the substitute; 0.165 M). The best rehardening properties were observed for low-viscous mucin- or CMC-containing saliva substitutes with xylitol (2% or 0.13 M; Saliva Orthana, Copenhagen, Denmark). The difference between xylitol and sorbitol was significant ($P<.01$).

Influence of Xylitol and Sucrose on Enamel Demineralization In Vivo

Fissure-like plaque retention grooves were created in human enamel blocks and demineralized. The blocks were mounted in prostheses of 11 subjects who used a 2.5% (0.164 M) xylitol, 2.5% (0.073 M) sucrose, or a water solution in a randomized cross-over design. During a period of 16 days, the subjects submerged the prosthesis twice a day in the test solutions for 5 minutes. Mineral loss and lesion depth were measured before and after the experiment by means of quantitative microradiography and polarized light microscopy. At the surface enamel, a significant reduction of enamel demineralization was found after the xylitol treatment. The lesion depth at the surface enamel increased 17 μm after sucrose and 7 μm after xylitol ($P< .05$) [134]. Similar enamel blocks were also treated for 28 days with 35% (2.3 M) xylitol toothpaste [135]. The results were in line with the above study.

Effect of Xylitol, Lactitol, and Other Sweeteners on Tooth Demineralization

Mixed cultures of dental plaque organisms were incubated for 24 hours in media containing different sweeteners. The attack of acids so generated was measured by Ca and phosphorus analyses. Demineralization was most severe with glucose and sucrose. Lactitol and xylitol showed extremely low enamel demineralization figures [136].

Effect of Fluoride-Polyol Gum on Remineralization

Maxillary acrylic appliances carrying carious enamel sections were worn by subjects who used Fluogum (0.113 mg F/stick) that also contains xylitol and sorbitol. After three days of chewing 15 sticks there was a significant reduction in both lesion depth and in the size of the body of the lesions by an average of 5% ($P< .05$) [137]. It is possible that both polyols contributed to remineralization.

Combined Effect of Xylitol and Fluoride on Enamel Demineralization in Vitro

Bovine enamel was exposed to a buffered lactic acid solution at pH 4.5 at +37°C. The selected enamel windows were treated with 2.63 M xylitol, 0.3 mM NaF, and their combination. Control enamel was not treated. The lesion depth was assayed by means of transversal microradiography. Xylitol and NaF had comparable effects on lesion reduction. The effects of F^- and xylitol were additive [138].

Effect of Sorbitol Chewing Gum on Enamel Lesion Remineralization

Subjects wore in situ appliances on which were mounted enamel sections containing artificial caries lesions [139]. Sorbitol-containing gum (the U.S. brand Orbit Extra) was chewed by volunteers for 20 minutes 5 times a day over a period of 7 weeks according to a crossover design. Microradiography showed that gum chewing caused 18.2% remineralization compared with the control's 12.1% remineralization (P = .07). However, a regular sucrose gum, used in an identical protocol, also caused significant remineralization (18.3% versus 10.8% in the control). The latter observation and the long chewing time used in this study suggest that any possible, specific polyol effect was possibly masked by a common "salivary effect". Hence, no proof was obtained for a specific remineralization-enhancing effect of sorbitol itself.

Effects of Sucralose, Xylitol, and Sorbitol on Remineralization of Rat Caries

Rats infected with Streptococcus sobrinus were first given drinking water containing 10% sucrose. A group of animals were thereafter fed either the same sucrose water, or received their nutrition by gavage and drank water containing 0.03% sucralose, 10% xylitol, or 17% sorbitol (added at a sweetness equivalent to 10% sucrose). The authors [140] reported that removal of the cariogenic challenge (sucrose) allowed remineralization to occur and that no sweetening agent was superior to another in this respect. The following details of the study deserve attention. (1) The intervention lasted only three weeks. (2) The rats were mono-infected with one bacterial species. (3) The authors' attempt to create almost isosweet drinking water for the animals resulted in significant differences in the chemical concentrations of the sweeteners used: sucrose 0.292 M, sorbitol 0.934 M, xylitol 0.658 M, and sucralose 0.876 mM. These characteristics of the study make it impossible to draw conclusions on the relative ability of the named sweeteners to facilitate remineralization or to reverse the process of "natural" caries.

Effects of Sorbitol and Sorbitol/Xylitol Chewing Gums on Human Enamel Remineralization

Intraoral remineralization of experimental caries-like lesions in human enamel was studied by Manning et al. [141]. Polarized light microscopy and quantitative microradiography showed that 20-min chewing of a sorbitol gum and a 3:1 sorbitol/xylitol gum 5 times per day over a period of 21 days promoted remineralization to a similar extent. The following features of the study deserve attention: (1) The long chewing time used may have abolished any possible, specific polyol effects. (2) Both chewing gums contained sorbitol as the only or as the clearly predominating polyol sweetener. Hence, it may be difficult to compare the remineralization-enhancing effect of xylitol and sorbitol in this study.

Xylitol-Induced Changes of Enamel Microhardness after Consumption of Xylitol Candy

Slabs of bovine enamel were inserted in cavities of children with rampant caries. The negative control subjects did not receive sweets whereas the treated subjects received 20 g xylitol daily in the form of candies. Predemineralized and non-demineralized enamel showed pronounced, statistically significant rehardening at exposure to xylitol ($P<.001$). Microradiography confirmed these findings [142].

Effect of Xylitol and Sorbitol in Chewing-Gums on Mineral Loss of Enamel

Human subjects with $>3 \times 10^5$ mutans streptococci per ml of saliva participated in a cross-over study involving the use of four different chewing-gums containing: (1) 70% xylitol, (2) 35% xylitol + 35% sorbitol, (3) 17.5% xylitol + 52.5% sorbitol, and (4) 70% sorbitol [143]. The subjects used 12 pieces of each gum (1.6 g) per day for 25 days. The wash-out periods lasted about 10 weeks. The subjects wore a removable palatinal plate containing two demineralized human enamel samples. The authors reported that the lesion depth and the mineral loss values, assessed microradiographically, did not differ significantly between groups. Increased concentration of xylitol in the gum resulted in a lower number of mutans streptococci in saliva and dental plaque. However, the pH drop in plaque measured in vivo after a 1-minute mouthrinse with a 10% sorbitol solution was least pronounced after the 70% xylitol gum and most pronounced after the 70% sorbitol gum period ($P<.01$). It is possible that the cross-over study design, the shorter wash-out periods, and shortness of the treatment affected the microradiographical data [144].

Stabilization (Remineralization) of Rampant Caries by the Use of Polyol Chewing Gums

The results from two cohort studies on arrest and non-progression of dentin caries after long-term usage of xylitol- and sorbitol-containing chewing gums by young subjects are summarized [145, 146]. The original research papers [61, 72] suggested that 40- and 24-month consumption of xylitol-containing chewing gum was superior to sorbitol-containing gum in the arrest of dentin and enamel caries. A re-examination 5 years after the 24-month study lent further support to this finding [83].

Remineralization of Dentin Caries Lesions of Primary Teeth by Means of Physical, Chemical, and Histologic Procedures

The Stann Creek study in Belize showed that habitual use of xylitol gum was associated with arrest of dental caries in young subjects [72]. After a 20–22-month intervention (when the children were 8 years old), a total of 23 primary teeth with extensive dentin caries lesions, whose surface in clinical examination was found to be totally remineralized, could be removed because the teeth were near their physiologic exfoliation time. The majority of the specimens had been remineralized from the surface by a non-cellular-mediated process. The topmost 20-μm layer of the lesions exhibited the highest Ca : P ratio. The rehardened surface layer (normally < 0.1 mm in thickness) was significantly ($P < .001$) harder than sound dentin and nearly as hard as sound enamel. The extracellular remineralization was most likely mediated by odontoblasts [147]. The bioinorganic and physicochemical mechanisms behind these effects are outlined [14, 15, 102, 104].

Timing of First Restorations in Relation to a Preventive Xylitol Trial

Re-examination 3 and 5 years after a 2-to 3-year xylitol gum program suggested that xylitol had “permanently prevented caries” [80]. An independent retrospective analysis showed that the need for restorations was significantly postponed in xylitol-receiving subjects [81].

Rehardening of Enamel Caries Lesions after a 16-Month Xylitol Gum Program that was preceded by a 40-month Sucrose Gum Program.

The 40-month Belize chewing gum program showed that the use of sucrose gum by subjects with high dietary sugar consumption was associated with high caries activity. After the termination of the 40-month trial, subjects of the sucrose gum group were recruited to an intense xylitol gum program (xylitol consumption level: 14 g per day and subject) for 16 months. The intensified xylitol gum usage was associated with a significant reduction in the mean DMFS score ($P = .0013$). The reduction of the score most likely resulted from the change of the D component of the index and possibly reflected rehardening of some caries lesions to a non-progressive carious state [82].

Influence of Xylitol and NaF on Dental Erosion In Vitro

Sectioned bovine incisors were treated with orange juice only or with juice containing 25% (1.64 M) xylitol and/or 0.5 ppm F. The samples were immersed in the test solutions six times daily for 5 min on each occasion for 24 days (total exposure: 12 hours). Mineral loss measurements showed that xylitol and F had an additive effect on the reduction of dental erosion by orange juice in vitro [148]. Using a similar study design, the same authors concluded that "tolerable levels of xylitol alone may not show significant caries inhibiting and remineralizing effect, but may act as a caries inhibitor additively with fluoride" [149].

Japanese Remineralization Experiments with Xylitol and Ca Hydrogenphosphate

Yanagisawa and his co-workers [150, 151] have provided crystallographic evidence on the restorative process of xylitol-associated remineralization. In their studies human enamel specimens were demineralized in acetate buffer at pH 4.0 and subsequently immersed at 37°C for two weeks in a remineralizing solution containing Ca, phosphate, F, and 20% xylitol (control solutions did not contain xylitol). Contact microradiography showed that, in the absence of xylitol, remineralization was mainly observed in the surface layers of the samples. In the presence of xylitol, somewhat less remineralization was observed in the outermost surface layers (up to 10 μm), whereas the middle and deeper layers exhibited pronounced remineralization. Multipurpose image processor studies confirmed this observation. It was concluded that xylitol acted as a carrier of Ca, maintaining a constant Ca level by introducing more mineral from the surface layers into the middle and deeper zones. Eventually, in the presence of xylitol, remineralization occurred over the entire demineralized zone [152, 153]. Further studies demonstrated that a combination of xylitol and Ca hydrogenphosphate was more effective than this same Ca salt supplemented with erythritol, sorbitol, maltitol, or palatinit [154]. When funoran (a sulphated polysaccharide of the seaweed Gloipeltis furcata) was added to the xylitol-Ca salt mixture, an increase in the extent of remineralization was observed [155]. Another study showed that the use of a chewing gum containing xylitol, Ca hydrogenphosphate, and funoran was more effective than a similar gum containing maltitol [156]. The inhibitory effect of funoran on the adherence and colonization of oral bacteria was further clarified by Saeki [157–159]. Recently, Japanese and Thai researchers teamed up for studies which supported the above concept, that is, the remineralization-associated influence of xylitol chewing gum containing funoran and calcium hydrogenphosphate [160].

Remineralization of Enamel Lesions by Various Forms of Ca in Mouth Rinse or Chewing Gum

Casein phosphopeptide and amorphous Ca phosphate nanocomplexes (CPP-ACP) have been intensively tested by Reynolds's research team [126, 161]. The CPP-ACP complex was tested in two different Lotte (Tokyo, Japan) xylitol gum formulas (one containing xylitol and $CaCO_3$, the other xylitol, $CaHPO_4/CaCO_3$, and funoran). Several CPP-ACP mouth rinse preparations were also tested. The xylitol-CPP-ACP and the sorbitol-CPP-ACP combinations prevented demineralization and promoted remineralization [126]. Xylitol and sorbitol gums did not differ [161]. These studies have employed a cross-over design with wash-out periods of only one week or somewhat longer between interventions. Owing to the fact that xylitol can exert long-term effects on dental caries [144], it is impossible to draw firm conclusions from these studies as to the relative ability of polyols to enhance remineralization.

Effect of Polyol-Containing Saliva Substitutes on Demineralized Dentin Specimens

Bovine dentin specimens were demineralized for 14 days at pH 5.5 and subsequently treated for 14 days with mucin- and CMC-based saliva substitutes that also contained 2% xylitol or sorbitol. Loss of mineral was assayed by means of transversal microradiography. For the dentin specimens, significant ($P < .05$) differences were observed between xylitol and sorbitol, the xylitol-containing saliva substitutes displaying lower mineral loss values. For enamel specimens, no significant differences were observed [162].

Effect of Xylitol + Ca Lactate on Remineralization

Artificial caries-like lesions were created in human enamel slabs which were subsequently worn by volunteers who used, in a three-leg study, over a period of two weeks, chewing gum containing either xylitol or xylitol + Ca lactate, or did not receive any gum [163]. X-ray spectrometry showed that the xylitol-Ca combination remineralized the lesions more effectively than the xylitol and the no-gum periods.

Caries Prevention with Xylitol-Containing Hard Candies among Disabled Pupils

A field study of 18 months' intervention was carried out in physically disabled school children, who received about 2.3 g xylitol in the form of hard candies in three daily episodes. The control subjects did not receive xylitol candies. The authors concluded that xylitol had "a strong and a clear remineralizing effect on caries" [57].

Effect of Xylitol + F Toothpaste on the Remineralization of Enamel In Vitro

Human enamel specimens, demineralized at pH 4.5, were treated twice a day for 14 days in a silicabased toothpaste slurry containing 500 ppm F, 500 ppm F + 5% xylitol, or no added F and xylitol. Quantitative lightinduced fluorescence studies showed that the combination of xylitol and F toothpaste was superior to the other treatments in occasioning remineralization of demineralized enamel [164].

Effect of Xylitol and Fluoride on Enamel Erosion In Vitro.

Since enamel erosion and tooth demineralization are chemically partly similar processes, one human enamel erosion study will be mentioned below. Human third molars were immersed for one min or five min in various test solutions four times a day for 14 days. Mineral loss was determined from lesion depth and surface hardness. Addition of xylitol, NaF, or a xylitol/NaF combination to an acidic drink (orange juice) reduced (but did not prevent) enamel erosion [165].

Effect of Isomalt on Enamel De- and Remineralization

In an in vitro study, subsurface bovine enamel lesions were subjected to 3-week pH-cycling involving 5-minute rinses with 10% isomalt (α-D-glucopyranosyl-1,6-sorbitol) solutions daily and 10% isomalt additions to re- or demineralizing solutions [166]. A 0.2 ppm fluoride "background" was used during the remineralization phase. In an in situ study, subsurface lesions were exposed for 2 months in vivo and brushed 3 times a day with a tooth paste containing 10% isomalt. In the in vitro study, 5-min rinses resulted in slightly increased remineralization, while continuous presence of 10% isomalt (in re- or demineralizing solutions) inhibited de- and/or demineralization. The in situ test confirmed enhancement of remineralization. It is possible that isomalt stabilizes the calcium phosphate system present in enamel surface.

ERYTHRITOL

The noncariogenicity of erythritol has been investigated in rats [167] and, based on the presence of various risk factors of caries (such as mutans streptococci and dental plaque), erythritol has been regarded as noncariogenic in humans [168–170]. All studies strongly support the idea of erythritol as a caries-reducing dietary polyol. Especially the way erythritol inhibited the growth of certain mutans streptococci isolates is interesting [170]; the results indicate that the mechanism of growth inhibition differs from that caused by xylitol. Partly based on such experiments, it is tempting to postulate that certain combinations of erythritol and xylitol will turn out to exert promising caries-limiting effects in humans. The combined effects may exceed or at least equal the separate effects of these polyols.

Also studies on erythritol have re-emphasized the inevitability of the existence of important differences between individual alditols; the alditols cannot be regarded as an entity with exactly identical molecular parameters and similar biological effects. Such a contention would be incongruent with accepted physical and chemical laws. It would be space-consuming to list all separate studies that have demonstrated the existence of selective alditol effects. In the case of erythritol, however, the following examples may serve a purpose: Sugar alcohols, especially erythritol, enhanced the fungicidal effect of benzethonium chloride toward in vitro candidal biofilms;

the difference with xylitol and D-glucitol (sorbitol) was significant [171]. De Cock and Bechert [90] emphasized the free-radical-quenching effect of erythritol, a property which may play a role in the "functionality" of erythritol-containing food products.

Several erythritol caries trials have been initiated in various parts of the world. Although there is still no valid clinical proof, the known molecular parameters of simple dietary alditols suggest that their caries-limiting efficacy may follow the simple homologous alditol series, will depend on the number of hydroxyl groups present in the alditol molecule, and will decrease as follows: erythritol ≥ xylitol > sorbitol. Combinations of erythritol and xylitol may have an edge over either of the alditols used separately: their microbiologic mechanism of action in caries prevention seems to differ and they are believed to exert a concerted and additive effect. However, no long-term human caries trial on erythritol has been completed. Hence, the possible difference between erythritol and xylitol in terms of carieslimiting ability will hopefully be elucidated in the near future. It will be essential to carry out trials where all of the above polyols (one tetritol, one pentitol, and one hexitol) will be simultaneously tested. The excellent gastrointestinal tolerability of peroral erythritol even in infants may promote its combination with xylitol in caries-limiting strategies.

ADDENDUM

The following research papers and statements have been published after the above text was completed. The effect of xylitol chewing gum on the acquisition pattern of 39 bacterial species (including mutans streptococci) was investigated in infants [172]. Mothers used xylitol or sorbitol gum (4.2 g/day) or no gum; gums were used 3 times a day for 9 months. The authors concluded that maternal use of xylitol gum did not result in statistically significant differences in the microbial plaque composition of 9- to 14-month-old infants. These results partly contradict those obtained in Finnish and Swedish mother-child studies (Table 1) where the counts of mutans streptococci decreased in children whose mothers had used xylitol. Perhaps the treatment period should have been extended beyond 14 months since the "window of infectivity" for mutans streptococci does remain open longer than 14 months. It is also possible that the polyol level in the tested gums was too low.

The following study completes caries trial information of Table 1: Pediatric topical oral xylitol syrup was administered in a group of 94 children aged 9 to 15 months for about 10.5 months. Parents administered syrup twice a day (2 xylitol 4 g doses and 1 sorbitol dose) (Xyl-2x group) or three times per day (3 xylitol 2.67 g doses) (Xyl-3x group) versus a control syrup (1 xylitol 2.67 g dose and 2 sorbitol doses). The data showed that oral xylitol syrup administered topically 2 or 3 times daily at a total daily dose of 8 g was effective in preventing early childhood caries [173]. These findings seem to be in congruence with previous field experience (Table 1) and also with the theory of the pharmacologic mode of action of xylitol in caries limitation [18].

The concept of the formation of oral biofilm has emerged during the last decades to facilitate the understanding of physiological oral processes. A recent study suggested that xylitol not only effectively inhibits acid production of cariogenic bacteria, but also prevents the formation of a multi-species bio-film [174]. In many instances oral bio-film is tantamount to dental plaque.

It is noteworthy that recently published anonymous dental journal editorials have emphasized the role of xylitol in caries prevention [175–177]. In an article in The Journal of American Dental Association [175] the potential role of xylitol-containing oral syrup in the prevention of childhood caries was discussed. The European Union also finally cleared xylitol for certain anti-caries claims [176]. The American Academy of Pediatric Dentistry's Council on Clinical Affairs issued its policy on the use of xylitol in caries prevention [177]. The resolutions and endorsements contained in the above texts are reflected in recent conclusions of leading experts on the use of polyol gums to prevent dental caries in general [178] and especially in early childhood [179, 180].

Finally, a recent research paper has shown that xylitol can indeed modify dental plaque, resulting in marked reduction in plaque acidogenicity (which was not detected using D-glucitol) [181], a finding that confirms similar results obtained almost 20 years earlier [182, 183]. Previously observed morphologic effects of xylitol on S. mutans [184] received partial support from a recent study [185] which showed that chewing xylitol gum over one year may negatively affect the synthesis of extracellular polysaccharides by S. mutans (because the adherence of colonies in xylitolreceiving subjects decreased). Related to these findings is the observation that both erythritol and xylitol can decrease polysaccharide-mediated cell adherence that contributes to plaque accumulation [186]. Since cooked starch can be regarded as potentially cariogenic, it is noteworthy that xylitol, along with maltotriitol ("gluco-gluco-sorbitol"; derived from maltotriose) and acarbose (a pseudotetrasaccharide containing an unsaturated cyclitol moiety; an α-glucosidase inhibitor) decreased the production of acid from starch by S. mutans and S. sobrinus [187].

CONCLUSIONS

Dental caries is a multi-factorial, diet-associated infectious disease that initiates as minor calcium-deficient lesions in tooth enamel. The repair (remineralization) of minor enamel defects is a normal physiological process that is well known to clinicians and researchers in dentistry and oral biology. This process can be facilitated by various dietary and oral hygiene procedures. The remineralization (rehardening) process may also concern dentin caries lesions. Consequently, the disease (dental caries) is reversible, if detected and treated sufficiently early. The scientific and clinical information available today indicates that habitual use of xylitol, a sugar alcohol of the pentitol type, can be associated with significant reduction in the incidence of dental caries and with remineralization of both enamel and dentin caries lesions. Suitable xylitol-containing "vehicles" include chewing gum, hard candies, gum arabic-based saliva stimulants, and dentifrice. Mouth rinses ("artificial saliva") and fluoride gels aimed at hyposalivation and at special dental caries patients, respectively, also serve as clinically effective oral health adjuvants. Other dietary polyols that can remarkably lower the incidence of caries include erythritol which is a tetritol-type alditol. Based on the molecular parameters of simple dietary alditols, it is conceivable to predict that their efficacy in caries reduction will follow the homologous series, that is, that the number of hydroxyl groups present in the alditol molecule will determine the efficacy as follows: erythritol $\geq$ xylitol $>$ sorbitol (the possible difference between erythritol and xylitol has not yet been adequately established). Most caries-related information available today is focused on the effects of xylitol and sorbitol. The present review examines the physical, bioinorganic, and biological chemistry of alditols from points of view that are believed to play a role in oral biology and caries prevention. The review also provides an account of tooth remineralization studies carried out with xylitol and sorbitol, as well as reports on recent caries-associated findings on erythritol.

REFERENCES

1. J. Head, "A study on saliva and its action on tooth enamel in reference to its hardening and softening," The Journal of the American Medical Association, vol. 59, no. 24, pp. 2118–2122, 1912.
2. T. Koulourides and W. Pigman, "Studies on rehardening of artificially softened enamel," Journal of Dental Research, vol. 39, no. 1, p. 198, 1960.
3. T. Kolourides, "Remineralization of enamel and dentin," in Dental Clinics of North America, E. Johansen and M. Shapiro, Eds., pp. 485–497, WB Saunders, Philadelphia, Pa, USA, 1962.
4. T. Koulourides, F. Feagin, and W. Pigman, "Remineralization of dental enamel by saliva in vitro," Annals of the New York Academy of Sciences, vol. 131, no. 2, pp. 751–757, 1965.
5. "Demineralization/remineralization—Working Group Consensus Report," Journal of Dental Research, vol. 65, pp. 1532– 1536, 1986.
6. S. Kashket, "Historical review of remineralization research," Journal of Clinical Dentistry, vol. 10, no. 2, pp. 56–64, 1999.
7. S. A. Leach, E. A. Agalamanyi, and R. M. Green, "Remineralization of the teeth by dietary means," in Remineralization of the Teeth, S. A. Leach and W. M. Edgar, Eds., pp. 51–73, IRL Press, Oxford, UK, 1983.
8. W. M. Edgar, "Diet, functional foods and oral health," in Functional Foods, Ageing and Degenerative Disease, C. Remarcle and B. Reusens, Eds., pp. 184–199, Woodhead, Cambridge, UK, 2004.
9. P. Fejerskov and A. Thylstrup, "Pathology of dental caries," in Textbook of Cariology, P. Fejerskov and A. Thylstrup, Eds., pp. 204–234, Munksgaard, Copenhagen, Denmark, 1st edition, 1986.
10. W. E. Herbert and W. A. Vale, Operative Dental Surgery, Edward Arnold, London, UK, 8th edition, 1962.
11. A. Scheinin and K. K. Makinen, "The e ¨ ffect of various sugars on the formation and chemical composition of dental plaque," The International Dental Journal, vol. 21, no. 3, pp. 302–321, 1971.
12. A. Scheinin and K. K. Makinen, "E ¨ ffect of sugars and sugar mixtures on dental plaque," Acta Odontologica Scandinavica, vol. 30, no. 2, pp. 235–257, 1972.

13. A. Scheinin and K. K. Makinen, "Turku sugar studies I-XXI," ¨ Acta Odontologica Scandinavica, vol. 33, supplement 70, pp. 1–351, 1975.
14. K. K. Makinen, "New biochemical aspects of sweeteners," ¨ The International Dental Journal, vol. 35, no. 1, pp. 23–35, 1985.
15. K. K. Makinen, "Latest dental studies on xylitol and mecha- ¨ nism of action of xylitol in caries limitation," in Progress in Sweeteners, T. H. Grenby, Ed., pp. 331–362, Elsevier, London, UK, 1989.
16. K. K. Makinen, "Prevention of dental caries by xylitol: issues ¨ relating to health claims," in America's Foods Health Messages and Claims, J. E. Tillotson, Ed., pp. 167–192, CRC Press, Boca Raton, Fla, USA, 1993.
17. K. K. Makinen, "Sugar alcohols," in ¨ Functional Foods, Designer Foods, Pharmafoods, Nutraceuticals, I. Goldberg, Ed., pp. 219–241, Chapman & Hall, New York, NY, USA, 1994.
18. K. K. Makinen, "Can the pentitol-hexitol theory explain the ¨ clinical observations made with xylitol?" Medical Hypotheses, vol. 54, no. 4, pp. 603–613, 2000.
19. K. K. Makinen, "The rocky road of xylitol to its clinical ¨ application," Journal of Dental Research, vol. 79, no. 6, pp. 1352–1355, 2000.
20. K. K. Makinen, "Sweeteners and dental health," in ¨ Functional Foods, Degenerative Disease, and Ageing, C. Remacle and B. Reusens, Eds., pp. 200–219, Woodhead, Cambridge, UK, 2004.
21. D. Birkhed, "Cariologic aspects of xylitol and its use in chewing gumml: a review," Acta Odontologica Scandinavica, vol. 52, no. 2, pp. 116–127, 1994.
22. J. Tanzer, "Xylitol chewing gum and dental caries," The International Dental Journal, vol. 45, supplement 1, pp. 65– 76, 1995.
23. L. Trahan, "Xylitol: a review of its action on mutans streptococci and dental plaque—its clinical significance," The International Dental Journal, vol. 45, supplement 1, pp. 77– 92, 1995.
24. W. M. Edgar, "Sugar substitutes, chewing gum and dental caries—a review," British Dental Journal, vol. 184, no. 1, pp. 29–32, 1998.
25. R. S. Levine, "Briefing paper: xylitol, caries and plaque," British Dental Journal, vol. 185, no. 10, p. 520, 1998.
26. C. Hayes, "The effect of non-cariogenic sweeteners on the prevention of dental caries: a review of the evidence," Journal of Dental Education, vol. 65, no. 10, pp. 1106–1109, 2001.

27. J. Peldyak and K. K. Makinen, "Xylitol for caries prevention," ¨ Journal of Dental Hygiene, vol. 76, no. 4, pp. 276–285, 2002.
28. A. Maguire and A. J. Rugg-Gunn, "Xylitol and caries prevention—is it a magic bullet?" British Dental Journal, vol. 194, no. 8, pp. 429–436, 2003.
29. B. A. Burt, "The use of sorbitol- and xylitol-sweetened chewing gum in caries control," Journal of the American Dental Association, vol. 137, no. 2, pp. 190–196, 2006.
30. K. K. Makinen, "Public use and recommendations of xylitol ¨ in the prevention of dental caries," Finnish Dental Journal, vol. 13, supplement 1, pp. 66–75, 2006.
31. K. K. Makinen, "Oral care gum products," in ¨ Food Constituents and Oral Health, M. Wilson, Ed., pp. 433–454, Woodhead, Cambridge, UK, 2009.
32. National Institutes of Health, "Consensus development conference statement. Diagnosis and management of dental caries through life," March 2002, http://nidcr.gov/news/ consensus.asp.
33. A. Deshpande and A. R. Jadad, "The impact of polyolcontaining chewing gums on dental caries: a systematic review of original randomized controlled trials and observational studies," Journal of the American Dental Association, vol. 139, no. 12, pp. 1602–1614, 2008.
34. S. Mickenautsch, S. C. Leal, V. Yengopal, A. C. Bezerra, and V. Cruvinel, "Sugar-free chewing gum and dental caries—a systematic review," Journal of Applied Oral Science, vol. 15, no. 2, pp. 83–88, 2007.
35. V. Machiulskiene, B. Nyvad, and V. Baelum, "Caries preventive effect of sugar-substituted chewing gum," Community Dentistry and Oral Epidemiology, vol. 29, no. 4, pp. 278–288, 2001.
36. C. Hayes, "Xylitol gum decreases the decayed, missing, and filled surfaces (DMFS) score over a 3-year period by an average of 1.9," Evidence-Based Dental Practice, vol. 2, pp. 14– 15, 2002.
37. K. A. Ly, P. Milgrom, and M. Rothen, "The potential of dental-protective chewing gum in oral health interventions," Journal of the American Dental Association, vol. 139, no. 5, pp. 553–563, 2008.
38. P. Milgrom, K. A. Ly, M. C. Roberts, M. Rothen, G. Mueller, and D. K. Yamaguchi, "Mutans streptococci dose response to xylitol chewing gum," Journal of Dental Research, vol. 85, no. 2, pp. 177–181, 2006.

39. P. Milgrom, M. Rothen, and L. Milgrom, "Developing public health interventions with xylitol for the US and USassociated territories and states," Finnish Dental Journal, vol. 13, supplement 1, pp. 28–37, 2006.
40. K. A. Ly, P. Milgrom, M. C. Roberts, D. K. Yamaguchi, M. Rothen, and G. Mueller, "Linear response of mutans streptococci to increasing frequency of xylitol chewing gum use: a randomized controlled trial," BMC Oral Health, vol. 6, article 6, 2006.
41. K. A. Ly, C. A. Riedy, P. Milgrom, M. Rothen, M. C. Roberts, and L. Zhou, "Xylitol gummy bear snacks: a school-based randomized clinical trial," BMC Oral Health, vol. 8, no. 1, article 20, 2008.
42. M. Svanberg and D. Birkhed, "Effect of dentifrices containing either xylitol and glycerol or sorbitol on mutans streptococci in saliva," Caries Research, vol. 25, no. 6, pp. 449–453, 1991.
43. L. G. Petersson, D. Birkhed, A. Gleerup, M. Johansson, and G. Jonsson, "Caries-preventive e ¨ ffect of dentifrices containing various types and concentrations of fluorides and sugar alcohols," Caries Research, vol. 25, no. 1, pp. 74–79, 1991.
44. J. L. Sintes, C. Escalante, B. Stewart, et al., "Enhanced anticaries efficacy of a 0.243% sodium fluoride/xylitol/silica dentifrice: 3-year clinical results," American Journal of Dentistry, vol. 8, no. 5, pp. 231–235, 1995.
45. J. L. Sintes, A. El´ıas-Boneta, B. Stewart, A. R. Volpe, and J. Lovett, "Anticaries efficacy of a sodium monofluorophosphate dentifrice containing xylitol in a dicalcium phosphate dihydrate base. A 30-month caries clinical study in Costa Rica," American Journal of Dentistry, vol. 15, no. 4, pp. 215– 219, 2002.
46. A. S. Aaltonen, J. T. Suhonen, J. Tenovuo, and I. Inkila-Saari, ¨ "Efficacy of a slow-release device containing fluoride, xylitol and sorbitol in preventing infant caries," Acta Odontologica Scandinavica, vol. 58, no. 6, pp. 285–292, 2000.
47. H. Maehara, Y. Iwami, H. Mayanagi, and N. Takahashi, "Synergistic inhibition by combination of fluoride and xylitol on glycolysis by mutans streptococci and its biochemical mechanism," Caries Research, vol. 39, no. 6, pp. 521–528, 2005.
48. V. G. Petin, J. K. Kim, R. O. Kritsky, and L. N. Komarova, "Mathematical description, optimization and prediction of synergistic interaction of fluoride and xylitol," Chemosphere, vol. 72, no. 5, pp. 844–849, 2008.

49. G. H. Hildebrandt and B. S. Sparks, "Maintaining mutans streptococci suppression: with xylitol chewing gum," Journal of the American Dental Association, vol. 131, no. 7, pp. 909– 916, 2000.
50. E.-M. Decker, G. Maier, D. Axmann, M. Brecx, and C. von Ohle, "Effect of xylitol/chlorhexidine versus xylitol or chlorhexidine as single rinses on initial biofilm formation of cariogenic streptococci," Quintessence International, vol. 39, no. 1, pp. 17–22, 2008.
51. P. Alanen, P. Isokangas, and K. Gutmann, "Xylitol candies in caries prevention: results of a field study in Estonian children," Community Dentistry and Oral Epidemiology, vol. 28, no. 3, pp. 218–224, 2000.
52. P. Alanen, M.-L. Holsti, and K. Pienihakkinen, "Sealants and ¨ xylitol chewing gum are equal in caries prevention," Acta Odontologica Scandinavica, vol. 58, no. 6, pp. 279–284, 2000.
53. G. Westergren, B. Krasse, D. Birkhed, and S. Edwardsson, "Genetic transfer of markers for sorbitol (D-glucitol) metabolism in oral streptococci," Archives of Oral Biology, vol. 26, no. 5, pp. 403–407, 1981.
54. S. Kalfas, G. Svensater, D. Birkhed, and S. Edwardsson, ¨ "Sorbitol adaptation of dental plaque in people with low and normal salivary-secretion rates," Journal of Dental Research, vol. 69, no. 2, pp. 442–446, 1990.
55. A. Nordblad, L. Suominen-Taipale, H. Murtomaa, E. Vartiainen, and K. Koskela, "Smart Habit Xylitol campaign, a new approach in oral health promotion," Community Dental Health, vol. 12, no. 4, pp. 230–234, 1995.
56. S. Honkala, E. Honkala, J. Tynjal¨ a, and L. Kannas, "Use of ¨ xylitol chewing gum among Finnish schoolchildren," Acta Odontologica Scandinavica, vol. 57, no. 6, pp. 306–309, 1999.
57. E. Honkala, S. Honkala, M. Shyama, and S. A. Al-Mutawa, "Field trial on caries prevention with xylitol candies among disabled school students," Caries Research, vol. 40, no. 6, pp. 508–513, 2006.
58. A. Trummler and W. Strubig, "Beeinflussung verschiedener ¨ speichelparameters nach taglicher verwendung von xylit- ¨ kaugummi in der schule," Oralprophylaxe Kinderzahnheilkunde, vol. 30, pp. 101–105, 2008.
59. W. Strubig, ¨Uber den Abbau von Zucker und Zuck- ¨eraustauschstoffen durch die Mischflora der Menschlichen Mundhohle ¨ , Quintessant,

Berlin, Germany, 1986.

60. W. Strubig, "Caries etiologic aspects of sugar and sugar ¨ substitutes," Zahnarztlicher Gesundheitsdienst ¨ , vol. 19, no. 2, pp. 10–13, 1989.
61. K. K. Makinen, C. A. Bennett, P. P. Hujoel, et al., "Xylitol ¨ chewing gums and caries rates: a 40-month cohort study," Journal of Dental Research, vol. 74, no. 12, pp. 1904–1913, 1995.
62. S. A. Leach and R. M. Green, "Effect of xylitol-supplemented diets on the progression and regression of fissure caries in the albino rat," Caries Research, vol. 14, no. 1, pp. 16–23, 1980.
63. K. W. Shyu and M. Y. Hsu, "The cariogenicity of xylitol, mannitol, sorbitol, and sucrose," Proceedings of the National Science Council, Republic of China, vol. 4, pp. 21–26, 1980.
64. R. Havenaar, J. H. J. Huis in't Veld, J. D. de Stoppelaar, and O. Backer Dirks, "Anti-cariogenic and remineralizing properties of xylitol in combination with sucrose in rats inoculated with Streptococcus mutans," Caries Research, vol. 18, no. 3, pp. 269– 277, 1984.
65. A. Scheinin, K. K. Makinen, and K. Ylitalo, "Turku sugar ¨ studies V. Final report on the effect of sucrose, fructose and xylitol diets on the caries incidence in man," Acta Odontologica Scandinavica, vol. 34, no. 4, pp. 179–216, 1976.
66. A. N. Galiullin, "Evaluation of the caries-preventive action of xylitol," Kazan Medical Journal, vol. 67, pp. 16–18, 1981 (Russian).
67. D. Kandelman, A. Bar, and A. Hefti, "Collaborative WHO ¨ xylitol field study in French Polynesia. I. Baseline prevalence and 32-month caries increment," Caries Research, vol. 22, no. 1, pp. 55–62, 1988.
68. A. Scheinin, J. Ban´ oczy, J. Sz ´ oke, et al., "Collaborative WHO ¨ xylitol field studies in Hungary. I. Three-year caries activity in institutionalized children," Acta Odontologica Scandinavica, vol. 43, no. 6, pp. 327–347, 1985.
69. A. Scheinin, K. Pienihakkinen, J. Tiekso, et al., "Collaborative ¨ WHO xylitol field studies in Hungary. VII. Two-year caries indicence in 976 institutionalized children," Acta Odontologica Scandinavica, vol. 43, no. 6, pp. 381–387, 1985.
70. D. Kandelman and G. Gagnon, "A 24-month clinical study of the incidence and progression of dental caries in relation to consumption of chewing gum containing xylitol in school preventive programs," Journal of Dental Research, vol. 69, no. 11, pp. 1771–1775, 1990.

71. P. Isokangas, P. Alanen, J. Tiekso, and K. K. Makinen, "Xylitol ¨ chewing gum in caries prevention: a field study in children," The Journal of the American Dental Association, vol. 117, no. 2, pp. 315–320, 1988.
72. K. K. Makinen, P. P. Hujoel, C. A. Bennett, K. P. Isotupa, P.- ¨ L. Makinen, and P. Allen, "Polyol chewing gums and caries ¨ rates in primary dentition: a 24-month cohort study," Caries Research, vol. 30, no. 6, pp. 408–417, 1996.
73. K. K. Makinen, D. Pemberton, P.-L. M ¨ akinen, et al., "Polyol- ¨ combinant saliva stimulants and oral health in veterans affairs patients—an exploratory study," Special Care in Dentistry, vol. 16, no. 3, pp. 104–115, 1996.
74. P. Isokangas, E. Soderling, K. Pienih ¨ akkinen, and P. Alanen, ¨ "Occurrence of dental decay in children after maternal consumption of xylitol chewing gum, a follow-up from 0 to 5 years of age," Journal of Dental Research, vol. 79, no. 11, pp. 1885–1889, 2000.
75. I. Thorild, B. Lindau, and S. Twetman, "Effect of maternal use of chewing gums containing xylitol, chlorhexidine or fluoride on mutans streptococci colonization in the mothers' infant children," Oral Health Preventive Dentistry, vol. 1, no. 1, pp. 53–57, 2003.
76. I. Thorild, B. Lindau, and S. Twetman, "Caries in 4-year-old children after maternal chewing of gums containing combinations of xylitol, sorbitol, chlorhexidine and fluoride," European Archives of Paediatric Dentistry, vol. 7, no. 4, pp. 241–245, 2006.
77. H. Hausen, L. Seppa, R. Poutanen, et al., "Noninvasive ¨ control of dental caries in children with active initial lesions: a randomized clinical trial," Caries Research, vol. 41, no. 5, pp. 384–391, 2007.
78. P. Isokangas, J. Tiekso, P. Alanen, and K. K. Makinen, ¨ "Long-term effect of xylitol chewing gum on dental caries," Community Dentistry and Oral Epidemiology, vol. 17, no. 4, pp. 200–203, 1989.
79. P. Isokangas, J. Tenovuo, E. Soderling, H. M ¨ annist ¨ o, and K. ¨ K. Makinen, "Dental caries and mutans streptococci in the ¨ proximal areas of molars affected by the habitual use of xylitol chewing gum," Caries Research, vol. 25, no. 6, pp. 444–448, 1991.
80. P. Isogangas, K. K. Makinen, J. Tiekso, and P. Alanen, "Long- ¨ term effect of xylitol chewing gum in the prevention of dental caries: a follow-up 5 years after termination of a prevention program," Caries Research, vol. 27, no. 6, pp. 495–498, 1993.

81. J. I. Virtanen, R. S. Bloigu, and M. A. Larmas, "Timing of first restorations before, during, and after a preventive xylitol trial," Acta Odontologica Scandinavica, vol. 54, no. 4, pp. 211– 216, 1996.
82. K. K. Makinen, P. P. Hujoel, C. A. Bennett, et al., "A ¨ descriptive report of the effects of a 16-month xylitol chewing-gum programme subsequent to a 40-month sucrose gum programme," Caries Research, vol. 32, no. 2, pp. 107– 112, 1998.
83. P. P. Hujoel, K. K. Makinen, C. A. Bennett, et al., "The ¨ optimum time to initiate habitual xylitol gum-chewing for obtaining long-term caries prevention," Journal of Dental Research, vol. 78, no. 3, pp. 797–803, 1999.
84. C. J. Carr and J. C. Krantz, "Metabolism of the sugar alcohols and their derivatives," Advances in Carbohydrate Chemistry, vol. 1, pp. 175–192, 1945.
85. R. L. Lohmar, "The polyols," in The Carbohydrates, Chemistry, Biochemistry, Physiology, W. Pigman, Ed., pp. 241–298, Academic Press, New York, NY, USA, 1962.
86. O. Touster and D. R. D. Shaw, "Biochemistry of the acyclic polyols," Physiological Reviews, vol. 42, pp. 181–225, 1962.
87. J. A. Mills, "Conformations of higher alditols," Australian Journal of Chemistry, vol. 27, pp. 1433–1446, 1974.
88. G. A. Scangos and A. M. Reiner, "Acquisition of ability to utilize xylitol: disadvantages of a constitutive catabolic pathway in Escherichia coli," Journal of Bacteriology, vol. 134, no. 2, pp. 501–505, 1978.
89. R. M. Corrales, L. Luo, E. Y. Chang, and S. C. Pflugfelder, "Effects of osmoprotectants on hyperosmolar stress in cultured human corneal epithelial cells," Cornea, vol. 27, no. 5, pp. 574–579, 2008.
90. P. de Cock and C.-L. Bechert, "Erythritol. Functionality in noncaloric functional beverages," Pure and Applied Chemistry, vol. 74, no. 7, pp. 1281–1289, 2002.
91. H. Bundgaard and C. Larsen, "The influence of carbohydrates and polyhydric alcohols on the stability of cephalosporins in aqueous solution," International Journal of Pharmaceutics, vol. 16, no. 3, pp. 319–325, 1983.
92. R. B. Killion Jr. and V. J. Stella, "The nucleophilicity of dextrose, sucrose, sorbitol, and mannitol with p-nitrophenyl esters in aqueous solution," International Journal of Pharmaceutics, vol. 66, no. 1–3, pp.

149–155, 1990.

93. K. Kiyosawa, "Volumetric properties of polyols (ethylene glycol, glycerol, meso-erythritol, xylitol and mannitol) in relation to their membrane permeability: group additivity and estimation of the maximum radius of their molecules," Biochimica et Biophysica Acta, vol. 1064, no. 2, pp. 251–255, 1991.
94. R. Schollner, D. Sieler, and E. Brettner, "Komplexbildung ¨ von geradkettigen Polyolen in waßriger L ¨ osung mit partiell ¨ hydratisierten K+ and Ca2+ Ionen in X- und Y-Zeolithen," Journal fur Praktische Chemie ¨ , vol. 337, pp. 567–575, 1995.
95. J. F. Back, D. Oakenfull, and M. B. Smith, "Increased thermal stability of proteins in the presence of sugars and polyols," Biochemistry, vol. 18, no. 23, pp. 5191–5196, 1979.
96. K. Kiyosawa, "Permeability of the Chara cell membrane for ethylene glycol, glycerol, meso-erythritol, xylitol and mannitol," Plant Physiology, vol. 88, pp. 366–371, 1993.
97. J. Chirife, G. Favetto, and C. Ferro Fontan, "Microbial growth ´ at reduced water activities: some physicochemical properties of compatible solutes," Journal of Applied Bacteriology, vol. 56, pp. 259–268, 1984.
98. J. K. Beattie and M. T. Kelso, "Equilibrium and dynamics of the binding of calcium ion to sorbitol (D-glucitol)," Australian Journal of Chemistry, vol. 34, pp. 2563–2568, 1981.
99. T. H. Grenby, A. H. Bashaarat, and K. H. Gey, "A clinical trial to compare the effects of xylitol and sucrose chewing-gums on dental plaque growth," British Dental Journal, vol. 152, no. 10, pp. 339–343, 1982.
100. H. Hurttia, V.-M. Multanen, K. K. Makinen, J. Tenovuo, and ¨ K. Paunio, "Effects on oral health of mouthrinses containing xylitol, sodium cyclamate and sucrose sweeteners in the absence of oral hygiene. III. Composition and bone resorbing potential of dental plaque," Proceedings of the Finnish Dental Society, vol. 80, no. 1, pp. 20–27, 1984.
101. L. M. Steinberg, F. Odusola, and I. D. Mandel, "Remineralizing potential, antiplaque and antigingivitis effects of xylitol and sorbitol sweetened chewing gum," Clinical Preventive Dentistry, vol. 14, no. 5, pp. 31–34, 1992.

102. K. K. Makinen and E. S ¨oderling, “Solubility of calcium salts, ¨enamel, and hydroxyapatite in aqueous solutions of simple carbohydrates,” Calcified Tissue International, vol. 36, no. 1, pp. 64–74, 1984.
103. E. Soderling and K. K. M ¨ akinen, “Aggregation of human sali- ¨ vary Ca-proteinates in the presence of simple carbohydrates in vitro,” Scandinavian Journal of Dental Research, vol. 94, no. 2, pp. 125–131, 1986.
104. K. K. Makinen, E. S ¨ oderling, D. R. Peacor, P.-L. M ¨ akinen, and ¨ L. M. Park, “Carbohydrate-controlled precipitation of apatite with coprecipitation of organic molecules in human saliva: stabilizing role of polyols,” Calcified Tissue International, vol. 44, pp. 258–268, 1989.
105. M. Carlevaro, E. R. Caffarena, and J. R. Grigera, “Hydration properties of xylitol: computer simulation,” International Journal of Biological Macromolecules, vol. 23, no. 2, pp. 149– 155, 1998.
106. K. Izumori and K. Yamanaka, “Selective inhibition of Klebsiella aerogenes growth on pentoses by pentitols,” Journal of Bacteriology, vol. 134, no. 3, pp. 713–717, 1978.
107. S. Schauder, K. H. Schneider, and F. Giffhorn, “Polyol metabolism of Rhodobacter sphaeroides: biochemical characterization of a short-chain sorbitol dehydrogenase,” Microbiology, vol. 14, part 8, pp. 1857–1863, 1995.
108. C. Kahle, K. H. Schneider, and F. Giffhorn, “Pentitol metabolism of Rhodobacter sphaeroides Si4: purification and characterization of a ribitol dehydrogenase,” Journal of General Microbiology, vol. 138, no. 6, pp. 1277–1281, 1992.
109. H. S. Singh, V. P. Singh, B. S. Arya, and G. R. Varma, “Kinetics and mechanism of oxidation of xylitol and galactitol by hexacyanoferrate(III) ion in aqueous alkaline medium,” Monatshefte fur Chemie ¨ , vol. 112, pp. 1253–1260, 1981.
110. C. J. Kleber, M. S. Putt, and J. C. Muhler, “Enamel dissolution by various food acidulants in a sorbitol candy,” Journal of Dental Research, vol. 57, no. 3, pp. 447–451, 1978.
111. V. Luostarinen, K. Paunio, J. Varrela, et al., “Turku sugar studies, XV. Vascular reactions in the hamster cheek pouch to human gingival exudate,” Acta Odontologica Scandinavica, vol. 33, supplement 70, pp. 287–291, 1975.

112. V. Luostarinen, K. K. Makinen, and P.-L. M ¨ akinen, "E ¨ ffects on oral health of mouthrinses containing xylitol, sodium cyclamate and sucrose sweeteners in the absence of oral hygiene. V. Response of hamster cheek pouch microcirculation to dental plaque," Proceedings of the Finnish Dental Society, vol. 80, no. 1, pp. 35–39, 1984.
113. J. Tenovuo, H. Mielityinen, and K. Paunio, "Effect of dental plaque grown in the presence of xylitol or sucrose on bone resorption in vitro," Pharmacology and Therapeutics in Dentistry, vol. 6, no. 1-2, pp. 35–43, 1981.
114. H. Mielityinen, J. Tenovuo, E. Soderling, and K. Paunio, ¨ "Effect of xylitol and sucrose plaque on release of lysosomal enzymes from bones and macrophages in vitro," Acta Odontologica Scandinavica, vol. 41, no. 3, pp. 173–180, 1983.
115. K. Paunio, H. Hurttia, J. Tenovuo, K. K. Makinen, and J. ¨ Tiekso, "Effects on oral health of mouthrinses containing xylitol, sodium cyclamate and sucrose sweeteners in the absence of oral hygiene. I. Clinical findings and analysis of gingival exudates," Proceedings of the Finnish Dental Society, vol. 80, no. 1, pp. 3–12, 1984.
116. U. Harjola and H. Liesmaa, "Effects of polyol and sucrose candies on plaque, gingivitis and lactobacillus index scores," Acta Odontologica Scandinavica, vol. 36, no. 4, pp. 237–242, 1978.
117. U. Pakkala, H. Liesmaa, and K. K. Makinen, "Use of xylitol ¨ in the control of oral hygiene in mentally retarded children: a clinical and biochemical study," Proceedings of the Finnish Dental Society, vol. 71, no. 5, pp. 271–277, 1981.
118. S. J. Han, S. Y. Jeong, Y. J. Nam, K. H. Yang, H. S. Lim, and J. Chung, "Xylitol inhibits inflammatory cytokine expression induced by lipopolysaccharide from Porphyromonas gingivalis," Clinical and Diagnostic Laboratory Immunology, vol. 12, no. 11, pp. 1285–1291, 2005.
119. A. M. Vacca Smith and W. H. Bowen, "In situ studies of pellicle formation on hydroxyapatite discs," Archives of Oral Biology, vol. 45, pp. 277–291, 2000.
120. H. Meyer-Lueckel, N. Umland, W. Hopfenmuller, and A. M. Kielbassa, "Effect of mucin alone and in combination with various dentifrices on in vitro remineralization," Caries Research, vol. 38, no. 5, pp. 478–483, 2004.

121. A. M. Kielbassa, U. Oeschger, J. Schulte-Monting, and H. Meyer-Lueckel, "Microradiographic study on the effects of salivary proteins on in vitro demineralization of bovine enamel," Journal of Oral Rehabilitation, vol. 32, no. 2, pp. 90– 96, 2005.
122. H. Meyer-Lueckel and A. M. Kielbassa, "Influence of calcium phosphates added to mucin-based saliva substitutes on bovine dentin," Quintessence International, vol. 37, no. 7, pp. 537–544, 2006.
123. R. K. Rose, G. H. Dibdin, and R. P. Shellis, "A quantitative study of calcium binding and aggregation in selected oral bacteria," Journal of Dental Research, vol. 72, no. 1, pp. 78– 84, 1993.
124. R. K. Rose, S. D. Hogg, and R. P. Shellis, "A quantitative study of calcium binding by isolated streptococcal cell walls and lipoteichoic acid: comparison with whole cells," Journal of Dental Research, vol. 73, no. 11, pp. 1742–1747, 1994.
125. E. C. Reynolds, "Calcium phosphate-based remineralization systems: scientific evidence?" Australian Dental Journal, vol. 53, pp. 268–273, 2008.
126. E. C. Reynolds, F. Cai, P. Shen, and G. D. Walker, "Retention in plaque and remineralization of enamel lesions by various forms of calcium in a mouthrinse or sugar-free chewing gum," Journal of Dental Research, vol. 82, no. 3, pp. 206–211, 2003.
127. I. Tarjan and L.- ′ A. Linden, "Investigation of enamel perme- ° ability with marked saccharose and xylit," Fogorvos Szle, vol. 74, pp. 235–238, 1981 (Hungarian).
128. A. Scheinin, K. K. Makinen, and K. Ylitalo, "Turku sugar ¨ studies. I. An intermediate report on the effect of sucrose, fructose and xylitol diets on the caries incidence in man," Acta Odontologica Scandinavica, vol. 32, no. 6, pp. 383–412, 1974.
129. A. Scheinin, K. K. Makinen, E. Tammisalo, and M. Rekola, ¨ "Turku sugar studies XVIII. Incidence of dental caries in relation to 1-year consumption of xylitol chewing gum," Acta Odontologica Scandinavica, vol. 33, supplement 70, pp. 307– 316, 1975.
130. M. Rekola, "Changes in buccal white spots during 2-year consumption of dietary sucrose or xylitol," Acta Odontologica Scandinavica, vol. 44, no. 5, pp. 285–290, 1986.
131. M. Rekola, "Approximal caries development during 2-year total substitution of dietary sucrose with xylitol," Caries Research, vol. 21,

no. 1, pp. 87–94, 1987.

132. J. D. B. Featherstone, T. W. Cutress, B. E. Rodgers, and P. J. Dennison, "Remineralization of artificial caries-like lesions in vivo by a self-administered mouthrinse or paste," Caries Research, vol. 16, pp. 235–242, 1982.
133. A. Vissink, E. J. S'Gravenmade, T. B. F. M. Gelhard, A. K. Panders, and M. H. Franken, "Properties of mucin- or CMCcontaining saliva substitutes on softened human enamel," Caries Research, vol. 19, pp. 212–218, 1985.
134. M. T. Smits and J. Arends, "Influence of extraoral xylitol and sucrose dippings on enamel demineralization in vivo," Caries Research, vol. 22, no. 3, pp. 160–165, 1988.
135. M. T. Smits, Xylitol and dental caries, Academic dissertation, University of Groningen, Amsterdam, The Netherlands, 1987.
136. T. H. Grenby, A. Phillips, and M. Mistry, "Studies of the dental properties of lactitol compared with five other bulk sweeteners in vitro," Caries Research, vol. 23, no. 5, pp. 315– 319, 1989.
137. F. N. Hattab, R. M. Green, K. M. Pang, and Y. C. Mok, "Effect of fluoride-containing chewing gum on remineralization of carious lesions and on fluoride uptake in man," Clinical Preventive Dentistry, vol. 11, no. 6, pp. 6–11, 1989.
138. J. Arends, M. Smits, J. L. Ruben, and J. Christoffersen, "Combined effect of xylitol and fluoride on enamel demineralization in vitro," Caries Research, vol. 24, no. 4, pp. 256–257, 1990.
139. S. L. Creanor, R. Strang, W. H. Gilmour, et al., "The effect of chewing gum use on in situ enamel lesion remineralization," Journal of Dental Research, vol. 71, pp. 1895–1900, 1992.
140. W. H. Bowen and S. K. Pearson, "The effects of sucralose, xylitol, and sorbitol on remineralization of caries lesions in rats," Journal of Dental Research, vol. 71, no. 5, pp. 1166–1168, 1992.
141. R. H. Manning, W. M. Edgar, and E. A. Agalamanyi, "Effects of chewing gums sweetened with sorbitol or a sorbitol/xylitol mixture on the remineralisation of human enamel lesions in situ," Caries Research, vol. 26, no. 2, pp. 104–109, 1992.
142. A. Scheinin, E. Soderling, U. Scheinin, R. L. Glass, and M.- ¨ L. Kallio, "Xylitol-induced changes of enamel microhardness paralleled by microradiographic observations," Acta Odontologica Scandinavica,

vol. 51, no. 4, pp. 241–246, 1993.

143. K. Wennerholm, J. Arends, D. Birkhed, J. Ruben, C. G. Emilson, and A. G. Dijkman, "Effect of xylitol and sorbitol in chewing-gums on mutans streptococci, plaque pH and mineral loss of enamel," Caries Research, vol. 28, no. 1, pp. 48–54, 1994.

144. K. K. Makinen, "An end to crossover designs for studies on ¨ the effect of sugar substitutes on caries?" Caries Research, vol. 43, pp. 331–333, 2009.

145. K. K. Makinen, P.-L. M ¨ akinen, H. R. Pape Jr., et al., ¨ "Stabilization of rampant caries: polyol gums and arrest of dentine caries in two long-term cohort studies in young subjects," International Dental Journal, vol. 45, pp. 93–107, 1995.

146. K. K. Makinen, P.-L. M ¨ akinen, H. R. Pape Jr., et al., "Conclu- ¨ sion and review of the 'Michigan Xylitol Programme' (1986– 1995) for the prevention of dental caries," International Dental Journal, vol. 46, pp. 22–34, 1996.

147. K. K. Makinen, D. J. Chiego Jr., P. Allen, et al., "Physical, ¨ chemical, and histologic changes in dentin caries lesions of primary teeth induced by regular use of polyol chewing gums," Acta Odontologica Scandinavica, vol. 56, pp. 148–156, 1998.

148. B. T. Amaechi, S. M. Higham, and W. M. Edgar, "The influence of xylitol and fluoride on dental erosion in vitro," Archives of Oral Biology, vol. 43, no. 2, pp. 157–161, 1998.

149. B. T. Amaechi, S. M. Higham, and W. M. Edgar, "Caries inhibiting and remineralizing effect of xylitol in vitro," Journal of Oral Science, vol. 41, no. 2, pp. 71–76, 1999.

150. T. Yanagisawa, "Ultrastructure of crystals in enamel carious lesions," Journal of Japanese Dental Association, vol. 46, pp. 1167–1176, 1994.

151. T. Yanagisawa, Y. Miake, Y. Saeki, and M. Takahashi, "Remineralization in enamel caries and restoration of carious lesions by enhanced remineralization induced by saliva and xylitol," Dentistry in Japan, vol. 39, pp. 208–215, 2003.

152. Y. Miake, M. Takahashi, Y. Saeki, and T. Yanagisawa, "Effect of xylitol on remineralization of demineralized enamel," The Shikwa Gakuho, vol. 99, pp. 393–399, 1999 (Japanese).

153. Y. Miake and T. Yanagisawa, "Effects of xylitol on remineralization of artificial demineralized enamel," Japanese Journal of Oral Biology,

vol. 42, pp. 580–589, 2000.

154. M. Takahashi, Y. Saeki, Y. Miake, and T. Yanagisawa, "Effects of sugar alcohols and calcium compounds on remineralization," Shikwa Gakuho, vol. 100, pp. 755–762, 2000 (Japanese).

155. Y. Saeki, M. Takahashi, S. Kamikawa, et al., "Remineralization effect of xylitol chewing gum containing Gloipeltis furcata extract and calcium hydrogenphosphate on initial caries-like enamel lesions," Japanese Journal of Oral Biology, vol. 42, pp. 590–600, 2000 (Japanese).

156. M. Takahashi, Y. Saeki, K. Fujimoto, H. Matsuzaki, Y. Miake, and T. Yanagisawa, "Remineralization effects of xylitol dragee gum containing Gloipeltis furcata extract and calcium hydrogenphosphate on initial caries-like enamel lesions in vivo," The Shikwa Gakuho, vol. 101, pp. 1033–1042, 2001 (Japanese).

157. Y. Saeki, "Effect of seaweed extracts on Streptococcus sobrinus adsorption to saliva-coated hydroxyapatite," The Bulletin of Tokyo Dental College, vol. 35, no. 1, pp. 9–15, 1994.

158. Y. Saeki, T. Kato, and K. Okuda, "Inhibitory effects of funoran on the adherence and colonization of oral bacteria," The Bulletin of Tokyo Dental College, vol. 37, no. 2, pp. 77–92, 1996.

159. Y. Saeki, T. Kato, Y. Naito, I. Takazoe, and K. Okuda, "Inhibitory effects of funoran on the adherence and colonization of mutans streptococci," Caries Research, vol. 30, no. 2, pp. 119–125, 1996.

160. S. Thaweboon, S. Nakornchai, Y. Miyake, et al., "Remineralizaron of enamel subsurface lesions by xylitol chewing gum containing funoran and calcium hydrogenphosphate," Southeast Asian Journal of Tropical Medicine and Public Health, vol. 40, no. 2, pp. 345–353, 2009.

161. P. Shen, F. Cai, A. Nowicki, J. Vincent, and E. C. Reynolds, "Remineralization of enamel subsurface lesions by sugar-free chewing gum containing casein phosphopeptide-amorphous calcium phosphate," Journal of Dental Research, vol. 80, no. 12, pp. 2066–2070, 2001.

162. H. Meyer-Lueckel, P. Tschoppe, W. Hopfenmuller, W.-R. Stenzel, and A. M. Kielbassa, "Effect of polymers used in saliva substitutes on demineralized bovine enamel and dentin," American Journal of Dentistry, vol. 19, no. 5, pp. 308– 312, 2006.

163. R. Suda, T. Suzuki, R. Takiguchi, K. Egawa, T. Sano, and K. Hasegawa, "The effect of adding calcium lactate to xylitol chewing gum on

remineralization of enamel lesions," Caries Research, vol. 40, no. 1, pp. 43–46, 2006.

164. H. Sano, S. Nakashima, Y. Songpaisan, and P. Phantumvanit, "Effect of a xylitol and fluoride containing toothpaste on the remineralization of human enamel in vitro," Journal of Oral Science, vol. 49, no. 1, pp. 67–73, 2007.

165. S. Chunmuang, S. S. Jitpukdeebodintra, C. Chuenarrom, and P. Benjakul, "Effect of xylitol and fluoride on enamel erosion in vitro," Journal of Oral Science, vol. 49, no. 4, pp. 293–297, 2007.

166. T. Takatsuka, R. A. Exterkate, and J. M. ten Cate, "Effects of isomalt on de- and remineralization, a combined in vitro pHcycling model and in situ study," Clinical Oral Investigations, vol. 12, pp. 173–177, 2008.

167. J. Kawanabe, M. Hirasawa, T. Takeuchi, T. Oda, and T. Ikeda, "Noncariogenicity of erythritol as a substrate," Caries Research, vol. 26, no. 5, pp. 358–362, 1992.

168. K. K. Makinen, K. P. Isotupa, T. Kivilompolo, P.-L. M ¨ akinen, ¨ J. Toivanen, and E. Soderling, "Comparison of erythritol and ¨ xylitol saliva stimulants in the control of dental plaque and mutans streptococci," Caries Research, vol. 35, no. 2, pp. 129– 135, 2001.

169. K. K. Makinen, K. P. Isotupa, T. Kivilompolo, et al., ¨ "The effect of polyol-combinant saliva stimulants on S. mutans levels in plaque and saliva of patients with mental retardation," Special Care in Dentistry, vol. 22, no. 5, pp. 187– 193, 2002.

170. K. K. Makinen, M. Saag, K. P. Isotupa, et al., "Similarity of the ¨ effects of erythritol and xylitol on some risk factors of dental caries," Caries Research, vol. 39, pp. 207–215, 2005.

171. T. Ichikawa, Y. Yano, Y. Fujita, T. Kashiwabara, and K. Nagao, "The enhancement effect of three sugar alcohols on the fungicidal effect of benzethonium chloride toward Candida albicans," Journal of Dentistry, vol. 36, no. 11, pp. 965–968, 2008.

172. M. Fontana, D. Catt, G. J. Eckert, et al., "Xylitol: effects on the acquisition of cariogenic species in infants," Pediatric Dentistry, vol. 31, no. 3, pp. 257–266, 2009.

173. P. Milgrom, K. A. Ly, O. K. Tut, et al., "Xylitol pediatric topical oral syrup to prevent dental caries: a double-blind randomized clinical trial of efficacy," Archives of Pediatrics and Adolescent Medicine, vol. 163, no. 7, pp. 601–607, 2009.

174. C. Badet, A. Furiga, and N. Thebaud, "E ′ ffect of xylitol on an in vitro model of oral biofilm," Oral Health & Preventive Dentistry, vol. 6, no. 4, pp. 337–341, 2008.
175. Anonymous, "Xylitol-containing oral syrup may prevent caries in children," The Journal of the American Dental Association, vol. 140, p. 972, 2009.
176. Anonymous, "Xylitol cleared for anti-caries health claims," British Dental Journal, vol. 206, no. 3, p. 123, 2009.
177. "American Academy on Pediatric Dentistry Council on Clinical Affairs," Pediatric Dental, vol. 30, supplement, pp. 36–37, 2008-2009.
178. S. Twetman, "Consistent evidence to support the use of xylitol- and sorbitol-containing chewing gum to prevent dental caries," The Journal of Evidence-Based Dental, vol. 10, pp. 10–11, 2009.
179. B. L. Edelstein, "Solving the problem of early childhood caries: a challenge for us all," Archives of Pediatrics and Adolescent Medicine, vol. 163, no. 7, pp. 667–668, 2009.
180. P. Milgrom, D. T. Zero, and J. M. Tanzer, "An examination of the advances in science technology of prevention of tooth decay in young children since the Surgeon General's report on oral health," Academic Pediatrics, vol. 9, no. 6, pp. 404– 409, 2009.
181. C. H. Splieth, M. Alkilzy, J. Schmitt, C. Berndt, and A. Welk, "E ffect of xylitol and sorbitol on plaque acidogenesis," Quintessence International, vol. 40, no. 4, pp. 279–285, 2009.
182. E. Soderling, K. K. M ¨ akinen, C.-Y. Chen, H. R. Pape Jr., ¨ W. Loesche, and P.-L. Makinen, "E ¨ ffect of sorbitol, xylitol, and xylitol/ sorbitol chewing gums on dental plaque," Caries Research, vol. 23, pp. 378–384, 1989.
183. O. Aguirre-Zero, D. T. Zero, and H. M. Proskin, "E ffect of chewing xylitol chewing gum on salivary flow rate and the acidogenic potential of dental plaque," Caries Research, vol. 27, no. 1, pp. 55–59, 1993.
184. H. Tuompo, J. H. Meurman, K. Lounatmaa, and J. Linkola, "E ffect of xylitol and other carbon sources on the cell wall of Streptococcus mutans," Scandinavian Journal of Dental Research, vol. 91, no. 1, pp. 17–25, 1983.
185. Y. E. Lee, Y. H. Choi, S. H. Jeong, H. S. Kim, S. H. Lee, and K. B. Song, "Morphological changes in Streptococcus mutans after chewing

gum containing xylitol for twelve months," Current Microbiology, vol. 58, pp. 332–337, 2009.

186. E. M. Soderling and A. M. Hietala-Lenkkeri, "Xylitol and ¨ erythritol decrease adherence of polysaccharide-producing oral streptococci," Current Microbiology, vol. 60, no. 1, pp. 25–29, 2010.

187. S. Aizawa, H. Miyasawa-Hori, K. Nakajo, et al., "E ffects of α -amylase and its inhibitors on acid production from cooked starch by oral streptococci," Caries Research, vol. 43, no. 1, pp. 17–24, 2009.

8

Fructose Metabolism and Relation to Atherosclerosis, Type 2 Diabetes, and Obesity

Astrid Kolderup[1], and Birger Svihus[2]

[1]Faculty of Public Health, Hedmark University College, P.O. Box 400, 2418 Elverum, Norway
[2]Norwegian University of Life Sciences, P.O. Box 5003, 1432 Aas, Norway

ABSTRACT

A high intake of sugars has been linked to diet-induced health problems. The fructose content in sugars consumed may also affect health, although the extent to which fructose has a particularly significant negative impact on health remains controversial. The aim of this narrative review is to describe the body's fructose management and to discuss the role of fructose as a risk factor for atherosclerosis, type 2 diabetes, and obesity. Despite some positive effects of fructose, such as high relative sweetness, high thermogenic effect, and low glycaemic index, a high intake of fructose, particularly when combined with glucose, can, to a larger extent than a similar glucose intake,

Citation: Fructose Metabolism and Relation to Atherosclerosis, Type 2 Diabetes, and Obesity. Astrid Kolderup and Birger Svihus. Hindawi Publishing Corporation Journal of Nutrition and Metabolism Volume 2015, Article ID 823081, 12 pages http://dx.doi.org/10.1155/2015/823081

lead to metabolic changes in the liver. Increased *de novo* lipogenesis (DNL), and thus altered blood lipid profile, seems to be the most prominent change. More studies with realistic consumption levels of fructose are needed, but current literature does not indicate that a normal consumption of fructose (approximately 50–60 g/day) increases the risk of atherosclerosis, type 2 diabetes, or obesity more than consumption of other sugars. However, a high intake of fructose, particularly if combined with a high energy intake in the form of glucose/starch, may have negative health effects via DNL.

INTRODUCTION

Sugars are important sources of energy in our diet, and a high intake of sugars has increasingly been identified as a considerable cause of major diet-induced health problems, namely, atherosclerosis, type 2 diabetes, and obesity [1–4]. An increased intake of sugar sweetened beverages in particular has been associated with these health problems in many epidemiological studies [5–7]. The main constituents of sugars are fructose and glucose, which can be present either alone or in combination, although most commonly fructose is mixed 50 : 50 with glucose. The ratio of glucose: fructose in the diet has been argued to be important for some of the effects of sugars [8, 9]. Thus, it is important to study the effects of fructose and glucose together, but also it is important to study the effects of fructose alone. Fructose has been claimed to be of benefit because it may aid glycemic control [10, 11], but also it has been claimed to be more harmful than other sugars, especially with regard to the development of atherosclerosis, type 2 diabetes, and obesity [8].

Fructose is found in a variety of foods. In table sugar, it is bound to glucose to form the disaccharide sucrose, whereas in honey it occurs in monosaccharide form. In fruit, berries, and vegetables, fructose occurs in both monosaccharide and disaccharide forms. Measured as intake from caloric sweeteners in USA, fructose intake was rather stable throughout the fifties and sixties but increased from the seventies until the end of the nineties, after which intake has declined [12]. The average American fructose intake was estimated to be 49 g per day in 2004 [13]. In Norway, the average daily intake of fructose can be estimated to be approximately 56 g/day, based on data from the Norwegian Directorate of Health's survey of consumption patterns [14] and composition data [15]. Globally, the main source of fructose is sucrose, which constitutes >90% of the energizing sweeteners used in the world [16]. In some countries, such as USA and Japan, high-fructose corn syrup (HFCS) is also an important source of fructose. HFCS is a mixture of fructose and glucose in different concentrations. It can contain up to 90% fructose [17], but the dominating concentration of fructose is 42% or 55% in commercial products [9]. In USA, consumption of HFCS increased sharply from 1970 until 1999, but it has declined since then [12, 18].

This review paper aims to describe how fructose, compared with other sugars, is managed by the body, thus clarifying the impact of fructose on atherosclerosis, type 2 diabetes, and obesity.

MATERIALS AND METHODS

The review is based on scientific peer-reviewed papers obtained using a nonsystematic search of the databases PubMed and Web of Science. The first step in the selection of literature was to identify the relevant keywords to search for in these databases. Various combinations of the following keywords were used: "fructose," "glucose," "sucrose," and "sugar" combined with "metabolism," "insulin resistance," "overweight," "obesity," "relative sweetness," "absorption capacity," "glycemic index," "de novo lipogenesis," "thermogenesis," "type 2 diabetes," and "appetite regulation."

The next step in the selection process consisted of inclusion or exclusion of papers based on the relevance to the study's aim. Through this initial selection process, both original and review papers were included. The reference lists of the papers were thoroughly studied to discover any possibly relevant papers that were not included. As far as possible, only original papers and reviews based on controlled trials and mechanistic studies were included. Unless otherwise specified, the results presented and discussed in this paper are statistically significant, and the experiments were conducted with a control group. Although animals may metabolize fructose differently from humans, animal studies have been included when found relevant, for example, due to lack of human studies. One possible limitation in this study lies in the nonsystematic nature of the search strategy, resulting in the possibility that relevant papers were not considered.

THE BODY'S FRUCTOSE MANAGEMENT

Absorption

Although there is some uncertainty about the mechanisms of fructose absorption, most of fructose seems to be absorbed by facilitated transport in the jejunum by the fructose transporter GLUT5 [18–20]. The body has limited capacity to absorb pure fructose, and intake of fructose can therefore lead to malabsorption [20–24]. Malabsorption of fructose results in bacterial fermentation, which leads to formation of short-chain fatty acids (acetate, propionate, and butyrate) and gases (hydrogen, methane, and carbon dioxide) [25, 26]. These processes can affect the motility of the intestine and cause various symptoms such as abdominal pain, bloating, and altered stool [27]. A large individual variation in absorption capacity for fructose has been observed. When fructose is consumed as a single oral dose, the maximum absorption capacity has been shown to vary between 5 and 50 g [28]. Several factors seem to affect this capacity, such as age and health [27, 29, 30], but of the dietary factors the presence of glucose is the most important [20, 23]. A significant increase in fructose absorption has been shown when fructose is coingested with equal amounts of glucose [21, 31]. There is still uncertainty about how glucose increases absorption capacity for fructose, but it may amongst others be due to an effect of glucose on the presence of fructose transporters [20]. Fructose does not appear in any fruit or vegetable without glucose [15]. Evolutionarily, this may explain why humans do not have the ability to absorb large amounts of pure fructose.

Metabolism

After absorption, fructose is transported by the portal vein to the liver, where it is effectively absorbed by liver cells [32], resulting in only small amounts entering the systemic circulation. The concentration of fructose in the blood is therefore only about 0.01 mmol/L, unlike that of glucose which is approximately 5.5 mmol/L [33]. Metabolism of fructose thus occurs primarily in the liver, but fructose may also be metabolized by enterocytes. In a study of pigs, it was shown that intestinal lactate production from fructose could account for 12% of absorbed fructose [34], but the functional significance of intestinal metabolism of fructose in humans remains unknown [18]. Bjorkman and Felig [35] also showed that infused fructose (48.6 g fructose in liquid form infused intravenously) was metabolized in the kidneys in humans who had fasted for 60 hours. Despite the artificially high blood fructose level, that study still shows that the kidneys have a relatively large capacity to metabolize fructose. It has been shown that GLUT5 is expressed in the membrane of fat, kidney, muscle, and brain cells [32, 36], but, due to very low levels of fructose in the blood, negligible amounts of fructose are probably metabolized in these tissues [9, 18].

As discussed above, the liver will metabolize a large majority of the ingested fructose, compared to only about 15–30% of ingested glucose [37, 38]. Most of the reactions in liver fructolysis are the same as those occurring in glycolysis, but fructose enters at a later stage in the glycolytic reaction chain than glucose [39]. Thus, fructose avoids the main control step in glycolysis, the phosphofructokinase step, which is tightly regulated by the energy status of the cell. The first step in fructolysis is the phosphorylation of fructose by fructokinase to fructose 1-phosphate. Unlike the phosphofructokinase in glycolysis, this enzyme is not inhibited by ATP [40]. The enzyme is considered virtually unregulated and, even when the liver's energy status (ATP) is high, the fructokinase will metabolize fructose to fructose 1-phosphate. Furthermore, fructose 1-phosphate is cleaved to dihydroxyacetone phosphate and glyceraldehyde by the enzyme aldolase B. Glyceraldehyde may be phosphorylated by triose kinase, with ATP as the phosphate donor, to form the glycolytic intermediate glyceraldehyde 3-phosphate [41]. After these steps, the carbon atoms from fructose follow the glycolytic steps.

Bypassing the phosphofructokinase step makes the flow of fructose carbon atoms through the biochemical pathways less controlled than for glucose. In this way, the liver will metabolize fructose in an unlimited way, as opposed to the case of glucose. This will influence the type and amount of metabolic products produced by the liver and is the main reason why fructose and glucose have different metabolic effects.

In the liver, fructose can enter metabolic pathways: it can be oxidized, converted to glucose (and glycogen), or converted to lactic acid, or enter *de novo* lipogenesis (DNL). After an overnight fast, approximately 50% of the fructose eaten as an oral dose of approximately 30–70 g is converted to glucose via gluconeogenesis [42]. Others have shown that after a similar fructose intake about 45% is oxidized; however, this includes both direct fructose oxidation and oxidation of glucose and lactic acid formed from fructose [43]. Another metabolic fate of fructose is to form lactic acid, but this seems to occur only at high fructose intakes [44–48]. The ingestion of 72 g fructose is the lowest amount that has been experimentally demonstrated to result in lactic acid formation [49]. Intake of fructose may also lead to formation of fatty acids via DNL. The diet composition may influence the distribution of fructose. Theytaz et al. [50] showed that coingestion of glucose with fructose decreased oxidation and gluconeogenesis from fructose carbon atoms. Intake of fructose together with glucose thus seems to affect the metabolic fate of fructose. To some degree, this effect may be due to higher insulin secretion after intake of glucose compared to fructose [51]. Insulin will, amongst others, decrease glucose production from fructose [52], and insulin will also stimulate DNL [53]. The extent to which fructose enters DNL is central to the health effects of fructose.

De Novo Lipogenesis (DNL)

DNL is the metabolic pathway that transforms surplus nonfat energy into fat by synthesis of fatty acids from acetyl-CoA [54]. The liver is the primary site of DNL, but DNL can also occur in lactating mammary glands and adipose tissue. Key lipogenic enzymes are present in adipose tissue, but to what degree adipose tissue contributes to the total DNL seems to be unclear. Although some data indicate that DNL in adipose tissue is of little importance [55–57], it has also been shown that it may play a more significant role [58], especially after ingestion of large amounts of carbohydrates [59, 60]. Some controversy exists regarding the capacity of the liver to carry out *de novo* lipogenesis in humans [60, 61], although a high capacity has been demonstrated in several experiments [62, 63].

As a result of the metabolic difference between glucose and fructose, a higher percentage of fructose compared to glucose can be converted to fat in the liver via DNL [63]. This has been shown in a number of animal and human studies, in which these sugars have been consumed in equal quantities under similar experimental conditions [63–66]. The greater potential of fructose to stimulate DNL is the main reason why fructose has been portrayed as particularly harmful. As described earlier, the liver absorbs most of the ingested fructose, and fructose metabolism bypasses the main control step in glycolysis, which means that a greater proportion of fructose, compared with glucose, is available for DNL. A normal diet, however, provides 3 to 5 times more glucose than fructose [12, 67], and this may influence the practical relevance of this metabolic difference. If, for example, glucose intake is very high, the liver may need to handle larger amounts of glucose than fructose. At two points in liver fructolysis, intermediates can enter lipid synthesis [9]. Fructose is converted to dihydroxyacetone phosphate, an intermediate in equilibrium with glycerol 3-phosphate, which forms the basis for the glycerol in triglycerides and phospholipids. Meanwhile, a large proportion of fructose carbons are metabolized directly to pyruvic acid and then acetyl-CoA. On metabolism of a large amount of fructose to acetyl-CoA, the amount of acetyl-CoA may exceed the citric acid cycle capacity of mitochondria. High levels of fructose may thus act as a nonregulated source of hepatic acetyl-CoA, which is a substrate that can enter DNL [9].

In addition, a high intake of fructose seems to stimulate gene expression and activity of lipogenic enzymes in the liver [40, 41, 66, 68]. Recently, fructose has also been shown to give a higher increase in fibroblast growth factor 21 (FGF21) than glucose. FGF21 is a hormone involved in glucose and lipid homeostasis [69], and high levels of FGF21 seem to be associated with metabolic disease [70].

It is clear that a high intake of fructose will cause a significant increase in DNL activity [71]. However, there is a paucity of knowledge about the effect of a normal fructose intake on this activity [72]. No research seems to have been conducted to assess the minimum amount or individual range of fructose that must be eaten to obtain a significant increase in DNL, which among others may be due to difficulties in quantifying DNL. In fact, measurement of hepatic DNL must be considered semiquantitative [54, 73]. Chong et al. [73] found that ^{13}C-labelled fructose contributed only 0.4% of the triglycerides in very-low-density lipoproteins (VLDLs) in men. This was measured 6 hours after eating a meal of 0.5 g fat/kg body weight and 0.75 g fructose/kg body weight after an overnight fast. The small increase in triglyceride levels in the blood after a fructose intake may be a result of delayed triglyceride secretion. Incorporation of carbon from ^{13}C-labelled fructose in VLDLs will not always reflect DNL, because some of the newly synthesized fatty acids will have delayed secretion [74, 75]. The level of DNL activity after fructose intake seems to vary significantly among individuals [76–78], and it also seems to vary during the day [79, 80].

Excessive intake of fructose, and hence increased DNL, may increase the risk of disease, because it may potentially cause both increased cholesterol levels in the blood and accumulation of fat in the liver [81]. In several animal studies, fat accumulation in the liver has been demonstrated after a high intake of fructose [82, 83], although most human studies have failed to demonstrate such an effect [84–86]. In fact, Chiu et al. [87] concluded, based on a systematic review with meta-analysis of controlled feeding trials, that fructose does not increase lipid content in liver when isocalorically exchanged for other carbohydrates. However, they pointed out that fructose ingested in large doses can raise liver lipid content, an effect that may be due to excess energy rather than fructose per se. Bravo et al. [88] conducted a study where participants consumed sucrose or HFCS for ten weeks at different levels of intake. In the study, fructose consumed as a part of a normal diet did not promote increased liver lipid content even at high intakes (90th percentile consumption level). In a study referred to by Rippe and

Angelopoulos [89], no increase in liver fat was observed after consumption of HFCS or sucrose at levels up to 30% of energy for ten weeks. However, Maersk et al. [90] found that intake of 1 L sucrose-sweetened soft drink per day for six months increased liver lipids in overweight participants compared with intake of same amount of milk, diet cola, and water. This may illustrate that the combination of fructose and glucose, as in sucrose, can lead to increased level of liver lipids. However, it is not possible to conclude that this is an effect of fructose, since there was no glucose control group in this study. Johnston et al. [91] found no difference in liver triglyceride level between overweight participants that ate a hypercaloric high-fructose or high-glucose diet. The authors concluded that this result indicates that the hypercaloric state rather than macronutrient composition is important for the accumulation of liver lipids. Carbohydrate-induced accumulation of fat in the liver can lead to nonalcoholic fatty liver disease (NAFLD) [92]. However, it is still unclear if fructose via increased DNL is particularly conducive to NAFLD [93]. The effect of fructose on lipid accumulation is thus unclear, but the effect of fructose on the blood lipid profile seems to be better documented.

ATHEROSCLEROSIS

It appears that high fructose intake can create an unfavorable lipid profile in blood via DNL [94]. The main product of DNL is palmitic acid [95], a fatty acid specifically shown to increase the risk of atherosclerosis [96]. Fatty acids formed by DNL will mainly be packed in VLDLs delivered into the bloodstream. This may, in turn, increase the level of low-density lipoproteins (LDLs) in the blood. In several studies, fructose has to a greater extent than glucose increased blood levels of triglycerides [51, 65, 97, 98] and LDLs [65, 99–103]. Aeberli et al. [104] showed that fructose increased the small dense LDLs, the type of LDLs that may in particular be linked to cardiovascular risk [105]. The level of high-density lipoproteins (HDLs) in blood does not seem to be affected by fructose [100, 101]. In most studies, an intake of fructose >100 g/day has been necessary to observe the adverse effects on lipid profiles [51, 65, 85, 98, 100, 106–108]. However, in a recent study by Aeberli et al. [109], a daily intake of about 77 g fructose and 34 g glucose for 3 weeks resulted in increased levels of total cholesterol and LDLs in the blood of healthy young men, compared with a daily intake of approximately 109 g glucose and 28 g fructose over the same time period.

The fact that both groups also ingested an unknown amount of starch and the fact that food intake was not controlled reduce the solidity of these results. Maersk et al. [90] found that intake of 50 g fructose per day together with 50 g glucose could have a negative effect on blood triglyceride level. The lack of a control group ingesting glucose makes it difficult to conclude that this is an effect of fructose. Conversely, Lowndes et al. [110] found no negative effect on lipid profile in overweight or obese individuals consuming HFCS or sucrose incorporated in a eucaloric diet for ten weeks at levels corresponding to the 25th and 50th percentiles of adult fructose consumption. Using the current knowledge, it does not appear that the consumption of moderate amounts of fructose (<50 g/day) alone will result in an unfavorable blood lipid profile [86, 111].

Due to the insignificant levels of fructose in peripheral blood, as described above, only glucose has the potential to be a substrate for DNL in adipose tissue. Although DNL in adipose tissue seems to be small as earlier discussed, glucose will, because of its presence in blood and by raising blood insulin level, probably to a higher extent than fructose, stimulate DNL in adipose tissue. Intake of glucose, in amounts that exceed the total capacity for glycogen storage and glucose oxidation, may thus increase DNL in adipose tissue more than the same amount of fructose. While fat formed in the liver has to be transported as lipoproteins in blood, this is avoided if the fat is formed directly in adipose tissue. Considering known negative health effects of lipoprotein residues, DNL occurring in adipose tissue may be preferable compared with DNL in the liver. This may illustrate a metabolic difference between glucose and fructose when consuming large amounts of sugars.

Another possible difference between fructose and glucose on risk factors for atherosclerosis is the effect of these sugars on blood uric acid level. Increased uric acid level has been associated with atherosclerosis in epidemiological studies, but the causality is uncertain [112–114]. Fructose appears to increase uric acid levels in the blood to a higher extent than glucose, especially at high intakes and when consumed as excess energy [86, 115, 116]. Intake of 0.5 g fructose/kg body weight is the lowest quantity shown to result in uric acid formation [117]. An increased blood level of uric acid can theoretically lead to elevated blood pressure because uric acid inhibits an enzyme in the endothelial cells of the arteries called endothelial nitric oxide synthase (eNOS). Activated eNOS leads to increased production of nitric oxide (NO), an important vasodilator. Thus, inhibition of eNOS may lead to vasoconstriction. Although only 0.5 g fructose/kg body weight has been shown to give uric acid formation and increased uric acid level theoretically could increase blood pressure, results from studies of the effect of fructose on blood pressure are very inconsistent [65, 72, 118, 119]. An average intake of fructose does not seem to lead to increased blood pressure [111, 120, 121]. The lack of causal link between uric acid level and atherosclerosis makes it difficult to draw conclusions on this effect of fructose.

TYPE 2 DIABETES

A high intake of sugar-sweetened beverages, with fructose as one of the major types of monosaccharides, has been associated with development of type 2 diabetes [5, 122]. Although this association does not prove causation, it is important to study the role of fructose in the development of type 2 diabetes. Central to the understanding of type 2 diabetes is the effect of nutrients on blood glucose homeostasis. Fructose must be converted to glucose in the liver to cause an increase in blood glucose level. As the conversion takes time and only a portion of the fructose will form glucose, fructose increases blood glucose less than similar levels of glucose [51]. Thus, the glycemic index for fructose is only 23 [10]. This, together with lack of stimulation of the pancreatic β cells [123], gives lower insulin secretion after intake of fructose compared with glucose [51, 124, 125]. These effects are positive because they contribute to blood glucose homeostasis. Additionally, moderate amounts of fructose have been shown to have positive effects on glycemic control [86, 126, 127]. However, it is claimed that fructose may also contribute negatively to blood glucose homeostasis by causing insulin resistance in the liver [9]. There is evidence that a high intake of fructose can cause insulin resistance in animals [128, 129], but several human studies have failed to demonstrate such an association [103, 130–132]. The operational definition of insulin resistance or sensitivity seems unclear, and many different methods have been used to measure it [133]. Thus, it is difficult to compare studies of insulin resistance. In human studies, in which fructose has been reported to cause insulin resistance, the daily intake of fructose has been as high as 110 g [109], approximately 250 g [134], 80 g [135], and 138 g [65]. This may indicate that the fructose intake must be high to potentially cause insulin resistance [86]. In the studies by Aeberli et al. [109], Stanhope et al. [65], and Beck-Nielsen et al. [134], total food intake was not controlled. Thus, the observed effect of fructose may also have been caused by differences in food intake between the control and experimental groups. In all the studies, in which insulin resistance has been shown, fructose was eaten together with glucose or starch, so the observations could also be the result of a combination of fructose and glucose. A number of hypotheses on how fructose can cause insulin resistance in the liver have been proposed. Lipid accumulation in the liver [136–138], metainflammation [83], and oxidative stress [139] are, either via inhibitory phosphorylation of the insulin receptor or the signaling molecules involved in insulin signaling, possible mechanisms for fructose-induced insulin resistance [9]. However, there are too few studies in humans and these are too divergent to be able to conclude firmly that there is a link between the consumption of fructose and insulin resistance. More long-term studies in which the daily intake of fructose is moderate are needed.

OBESITY

It is debatable whether fructose is less satiating than other sugars and thus can contribute to obesity through a high food intake. In a study by Page et al. [92], magnetic resonance (MR) images were taken of human brains after giving 75 g of a fructose or glucose drink. Glucose, but not fructose, reduced activity (regional cerebral blood flow) in the hypothalamus in the areas involved in energy regulation and reward systems; this is probably an indicator of satiety and may indicate that fructose is less satiating than glucose. Fructose will also to a lesser extent than glucose increase blood levels of insulin [51], leptin [51, 140], gastric inhibitory polypeptide [141], and glucagon-like peptide-1 [92], while at the same time it will attenuate levels of ghrelin less [51]. Although these hormonal effects may indicate that fructose is less satiating than glucose, this has not been confirmed in studies of the ability of fructose to satiate. In such studies, it has been shown that fructose has a greater appetite-reducing effect than glucose, when intake occurs before a meal [142, 143], or that there is no difference in the effects on appetite between fructose and glucose [144, 145]. Therefore, the effect of fructose on appetite remains unclear.

Although it is conceivable that fructose, via lack of stimulation of satiety signals, could contribute to obesity, fructose has several properties that act against obesity. As previously mentioned, the small intestine has a limited capacity to absorb fructose. This can lead to malabsorption at least if large amounts are consumed and consumption occurs without glucose-providing nutrients. Malabsorption of fructose will make less fructose enter the bloodstream and thus less energy will be available to the cells. In this way the malabsorption will act against obesity. It has also been shown in numerous studies that fructose has a greater thermogenic effect than glucose

[46, 146–148]. This means that the body uses more energy after eating fructose rather than glucose, so less energy will be available to be stored as fat. The relative sweetness of fructose is also greater than for glucose and sucrose [149, 150]. Although this will decrease with increasing temperature [151, 152], the high relative sweetness allows smaller amounts of fructose than glucose and sucrose to be used to achieve a particular sweetness in most applications. On the basis of these properties, it does not appear that fructose is more fattening than other sugars. This also agrees with experimental studies of the relationship between fructose intake and obesity in animals [153, 154] and humans [65, 86, 106, 155].

SUBSTRATE OXIDATION

It has been proposed that fructose can inhibit lipid oxidation [146]. For liver lipid oxidation this is logical, because the liver acquires energy from fructose and thus does not need to oxidize fat. Fructose can also increase DNL. It would not be expedient to form and break down fat simultaneously in the liver, and increased levels of malonyl-CoA (due to active DNL) will inhibit β-oxidation [41, 156]. It would also be logical for fructose to increase total body lipid oxidation more than glucose due to fructose's smaller contribution as an extrahepatic energy source. In some studies, however, fructose has been shown to increase the respiration quotient (RQ), the ratio of CO_2 exhaled to O_2 consumed, [157] more than glucose. This indicates that fructose to a higher degree than glucose reduces total body lipid oxidation and increases total body carbohydrate oxidation. Blaak and Saris [158] conducted a study in which participants ate 75 g fructose, starch, or glucose after a 12-hour fast in a crossover study. Fructose resulted in a significantly larger increase in the RQ, measured 6 hours after ingestion, than both glucose and starch. Tappy et al. [146] also showed a greater increase in RQ 4 hours after ingestion of 75 g fructose, compared with similar healthy participants eating 75 g glucose. Schwarz et al. [46] conducted a similar study (75 g fructose/glucose) with the same result. These results may be explained by the fact that fructose, more than glucose, enters DNL under specific conditions. Due to the fact that RQ varies for different substrates (e.g., 1 for carbohydrates and 0.7 for fats), RQ is used to determine the source of substrate oxidation [159]. However, during active DNL, CO_2 is produced without using O_2. Simultaneous occurrence of DNL and carbohydrate oxidation can lead to RQ values greater than 1 [159]. An increased RQ caused by DNL can, therefore, be misinterpreted as reduced lipid oxidation and increased carbohydrate oxidation. Such a misinterpretation may have occurred in the studies described above. Thus the effect of fructose on total body substrate oxidation remains unclear.

DISCUSSION

The distribution of fructose into metabolic pathways, especially DNL, is of key importance to the health effects of fructose. The distribution varies with the amount of fructose consumed, the duration of fructose exposure, the composition of diet/meal, and whether the measurement took place postprandially, after absorption, or under fasting conditions. Individual physiological, enzymatic, and endocrine factors are also important. Diet composition and the amount of fructose eaten and absorbed will be the focus of this discussion.

Malabsorption of fructose will affect the amount of fructose absorbed and can thus be an important confounding factor in studies in which factors that affect absorption capacity have not been taken into account [39]. Truswell et al. [21] showed that intake of 50 g fructose led to malabsorption in over 50% of the study participants. The results of studies in which fructose is malabsorbed can thus be inaccurate due to individual differences in the absorption capacity of fructose. As the small intestine has a large absorption capacity for glucose and a limited one for fructose, it is problematic to compare fructose with glucose as the sole carbohydrate source. In future studies, this should be controlled for, for example, by using the hydrogen breath test to assess fructose malabsorption.

The composition of diet and especially the amount of glucose/starch may have influence on the health effects of fructose. As fructose is present with glucose in most food products, it is more practical and relevant to look at the effects of fructose and glucose together than the effects of fructose alone. A larger increase in DNL after eating fructose and glucose together (50 : 50 glucose : fructose) rather than the same amount as pure glucose has been shown [63]. Eating glucose with fructose is likely to affect fructose's health effects by stimulating the flow of fructose to DNL [50]. This effect could be due to both increased absorption capacity for fructose when coingested with glucose and therefore greater availability of fructose carbon atoms going towards DNL and increased blood insulin levels when glucose is present in the diet. Insulin stimulates DNL directly and indirectly by inhibiting other important metabolic pathways for fructose, such as gluconeogenesis. It is also plausible that glucose will compete with fructose as an energy source for enterocytes and hepatic cells, thus making more fructose available for DNL. Coingestion of glucose with fructose has been shown to decrease oxidation and gluconeogenesis from fructose carbon atoms [50]. Thus, it appears that the combination of fructose and glucose is particularly unfortunate, although fructose is not a prerequisite for DNL. It is unclear how much glucose/starch must be included with fructose for this effect to be significant.

The amount of fructose consumed seems to be of great importance to the effects of fructose on health. The negative health effects of fructose have mainly been demonstrated at high intakes, and several studies have found an average intake of fructose to cause no health problems [12, 110, 111, 160]. Although the average daily intake of fructose is 50–60 g/day, some of the population will consume larger amounts [161, 162], so the negative health effects of fructose may be relevant at least for a proportion of the population. Many of the current studies are poorly suited to determine health risks of fructose because (a) the fructose intakes are unrealistically high, (b) fructose is given in isolation and not mixed with other carbohydrates as in practice, and/or (c) the studies are conducted on animals. Differences between human and animal physiology limit their applicability to humans. There is a need for more human studies under conditions more similar to the way fructose is normally consumed. Such studies, particularly related to DNL, will be necessary to understand the effects of a normal fructose intake. The effects of fructose on triglyceride and cholesterol levels in the blood, fat accumulation in liver, and insulin signaling all seem to be linked to the extent to which fructose enters DNL. Surprisingly, there have been no satisfactory studies to assess the proportion of fructose that enters DNL at different levels of intake. Such studies should be carried out and should include intake of both pure fructose and fructose together with glucose.

It is also important to note that, despite the metabolic difference between glucose and fructose, glucose consumption far exceeds fructose consumption in the human diet [12]. This quantitative aspect must be considered when comparing the health effects of glucose and fructose.

CONCLUSION

Although there is a paucity of published literature regarding physiological effects of fructose in humans, current literature does not indicate that a normal consumption of fructose (approximately 50–60 g/day) increases the risk of atherosclerosis, type 2 diabetes, or obesity more than consumption of other sugars. However, a high intake of fructose, particularly if combined with a high energy intake in the form of glucose/starch, may have negative health effects via DNL. More studies are clearly needed, particularly studies under more realistic consumption levels of fructose.

REFERENCES

1. Cohen A. M., Teitelbaum A., Balogh M., Groen J. J. Effect of interchanging bread and sucrose as main source of carbohydrate in a low fat diet on the glucose tolerance curve of healthy volunteer subjects. *The American Journal of Clinical Nutrition*. 1966;19(1):59–62.
2. Storlien L. H., Kraegen E. W., Jenkins A. B., Chisholm D. J. Effects of sucrose vs starch diets on *in vivo* insulin action, thermogenesis, and obesity in rats. *The American Journal of Clinical Nutrition*. 1988;47(3):420–427.
3. Giaccari A., Sorice G., Muscogiuri G. Glucose toxicity: the leading actor in the pathogenesis and clinical history of type 2 diabetes—mechanisms and potentials for treatment. *Nutrition, Metabolism and Cardiovascular Diseases*. 2009;19(5):365–377. doi: 10.1016/j.numecd.2009.03.018.
4. Laughlin M. R. Normal roles for dietary fructose in carbohydrate metabolism. *Nutrients*. 2014;6(8):3117–3129. doi: 10.3390/nu6083117.
5. Malik V. S., Popkin B. M., Bray G. A., Després J.-P., Willett W. C., Hu F. B. Sugar-sweetened beverages and risk of metabolic syndrome and type 2 diabetes: a meta-analysis. *Diabetes Care*. 2010;33(11):2477–2483. doi: 10.2337/dc10-1079.
6. Dhingra R., Sullivan L., Jacques P. F., et al. Soft drink consumption and risk of developing cardiometabolic risk factors and the metabolic syndrome in middle-aged adults in the community. *Circulation*. 2007;116(5):480–488.
7. Høstmark A. T. The Oslo Health Study: soft drink intake is associated with the metabolic syndrome. *Applied Physiology, Nutrition and Metabolism*. 2010;35(5):635–642. doi: 10.1139/h10-059.
8. Lustig R. H. Fructose: metabolic, hedonic, and societal parallels with ethanol. *Journal of the American Dietetic Association*. 2010;110(9):1307–1321. doi: 10.1016/j.jada.2010.06.008.
9. Havel P. J. Dietary fructose: implications for dysregulation of energy homeostasis and lipid/carbohydrate metabolism. *Nutrition Reviews*. 2005;63(5):133–137. doi: 10.1301/nr.2005.may.133-157.
10. Segal M. S., Gollub E., Johnson R. J. Is the fructose index more relevant with regards to cardiovascular disease than the glycemic

index? *European Journal of Nutrition*. 2007;46(7):406–417. doi: 10.1007/s00394-007-0680-9.

11. Sánchez-Lozada L. G., Le M., Segal M., Johnson R. J. How safe is fructose for persons with or without diabetes? *The American Journal of Clinical Nutrition*. 2008;88(5):1189–1190. doi: 10.3945/ajcn.2008.26812.
12. White J. S. Challenging the fructose hypothesis: new perspectives on fructose consumption and metabolism. *Advances in Nutrition*. 2013;4(2):246–256. doi: 10.3945/an.112.003137.
13. Marriott B. P., Cole N., Lee E. National estimates of dietary fructose intake increased from 1977 to 2004 in the United States. *The Journal of Nutrition*. 2009;139(6):S1228–S1235. doi: 10.3945/jn.108.098277.
14. Helsedirektoratet. Utviklingen i norsk kosthold. Matforsyningsstatistikk. 2013, (Norwegian), https://helsedirektoratet.no/Lists/Publikasjoner/Attachments/370/Utviklingen-i-norsk-kosthold-2013-matforsyningsstatistikk-IS-2116.pdf.
15. United States Department of Argiculture. *National Nutrient Database for Standard Reference*. United States Department of Argiculture; 2012. http://ndb.nal.usda.gov/ndb/search/list?fg=&man=&lfacet=&count=&max=&sort=&qlookup=&offset=&format=Full&new=
16. White J. S. Straight talk about high-fructose corn syrup: what it is and what it ain't. *The American Journal of Clinical Nutrition*. 2008;88(6):1716S–1721S. doi: 10.3945/ajcn.2008.25825b.
17. Ferder L., Ferder M. D., Inserra F. The role of high-fructose corn syrup in metabolic syndrome and hypertension. *Current Hypertension Reports*. 2010;12(2):105–112. doi: 10.1007/s11906-010-0097-3.
18. Tappy L., Le K.-A. Metabolic effects of fructose and the worldwide increase in obesity. *Physiological Reviews*. 2010;90(1):23–46. doi: 10.1152/physrev.00019.2009.
19. Madero M., Perez-Pozo S. E., Jalal D., Johnson R. J., Sánchez-Lozada L. G. Dietary fructose and hypertension. *Current Hypertension Reports*. 2011;13(1):29–35. doi: 10.1007/s11906-010-0163-x.
20. Jones H. F., Butler R. N., Brooks D. A. Intestinal fructose transport and malabsorption in humans. *The American Journal of Physiology—Gastrointestinal and Liver Physiology*. 2011;300(2):G202–G206. doi: 10.1152/ajpgi.00457.2010.

21. Truswell A. S., Seach J. M., Thorburn A. W. Incomplete absorption of pure fructose in healthy subjects and the facilitating effect of glucose. *The American Journal of Clinical Nutrition*. 1988;48(6):1424–1430.
22. Riby J. E., Fujisawa T., Kretchmer N. Fructose absorption. *American Journal of Clinical Nutrition*. 1993;58(5):S748–S753.
23. Corpe C. P., Burant C. F., Hoekstra J. H. Intestinal fructose absorption: clinical and molecular aspects. *Journal of Pediatric Gastroenterology and Nutrition*. 1999;28(4):364–374. doi: 10.1097/00005176-199904000-00004.
24. Beyer P. L., Caviar E. M., McCallum R. W. Fructose intake at current levels in the United States may cause gastrointestinal distress in normal adults. *Journal of the American Dietetic Association*. 2005;105(10):1559–1566. doi: 10.1016/j.jada.2005.07.002.
25. Pimentel M., Lin H. C., Enayati P., et al. Methane, a gas produced by enteric bacteria, slows intestinal transit and augments small intestinal contractile activity. *American Journal of Physiology—Gastrointestinal and Liver Physiology*. 2006;290(6):G1089–G1095. doi: 10.1152/ajpgi.00574.2004.
26. Wong J. M. W., de Souza R., Kendall C. W. C., Emam A., Jenkins D. J. A. Colonic health: fermentation and short chain fatty acids. *Journal of Clinical Gastroenterology*. 2006;40(3):235–243. doi: 10.1097/00004836-200603000-00015.
27. Gibson P. R., Newnham E., Barrett J. S., Shepherd S. J., Muir J. G. Review article: fructose malabsorption and the bigger picture. *Alimentary Pharmacology & Therapeutics*. 2007;25(4):349–363. doi: 10.1111/j.1365-2036.2006.03186.x.
28. Madsen J. L., Linnet J., Rumessen J. J. Effect of nonabsorbed amounts of a fructose-sorbitol mixture on small intestinal transit in healthy volunteers. *Digestive Diseases and Sciences*. 2006;51(1):147–153. doi: 10.1007/s10620-006-3100-8.
29. Dyer J., Wood I. S., Palejwala A., Ellis A., Shirazi-Beechey S. P. Expression of monosaccharide transporters in intestine of diabetic humans. *American Journal of Physiology: Gastrointestinal and Liver Physiology*. 2002;282(2):G241–G248. doi: 10.1152/ajpgi.00310.2001.
30. Kellett G. L., Brot-Laroche E. Apical GLUT2—a major pathway of intestinal sugar absorption. *Diabetes*. 2005;54(10):3056–3062. doi:

10.2337/diabetes.54.10.3056.

31. Kneepkens C. M. F., Vonk R. J., Fernandes J. Incomplete intestinal absorption of fructose. *Archives of Disease in Childhood*. 1984;59(8):735–738. doi: 10.1136/adc.59.8.735.
32. Douard V., Ferraris R. P. Regulation of the fructose transporter GLUT5 in health and disease. *American Journal of Physiology—Endocrinology and Metabolism*. 2008;295(2):E227–E237. doi: 10.1152/ajpendo.90245.2008.
33. Bray G. A. How bad is fructose? *The American Journal of Clinical Nutrition*. 2007;86(4):895–896.
34. Bjorkman O., Crump M., Phillips R. W. Intestinal metabolism of orally administered glucose and fructose in Yucatan miniature swine. *The Journal of Nutrition*. 1984;114(8):1413–1420.
35. Bjorkman O., Felig P. Role of the kidney in the metabolism of fructose in 60-hour fasted humans. *Diabetes*. 1982;31(6):516–520. doi: 10.2337/diab.31.6.516.
36. Litherland G. J., Hajduch E., Gould G. W., Hundal H. S. Fructose transport and metabolism in adipose tissue of Zucker rats: diminished GLUT5 activity during obesity and insulin resistance. *Molecular and Cellular Biochemistry*. 2004;261(1):23–33. doi: 10.1023/b:mcbi.0000028734.77867.d2.
37. Lam P., Ng K., Stanhope K. L., et al. Effects of consuming dietary fructose versus glucose on de novo lipogenesis in overweight and obese human subjects. *Berkeley Scientific Journal*. 2011;15(2)
38. Tappy L., Lê K.-A. Does fructose consumption contribute to non-alcoholic fatty liver disease? *Clinics and Research in Hepatology and Gastroenterology*. 2012;36(6):554–560. doi: 10.1016/j.clinre.2012.06.005.
39. Stanhope K. L. Role of fructose-containing sugars in the epidemics of obesity and metabolic syndrome. *Annual Review of Medicine*. 2012;63:329–343. doi: 10.1146/annurev-med-042010-113026.
40. Samuel V. T. Fructose induced lipogenesis: from sugar to fat to insulin resistance. *Trends in Endocrinology and Metabolism*. 2011;22(2):60–65.
41. Mayes P. A. Intermediary metabolism of fructose. *American Journal of Clinical Nutrition*. 1993;58(5):S754–S765.

42. Delarue J., Normand S., Pachiaudi C., Beylot M., Lamisse F., Riou J. P. The contribution of naturally labelled 13C fructose to glucose appearance in humans. *Diabetologia*. 1993;36(4):338–345. doi: 10.1007/BF00400238.
43. Sun S. Z., Empie M. W. Fructose metabolism in humans—what isotopic tracer studies tell us. *Nutrition& Metabolism*. 2012;9, article 89 doi: 10.1186/1743-7075-9-89.
44. Smith L. H., Jr., Ettinger R. H., Seligson D. A comparison of the metabolism of fructose and glucose in hepatic. *The Journal of Clinical Investigation*. 1953;32(4):273–282. doi: 10.1172/jci102736.
45. Sahebjami H., Scalettar R. Effects of fructose infusion on lactate and uric acid metabolism. *The Lancet*. 1971;297(7695):366–369.
46. Schwarz J.-M., Schutz Y., Froidevaux F., et al. Thermogenesis in men and women induced by fructose vs glucose added to a meal. *American Journal of Clinical Nutrition*. 1989;49(4):667–674.
47. Couchepin C., Kim-Anne L. Ê., Bortolotti M., et al. Markedly blunted metabolic effects of fructose in healthy young female subjects compared with male subjects. *Diabetes Care*. 2008;31(6):1254–1256. doi: 10.2337/dc07-2001.
48. Lecoultre V., Benoit R., Carrel G., et al. Fructose and glucose co-ingestion during prolonged exercise increases lactate and glucose fluxes and oxidation compared with an equimolar intake of glucose. *The American Journal of Clinical Nutrition*. 2010;92(5):1071–1079. doi: 10.3945/ajcn.2010.29566.
49. Kelsay J. L., Behall K. M., Clark W. M. Glucose, fructose, lactate and pyruvate in blood, and lactate and pyruvate in parotid saliva in response to sugars with and without other foods. *American Journal of Clinical Nutrition*. 1974;27(8):819–825.
50. Theytaz F., de Giorgi S., Hodson L., et al. Metabolic fate of fructose ingested with and without glucose in a mixed meal. *Nutrients*. 2014;6(7):2632–2649. doi: 10.3390/nu6072632.
51. Teff K. L., Elliott S. S., Tschöp M., et al. Dietary fructose reduces circulating insulin and leptin, attenuates postprandial suppression of ghrelin, and increases triglycerides in women. *The Journal of Clinical Endocrinology & Metabolism*. 2004;89(6):2963–2972. doi: 10.1210/jc.2003-031855.
52. Girard J. The inhibitory effects of insulin on hepatic glucose production

are both direct and indirect. *Diabetes*. 2006;55(2):S65–S69. doi: 10.2337/db06-S009.

53. Boden G., Salehi S., Cheung P., et al. Comparison of in vivo effects of insulin on SREBP-1c activation and INSIG-1/2 in rat liver and human and rat adipose tissue. *Obesity*. 2013;21(6):1208–1214. doi: 10.1002/oby.20134.
54. Hellerstein M. K., Schwarz J.-M., Neese R. A. Regulation of hepatic de novo lipogenesis in humans. *Annual Review of Nutrition*. 1996;16:523–557. doi: 10.1146/annurev.nu.16.070196.002515.
55. Shrago E., Glennon J. A., Gordon E. S. Comparative aspects of lipogenesis in mammalian tissues. *Metabolism: Clinical and Experimental*. 1971;20(1):54–62. doi: 10.1016/0026-0495(71)90059-x.
56. Guo Z. K., Cella L. K., Baum C., Ravussin E., Schoeller D. A. De nova lipognesis in adipose tissue of lean and obese women: application of deuterated water and isotope ratio mass spectrometry. *International Journal of Obesity*. 2000;24(7):932–937. doi: 10.1038/sj.ijo.0801256.
57. Diraison F., Yankah V., Letexier D., Dusserre E., Jones P., Beylot M. Differences in the regulation of adipose tissue and liver lipogenesis by carbohydrates in humans. *Journal of Lipid Research*. 2003;44(4):846–853. doi: 10.1194/jlr.m200461-jlr200.
58. Strawford A., Antelo F., Christiansen M., Hellerstein M. K. Adipose tissue triglyceride turnover, de novo lipogenesis, and cell proliferation in humans measured with ^2H_2O. *American Journal of Physiology—Endocrinology & Metabolism*. 2004;286(4):E577–E588. doi: 10.1152/ajpendo.00093.2003.
59. Chascione C., Elwyn D. H., Davila M., Gil K. M., Askanazi J., Kinney J. M. Effect of carbohydrate intake on de novo lipogenesis in human adipose tissue. *American Journal of Physiology: Endocrinology and Metabolism*. 1987;253(6):E664–E669.
60. Aarsland A., Chinkes D., Wolfe R. R. Hepatic and whole-body fat synthesis in humans during carbohydrate overfeeding. *The American Journal of Clinical Nutrition*. 1997;65(6):1774–1782.
61. Stubbs R. J., Mazlan N., Whybrow S. Carbohydrates, appetite and feeding behavior in humans. *The Journal of Nutrition*. 2001;131(10):2775S–2781S.

62. Acheson K. J., Schutz Y., Bessard T., Anantharaman K., Flatt J.-P., Jequier E. Glycoprotein storage capacity and de novo lipogenesis during massive carbohydrate overfeeding in man. *American Journal of Clinical Nutrition*. 1988;48(2):240–247.

63. Parks E. J., Skokan L. E., Timlin M. T., Dingfelder C. S. Dietary sugars stimulate fatty acid synthesis in adults. *Journal of Nutrition*. 2008;138(6):1039–1046.

64. Park O.-J., Cesar D., Faix D., Wu K., Shackleton C. H. L., Hellerstein M. K. Mechanisms of fructose-induced hypertriglyceridaemia in the rat. Activation of hepatic pyruvate dehydrogenase through inhibition of pyruvate dehydrogenase kinase. *Biochemical Journal*. 1992;282(3):753–757.

65. Stanhope K. L., Schwarz J. M., Keim N. L., et al. Consuming fructose-sweetened, not glucose-sweetened, beverages increases visceral adiposity and lipids and decreases insulin sensitivity in overweight/obese humans. *The Journal of Clinical Investigation*. 2009;119(5):1322–1334. doi: 10.1172/jci37385.

66. Crescenzo R., Bianco F., Falcone I., Coppola P., Liverini G., Iossa S. Increased hepatic de novo lipogenesis and mitochondrial efficiency in a model of obesity induced by diets rich in fructose. *European Journal of Nutrition*. 2013;52(2):537–545. doi: 10.1007/s00394-012-0356-y.

67. Carden T. J., Carr T. P. Food availability of glucose and fat, but not fructose, increased in the US between 1970 and 2009: analysis of the USDA food availability data system. *Nutrition Journal*. 2013;12, article 130 doi: 10.1186/1475-2891-12-130.

68. Hirahatake K. M., Meissen J. K., Fiehn O., Adams S. H. Comparative effects of fructose and glucose on lipogenic gene expression and intermediary metabolism in HepG2 liver cells. *PLoS ONE*. 2011;6(11) doi: 10.1371/journal.pone.0026583.e26583

69. Dushay J. R., Toschi E., Mitten E. K., Fisher F. M., Herman M. A., Maratos-Flier E. Fructose ingestion acutely stimulates circulating FGF21 levels in humans. *Molecular Metabolism*. 2014;4(1):51–57. doi: 10.1016/j.molmet.2014.09.008.

70. Bobbert T., Schwarz F., Fischer-Rosinsky A., et al. Fibroblast growth factor 21 predicts the metabolic syndrome and type 2 diabetes in Caucasians. *Diabetes Care*. 2013;36(1):145–149. doi: 10.2337/dc12-0703.

71. Faeh D., Minehira K., Schwarz J.-M., Periasami R., Seongsu P., Tappy L. Effect of fructose overfeeding and fish oil administration on hepatic de novo lipogenesis and insulin sensitivity in healthy men. *Diabetes*. 2005;54(7):1907–1913. doi: 10.2337/diabetes.54.7.1907.
72. Ha V., Sievenpiper J. L., De Souza R. J., et al. Effect of fructose on blood pressure: a systematic review and meta-analysis of controlled feeding trials. *Hypertension*. 2012;59(4):787–795. doi: 10.1161/hypertensionaha.111.182311.
73. Chong M. F.-F., Fielding B. A., Frayn K. N. Mechanisms for the acute effect of fructose on postprandial lipemia. *The American Journal of Clinical Nutrition*. 2007;85(6):1511–1520.
74. Vedala A., Wang W., Neese R. A., Christiansen M. P., Hellerstein M. K. Delayed secretory pathway contributions to VLDL-triglycerides from plasma NEFA, diet, and de novo lipogenesis in humans. *Journal of Lipid Research*. 2006;47(11):2562–2574. doi: 10.1194/jlr.m600200-jlr200.
75. Stanhope K. L., Havel P. J. Fructose consumption: potential mechanisms for its effects to increase visceral adiposity and induce dyslipidemia and insulin resistance. *Current Opinion in Lipidology*. 2008;19(1):16–24. doi: 10.1097/mol.0b013e3282f2b24a.
76. Marques-Lopes I., Ansorena D., Astiasaran I., Forga L., Martínez J. A. Postprandial de novo lipogenesis and metabolic changes induced by a high-carbohydrate, low-fat meal in lean and overweight men. *The American Journal of Clinical Nutrition*. 2001;73(2):253–261.
77. Horton J. D., Goldstein J. L., Brown M. S. SREBPs: activators of the complete program of cholesterol and fatty acid synthesis in the liver. *Journal of Clinical Investigation*. 2002;109(9):1125–1131. doi: 10.1172/jci200215593.
78. Tran C., Jacot-Descombes D., Lecoultre V., et al. Sex differences in lipid and glucose kinetics after ingestion of an acute oral fructose load. *British Journal of Nutrition*. 2010;104(8):1139–1147. doi: 10.1017/S000711451000190X.
79. Hudgins L. C., Hellerstein M. K., Seidman C. E., Neese R. A., Tremaroli J. D., Hirsch J. Relationship between carbohydrate-induced hypertriglyceridemia and fatty acid synthesis in lean and obese subjects. *Journal of Lipid Research*. 2000;41(4):595–604.

80. Timlin M. T., Parks E. J. Temporal pattern of de novo lipogenesis in the postprandial state in healthy men. *American Journal of Clinical Nutrition*. 2005;81(1):35–42.

81. Le K.-A., Ith M., Kreis R., et al. Fructose overconsumption causes dyslipidemia and ectopic lipid deposition in healthy subjects with and without a family history of type 2 diabetes. *The American Journal of Clinical Nutrition*. 2009;89(6):1760–1765. doi: 10.3945/ajcn.2008.27336.

82. Allen R. J. L., Leahy J. S. Some effects of dietary dextrose, fructose, liquid glucose and sucrose in the adult male rat. *British Journal of Nutrition*. 1966;20(2):339–347. doi: 10.1079/bjn19660034.

83. Bergheim I., Weber S., Vos M., et al. Antibiotics protect against fructose-induced hepatic lipid accumulation in mice: role of endotoxin. *Journal of Hepatology*. 2008;48(6):983–992.

84. Lê K.-A., Faeh D., Stettler R., et al. A 4-wk high-fructose diet alters lipid metabolism without affecting insulin sensitivity or ectopic lipids in healthy humans. *American Journal of Clinical Nutrition*. 2006;84(6):1374–1379.

85. Silbernagel G., MacHann J., Unmuth S., et al. Effects of 4-week very-high-fructose/glucose diets on insulin sensitivity, visceral fat and intrahepatic lipids: an exploratory trial. *British Journal of Nutrition*. 2011;106(1):79–86. doi: 10.1017/s000711451000574x.

86. Sievenpiper J. L., de Souza R. J., Cozma A. I., Chiavaroli L., Ha V., Mirrahimi A. Fructose vs. glucose and metabolism: do the metabolic differences matter? *Current Opinion in Lipidology*. 2014;25(1):8–19. doi: 10.1097/mol.0000000000000042.

87. Chiu S., Sievenpiper J. L., de Souza R. J., et al. Effect of fructose on markers of non-alcoholic fatty liver disease (NAFLD): a systematic review and meta-analysis of controlled feeding trials. *European Journal of Clinical Nutrition*. 2014;68(4):416–423. doi: 10.1038/ejcn.2014.8.

88. Bravo S., Lowndes J., Sinnett S., Yu Z., Rippe J. Consumption of sucrose and high-fructose corn syrup does not increase liver fat or ectopic fat deposition in muscles. *Applied Physiology, Nutrition and Metabolism*. 2013;38(6):681–688. doi: 10.1139/apnm-2012-0322.

89. Rippe J. M., Angelopoulos T. J. Sucrose, high-fructose corn syrup, and fructose, their metabolism and potential health effects: what do we really know? *Advances in Nutrition*. 2013;4(2):236–245. doi: 10.3945/

an.112.002824.

90. Maersk M., Belza A., Stødkilde-Jørgensen H., et al. Sucrose-sweetened beverages increase fat storage in the liver, muscle, and visceral fat depot: a 6-mo randomized intervention study. *The American Journal of Clinical Nutrition*. 2012;95(2):283–289. doi: 10.3945/ajcn.111.022533.
91. Johnston R. D., Stephenson M. C., Crossland H., et al. No difference between high-fructose and high-glucose diets on liver triacylglycerol or biochemistry in healthy overweight men. *Gastroenterology*. 2013;145(5):1016–1025. doi: 10.1053/j.gastro.2013.07.012.
92. Page K. A., Chan O., Arora J., et al. Effects of fructose vs glucose on regional cerebral blood flow in brain regions involved with appetite and reward pathways. *The Journal of the American Medical Association*. 2013;309(1):63–70. doi: 10.1001/jama.2012.116975.
93. Moore J. B., Gunn P. J., Fielding B. A. The role of dietary sugars and de novo lipogenesis in non-alcoholic fatty liver disease. *Nutrients*. 2014;6(12):5679–5703. doi: 10.3390/nu6125679.
94. Rizkalla S. W. Health implications of fructose consumption: a review of recent data. *Nutrition & Metabolism*. 2010;7, article 82 doi: 10.1186/1743-7075-7-82.
95. Aarsland A., Wolfe R. R. Hepatic secretion of VLDL fatty acids during stimulated lipogenesis in men. *Journal of Lipid Research*. 1998;39(6):1280–1286.
96. Connor W. E. Harbingers of coronary heart disease: dietary saturated fatty acids and cholesterol. Is chocolate benign because of its stearic acid content? *The American Journal of Clinical Nutrition*. 1999;70(6):951–952.
97. Cohen J. C., Schall R. Reassessing the effects of simple carbohydrates on the serum triglyceride responses to fat meals. *American Journal of Clinical Nutrition*. 1988;48(4):1031–1034.
98. Teff K. L., Grudziak J., Townsend R. R., et al. Endocrine and metabolic effects of consuming fructose- and glucose-sweetened beverages with meals in obese men and women: influence of insulin resistance on plasma triglyceride responses. *The Journal of Clinical Endocrinology & Metabolism*. 2009;94(5):1562–1569. doi: 10.1210/jc.2008-2192.

99. Hallfrisch J., Reiser S., Prather E. S. Blood lipid distribution of hyperinsulinemic men consuming three levels of fructose. *The American Journal of Clinical Nutrition*. 1983;37(5):740–748.

100. Reiser S., Powell A. S., Scholfield D. J., Panda P., Ellwood K. C., Canary J. J. Blood lipids, lipoproteins, apoproteins, and uric acid in men fed diets containing fructose or high-amylose cornstarch. *The American Journal of Clinical Nutrition*. 1989;49(5):832–839.

101. Bantle J. P., Swanson J. E., Thomas W., Laine D. C. Metabolic effects of dietary fructose in diabetic subjects. *Diabetes Care*. 1992;15(11):1468–1476. doi: 10.2337/diacare.15.11.1468.

102. Swanson J. E., Laine D. C., Thomas W., Bantle J. P. Metabolic effects of dietary fructose in healthy subjects. *American Journal of Clinical Nutrition*. 1992;55(4):851–856.

103. Stanhope K. L., Bremer A. A., Medici V., et al. Consumption of fructose and high fructose corn syrup increase postprandial triglycerides, LDL-cholesterol, and apolipoprotein-B in young men and women. *Journal of Clinical Endocrinology and Metabolism*. 2011;96(10):E1596–E1605. doi: 10.1210/jc.2011-1251.

104. Aeberli I., Gerber P. A., Hochuli M., et al. Low to moderate sugar-sweetened beverage consumption impairs glucose and lipid metabolism and promotes inflammation in healthy young men: a randomized controlled trial. *The American Journal of Clinical Nutrition*. 2011;94(2):479–485. doi: 10.3945/ajcn.111.013540.

105. Diffenderfer M. R., Schaefer E. J. The composition and metabolism of large and small LDL. *Current Opinion in Lipidology*. 2014;25(3):221–226. doi: 10.1097/MOL.0000000000000067.

106. Livesey G., Taylor R. Fructose consumption and consequences for glycation, plasma triacylglycerol, and body weight: meta-analyses and meta-regression models of intervention studies. *The American Journal of Clinical Nutrition*. 2008;88(5):1419–1437. doi: 10.3945/ajcn.2007.25700.

107. Swarbrick M. M., Stanhope K. L., Elliott S. S., et al. Consumption of fructose-sweetened beverages for 10 weeks increases postprandial triacylglycerol and apolipoprotein-B concentrations in overweight and obese women. *British Journal of Nutrition*. 2008;100(5):947–952. doi: 10.1017/S0007114508968252.

108. David Wang D., Sievenpiper J. L., De Souza R. J., et al. Effect of fructose

on postprandial triglycerides: a systematic review and meta-analysis of controlled feeding trials. *Atherosclerosis*. 2014;232(1):125–133. doi: 10.1016/j.atherosclerosis.2013.10.019.

109. Aeberli I., Hochuli M., Gerber P. A., et al. Moderate amounts of fructose consumption impair insulin sensitivity in healthy young men: a randomized controlled trial. *Diabetes Care*. 2013;36(1):150–156. doi: 10.2337/dc12-0540.
110. Lowndes J., Sinnett S., Pardo S., et al. The effect of normally consumed amounts of sucrose or high fructose corn syrup on lipid profiles, body composition and related parameters in overweight/obese subjects. *Nutrients*. 2014;6(3):1128–1144. doi: 10.3390/nu6031128.
111. Rippe J. M. The metabolic and endocrine response and health implications of consuming sugar-sweetened beverages: findings from recent randomized controlled trials. *Advances in Nutrition*. 2013;4(6):677–686. doi: 10.3945/an.113.004580.
112. Hashemi M., Yavari M., Amiri N., et al. Uric acid: a risk factor for coronary atherosclerosis? *Cardiovascular Journal of South Africa*. 2007;18(1):16–19.
113. Rodrigues T. C., Maahs D. M., Johnson R. J., et al. Serum uric acid predicts progression of subclinical coronary atherosclerosis in individuals without renal disease. *Diabetes Care*. 2010;33(11):2471–2473. doi: 10.2337/dc10-1007.
114. Feig D. I., Kang D.-H., Johnson R. J. Uric acid and cardiovascular risk. *The New England Journal of Medicine*. 2008;359(17):1811–1821. doi: 10.1056/nejmra0800885.
115. Cox C. L., Stanhope K. L., Schwarz J. M., et al. Consumption of fructose-but not glucose-sweetened beverages for 10 weeks increases circulating concentrations of uric acid, retinol binding protein-4, and gamma-glutamyl transferase activity in overweight/obese humans. *Nutrition & Metabolism*. 2012;9, article 68 doi: 10.1186/1743-7075-9-68.
116. Wang D. D., Sievenpiper J. L., de Souza R. J., et al. The effects of fructose intake on serum uric acid vary among controlled dietary trials. *The Journal of Nutrition*. 2012;142(5):916–923. doi: 10.3945/jn.111.151951.

117. Perheentupa J., Raivio K. Fructose-induced hyperuricaemia. *The Lancet*. 1967;2(7515):528–531.
118. Brown C. M., Dulloo A. G., Yepuri G., Montani J.-P. Fructose ingestion acutely elevates blood pressure in healthy young humans. *The American Journal of Physiology—Regulatory Integrative and Comparative Physiology*. 2008;294(3):R730–R737. doi: 10.1152/ajpregu.00680.2007.
119. Perez-Pozo S. E., Schold J., Nakagawa T., Sánchez-Lozada L. G., Johnson R. J., Lillo J. L. Excessive fructose intake induces the features of metabolic syndrome in healthy adult men: role of uric acid in the hypertensive response. *International Journal of Obesity*. 2010;34(3):454–461. doi: 10.1038/ijo.2009.259.
120. Jayalath V. H., Sievenpiper J. L., de Souza R. J., et al. Total fructose intake and risk of hypertension: a systematic review and meta-analysis of prospective cohorts. *Journal of the American College of Nutrition*. 2014;33(4):328–339. doi: 10.1080/07315724.2014.916237.
121. Ha V., Jayalath V. H., Cozma A. I., Mirrahimi A., de Souza R. J., Sievenpiper J. L. Fructose-containing sugars, blood pressure, and cardiometabolic risk: a critical review. *Current Hypertension Reports*. 2013;15(4):281–297. doi: 10.1007/s11906-013-0364-1.
122. Wang M., Yu M., Fang L., Hu R. Association between sugar-sweetened beverages and type 2 diabetes: a meta-analysis. *Journal of Diabetes Investigation*. 2015;6(3):360–366. doi: 10.1111/jdi.12309.
123. Kong M.-F., Chapman I., Goble E., et al. Effects of oral fructose and glucose on plasma GLP-1 and appetite in normal subjects. *Peptides*. 1999;20(5):545–551. doi: 10.1016/S0196-9781(99)00006-6.
124. Grant A. M., Christie M. R., Ashcroft S. J. H. Insulin release from human pancreatic islets in vitro. *Diabetologia*. 1980;19(2):114–117. doi: 10.1007/bf00421856.
125. Curry D. L. Effects of mannose and fructose on the synthesis and secretion of insulin. *Pancreas*. 1989;4(1):2–9. doi: 10.1097/00006676-198902000-00002.
126. Cozma A. I., Sievenpiper J. L., de Souza R. J., et al. Effect of fructose on glycemic control in diabetes: a systematic review and meta-analysis of controlled feeding trials. *Diabetes Care*. 2012;35(7):1611–1620. doi: 10.2337/dc12-0073.

127. Sievenpiper J. L., Chiavaroli L., De Souza R. J., et al. 'Catalytic' doses of fructose may benefit glycaemic control without harming cardiometabolic risk factors: A small meta-analysis of randomised controlled feeding trials. *British Journal of Nutrition*. 2012;108(3):418–423. doi: 10.1017/s000711451200013x.

128. Hwang I. S., Ho H., Hoffman B. B., Reaven G. M. Fructose-induced insulin resistance and hypertension in rats. *Hypertension*. 1987;10(5):512–516. doi: 10.1161/01.hyp.10.5.512.

129. Huang Y.-J., Fang V. S., Juan C.-C., Chou Y.-C., Kwok C.-F., Ho L.-T. Amelioration of insulin resistance and hypertension in a fructose-fed rat model with fish oil supplementation. *Metabolism: Clinical and Experimental*. 1997;46(11):1252–1258. doi: 10.1016/s0026-0495(97)90226-2.

130. Crapo P. A., Kolterman O. G. The metabolic effects of 2-week fructose feeding in normal subjects. *The American Journal of Clinical Nutrition*. 1984;39(4):525–534.

131. Grigoresco G., Rizkalla S. W., Halfon P., et al. Lack of detectable deleterious effects on metabolic control of daily fructose ingestion for 2-mo in NIDDM patients. *Diabetes Care*. 1988;11(7):546–550. doi: 10.2337/diacare.11.7.546.

132. Sunehag A. L., Toffolo G., Treuth M. S., et al. Effects of dietary macronutrient content on glucose metabolism in children. *Journal of Clinical Endocrinology & Metabolism*. 2002;87(11):5168–5178. doi: 10.1210/jc.2002-020674.

133. Feinman R. D., Fine E. J. Fructose in perspective. *Nutrition & Metabolism*. 2013;10(1, article 45) doi: 10.1186/1743-7075-10-45.

134. Beck-Nielsen H., Pedersen O., Lindskov H. O. Impaired cellular insulin binding and insulin sensitivity induced by high-fructose feeding in normal subjects. *The American Journal of Clinical Nutrition*. 1980;33(2):273–278.

135. Hallfrisch J., Ellwood K. C., Michaelis O. E., Reiser S., O'Dorisio T. M., Prather E. S. Effects of dietary fructose on plasma-glucose and hormone responses in normal and hyperinsulinemic men. *The Journal of Nutrition*. 1983;113(9):1819–1826.

136. Eiffert K. C., McDonald R. B., Stern J. S. High sucrose diet and exercise: effects on insulin-receptor function of 12- and 24-mo-old Sprague-Dawley rats. *Journal of Nutrition*. 1991;121(7):1081–1089.

137. Wei Y., Pagliassotti M. J. Hepatospecific effects of fructose on c-jun NH2-terminal kinase: implications for hepatic insulin resistance. *American Journal of Physiology—Endocrinology and Metabolism*. 2004;287(5):E926–E933. doi: 10.1152/ajpendo.00185.2004.
138. Nagai Y., Yonemitsu S., Erion D. M., et al. The role of peroxisome proliferator-activated receptor gamma coactivator-1 in the pathogenesis of fructose-induced insulin resistance. *Cell Metabolism*. 2009;9(3):252–264. doi: 10.1016/j.cmet.2009.01.011.
139. Delbosc S., Paizanis E., Magous R., et al. Involvement of oxidative stress and NADPH oxidase activation in the development of cardiovascular complications in a model of insulin resistance, the fructose-fed rat. *Atherosclerosis*. 2005;179(1):43–49. doi: 10.1016/j.atherosclerosis.2004.10.018.
140. Stanhope K. L., Griffen S. C., Bair B. R., Swarbrick M. M., Keim N. L., Havel P. J. Twenty-four-hour endocrine and metabolic profiles following consumption of high-fructose corn syrup-, sucrose-, fructose-, and glucose-sweetened beverages with meals. *The American Journal of Clinical Nutrition*. 2008;87(5):1194–1203.
141. Tran L. T., Yuen V. G., McNeill J. H. The fructose-fed rat: a review on the mechanisms of fructose-induced insulin resistance and hypertension. *Molecular and Cellular Biochemistry*. 2009;332(1-2):145–159. doi: 10.1007/s11010-009-0184-4.
142. Rodin J., Reed D., Jamner L. Metabolic effects of fructose and glucose: implications for food intake. *American Journal of Clinical Nutrition*. 1988;47(4):683–689.
143. Rodin J. Comparative effects of fructose, aspartame, glucose, and water preloads on calorie and macronutrient intake. *The American Journal of Clinical Nutrition*. 1990;51(3):428–435.
144. Spitzer L., Rodin J. Effects of fructose and glucose preloads on subsequent food-intake. *Appetite*. 1987;8(2):135–145. doi: 10.1016/s0195-6663(87)80006-5.
145. Warwick Z. S., Weingarten H. P. Dynamics of intake suppression after a preload: role of calories, volume, and macronutrients. *The American Journal of Physiology—Regulatory Integrative and Comparative Physiology*. 1994;266(4):R1314–R1318.
146. Tappy L., Randin J.-P., Felber J.-P., et al. Comparison of thermogenic

effect of fructose and glucose in normal humans. *The American Journal of Physiology—Endocrinology and Metabolism*. 1986;250(6):E718–E724.

147. Simonson D. C., Tappy L., Jequier E., Felber J.-P., DeFronzo R. A. Normalization of carbohydrate-induced thermogenesis by fructose in insulin-resistant states. *The American Journal of Physiology—Endocrinology and Metabolism*. 1988;254(2):E201–E207.
148. Schwarz J.-M., Acheson K. J., Tappy L., et al. Thermogenesis and fructose metabolism in humans. *American Journal of Physiology: Endocrinology and Metabolism*. 1992;262(5):E591–E598.
149. Hugenholtz J. The lactic acid bacterium as a cell factory for food ingredient production. *International Dairy Journal*. 2008;18(5):466–475. doi: 10.1016/j.idairyj.2007.11.015.
150. Gwak M.-J., Chung S.-J., Kim Y. J., Lim C. S. Relative sweetness and sensory characteristics of bulk and intense sweeteners. *Food Science and Biotechnology*. 2012;21(3):889–894. doi: 10.1007/s10068-012-0115-0.
151. Shallenberger R. S. Intrinsic chemistry of fructose. *Pure and Applied Chemistry*. 1978;50(11-12):1409–1420. doi: 10.1351/pac197850111409.
152. Wrolstad R. E. *Food Carbohydrate Chemistry*. 1st. New York, NY, USA: John Wiley & Sons; 2012.
153. Blakely S. R., Hallfrisch J., Reiser S., Prather E. S. Long-term effects of moderate fructose feeding on glucose-tolerance parameters in rats. *The Journal of Nutrition*. 1981;111(2):307–314.
154. Stanhope K. L., Havel P. J. Endocrine and metabolic effects of consuming beverages sweetened with fructose, glucose, sucrose, or high-fructose corn syrup. *The American Journal of Clinical Nutrition*. 2008;88(6):1733S–1737S. doi: 10.3945/ajcn.2008.25825d.
155. Sievenpiper J. L., de Souza R. J., Mirrahimi A., et al. Effect of fructose on body weight in controlled feeding trials: a systematic review and meta-analysis. *Annals of Internal Medicine*. 2012;156(4):291–304. doi: 10.7326/0003-4819-156-4-201202210-00007.
156. Schwarz J.-M., Linfoot P., Dare D., Aghajanian K. Hepatic de novo lipogenesis in normoinsulinemic and hyperinsulinemic subjects consuming high-fat, low-carbohydrate and low-fat, high-carbohydrate isoenergetic diets. *The American Journal of Clinical*

Nutrition. 2003;77(1):43–50.

157. Frayn K. N., Lund P., Walker M. Interpretation of oxygen and carbon dioxide exchange across tissue beds in vivo. *Clinical Science*. 1993;85(4):373–384.
158. Blaak E. E., Saris W. H. M. Postprandial thermogenesis and substrate utilization after ingestion of different dietary carbohydrates. *Metabolism: Clinical and Experimental*. 1996;45(10):1235–1242. doi: 10.1016/s0026-0495(96)90241-3.
159. Ferrannini E. The theoretical bases of indirect calorimetry: a review. *Metabolism: Clinical and Experimental*. 1988;37(3):287–301.
160. Tappy L., Mittendorfer B. Fructose toxicity: is the science ready for public health actions? *Current Opinion in Clinical Nutrition and Metabolic Care*. 2012;15(4):357–361. doi: 10.1097/mco.0b013e328354727e.
161. Vos M. B., Kimmons J. E., Gillespie C., Welsh J., Blank H. M. Dietary fructose consumption among US children and adults: the Third National Health and Nutrition Examination Survey. *The Medscape Journal of Medicine*. 2008;10(7, article 160)
162. Sland M., Haugen M., Eriksen F.-L., et al. High sugar consumption and poor nutrient intake among drug addicts in Oslo, Norway. *British Journal of Nutrition*. 2011;105(4):618–624. doi: 10.1017/S0007114510003971.

9

Dietary Carbohydrates Modulate Candida albicans Biofilm Development on the Denture Surface

Ivone Lima Santana[1], Letícia Machado Gonçalves[2], Andréa Araújo de Vasconcellos[2], Wander José da Silva[2], Jaime Aparecido Cury[3] and Altair Antoninha Del Bel Cury[2]

[1] Department of Dentistry I, School of Dentistry, Federal University of Maranhão, São Luis, Maranhão, Brazil,
[2] Department of Prosthodontics and Periodontology, Piracicaba Dental School, University of Campinas, Piracicaba, São Paulo, Brazil,
[3] Department of Physiological Sciences, Piracicaba Dental School, University of Campinas, Piracicaba, São Paulo, Brazil,

ABSTRACT

The purpose of this study was to investigate whether dietary carbohydrates can modulate the development of *Candida albicans* biofilms on the denture material surface. Poly (methyl methacrylate) acrylic resin discs were fabricated and had their surface roughness measured. Biofilms

Citation: Dietary Carbohydrates Modulate Candida albicans Biofilm Development on the Denture Surface. Ivone Lima Santana, Letı´cia Machado Gonc¸alves, Andre´a Arau´jo de Vasconcellos, Wander Jose´ da Silva, Jaime Aparecido Cury, Altair Antoninha Del Bel Cury. (2013) Dietary Carbohydrates Modulate Candida albicans Biofilm Development on the Denture Surface. PLoS ONE 8(5): e64645. doi:10.1371/journal.pone.0064645 of *C. albicans* ATCC 90028 were developed on saliva-coated specimens in culture medium without (control) or with carbohydrate supplementation by starch, starch+sucrose, glucose, or sucrose for 72 h. The cell count, metabolic activity, biovolume, average thickness, and roughness coefficient were evaluated at the adhesion phase (1.5 h) and after 24, 48, and 72 h. The secretion of proteinases and phospholipases, cell surface energy, and production of extra/intracellular polysaccharides were analyzed after 72 h of biofilm development. Data were analyzed by one- and two-way ANOVA followed by Tukey's test at 5% significance level. In the early stages of colonization (adhesion and 24 h), the glucose group showed the highest cell counts and metabolic activity among the groups ($p<0.05$). After maturation (48 and 72 h), biofilms exposed to glucose, sucrose, or starch+sucrose showed higher cell counts and metabolic activity than the control and starch groups ($p<0.001$). Compared to the control group, biofilms developed on starch or starch+sucrose had more proteinase activity ($p<0.001$), whereas biofilms developed on glucose or sucrose had more phospholipase activity ($p<0.05$). Exposure to starch+sucrose increased the production of extracellular and intracellular polysaccharides ($p<0.05$). Biofilms developed on starch or without carbohydrate supplementation presented cells with more hydrophobic behavior compared to the other groups. Confocal images showed hyphae forms on biofilms exposed to starch or starch+sucrose. Within the conditions studied, it can be concluded that dietary carbohydrates can modulate biofilm development on the denture surface by affecting virulence factors and structural features.

INTRODUCTION

The availability of nutrient sources plays an important role in the pathogenesis of fungal infections [1], [2]. Carbohydrates consumed in the diet are the primary and preferred nutrient sources for *Candida albicans* [3]–[5], the main pathogen related to *Candida*-associated denture stomatitis (CADS) [6]. Additionally, denture wearers tend to choose a carbohydrate-rich diet because of difficulty in chewing [7]. The constant supply of sugars such as glucose, sucrose, and starch may create an environment conductive to *C. albicans* colonization [3], [5], [8]. Thus, understanding how *C. albicans* biofilms behave under exposure to dietary carbohydrates is of outmost relevance, although little data are available so far.

The first step in the dynamics of CADS pathogenesis is the adhesion of *C. albicans* to the denture surface and epithelial cells, followed by cell multiplication, organization, and secretion of the extracellular matrix (ECM). This process results in the formation of a three-dimensional (3D) structure known as biofilm [9]. During biofilm formation, the abilities of cells to adhere and co-aggregate are influenced by the cell surface hydrophobicity, an important virulence factor [10]. In this respect, it is possible that exposure to dietary carbohydrates could alter the components of fungal cells, affecting their hydrophobicity [8], [11], [12] and, consequently, their capacity for biofilm formation [5]. Additionally, the production of organic acids during carbohydrate hydrolysis [13] could activate other virulence factors, such as the yeast-hyphae transition and the secretion of hydrolytic enzymes (*e.g.*, phospholipases and proteases) [10], [14]. These virulence factors are mainly related to the invasion of host tissues, which aggravates the inflammatory response of the oral mucosa [15], [16].

Also, microbial communities adapt to environmental conditions by altering their structural organization [17]. Thus, it is expected that dietary carbohydrates could provoke structural changes in *C. albicans* biofilms. Fermentable carbohydrates, such as glucose, sucrose, and starch, have been observed to serve as substrates for ECM synthesis in bacterial biofilm models [18]. Additionally, sucrose and starch are consumed simultaneously in modern societies [19]. This combination could enhance the pathogenicity of biofilms by altering the organization of cells and leading to the formation of a structurally distinct ECM relative to other carbon sources [20]. However, it remains unclear how these carbohydrates affect the structure of *C. albicans* biofilms. Thus, this study investigated whether dietary carbohydrates or their combinations can modulate the development of *C. albicans* biofilms on the surface of denture material.

MATERIALS AND METHODS

Study Design

This *in vitro* study had a randomized and blinded design. Discs were fabricated using a water bath poly (methyl methacrylate) (PMMA) acrylic resin. The surface roughness of all specimens and the surface free energy for the salivary-coated specimens were measured. Biofilms of *C. albicans* ATCC 90028 were developed on saliva-coated specimens in culture medium without or with carbohydrate supplementation by starch, starch+sucrose, glucose, or sucrose for 72 h. The cell count, metabolic activity, biovolume, average thickness, and roughness coefficient were evaluated at the adhesion phase (1.5 h) and after 24, 48, and 72 h of biofilm development. The secretion of proteinases and phospholipases, cell surface energy, and production of extra/intracellular polysaccharides were analyzed after 72 h of biofilm development. All of the experiments were performed in four replicates of three independent experiments on different days (n=12). Data were analyzed by one-way and two-way ANOVA followed by Tukey's test at the 5% significance level.

Fabrication of Specimens

Discs (10-mm diameter, 2-mm thickness) were fabricated using a water bath PMMA acrylic resin (QC-20; Dentsply Ltd., Weybridge, England) according to manufacturers' directions (25±1°C and 50% ±5% relative humidity). Prior to polymerization for 20 min at 100°C, the resin mass in the plastic phase was packed into a metal mold to standardize the specimen dimensions. Processed specimens were immersed in distilled water for 48 h at 35°C to release residual monomer [21]. To simulate the inner side of a denture, the specimen surfaces were ground in an horizontal polisher (model APL-4; Arotec, São Paulo, Brazil) by using progressively smoother aluminum oxide papers (320, 400, and 600 grit). The surface roughness of all of the specimens was analyzed by a profilometer (Surfcoder SE 1700; Kosaka Laboratory Ltd., Kosaka, Japan) accurate to the nearest 0.01 μm and calibrated at a specimen length of 0.8 mm, 2.4 mm percussion of measure, and 0.5 mm/s. The mean of three measurements for each specimen was calculated, and the surface roughness was standardized at 0.30±0.04 μm [22]. The surface free energy of the saliva-coated specimens was measured with a goniometer (Ramé-Hart Instrument Co., NJ, USA) by an acid-base technique [23]. The total surface free energy was 40.05±1.00 mJ/m^2.

After these measurements had been made, the specimens were ultrasonically cleansed (Thornton T 740; Thornton-Inpec Eletronica Ltd., Vinhedo, São Paulo, Brazil) in purified water for 20 min to remove any contaminants and artifacts.

Preparation of Candida Suspensions

C. albicans (ATCC 90028, reference strain) was aerobically cultured from original broth by incubation in Sabouraud Dextrose Agar (SDA; Difco, Detroit, MI, USA) for 24 h at 35°C. A loop of yeast cells was inoculated into Yeast Nitrogen Base culture medium (YNB; Difco) supplemented with 50 mM glucose and incubated aerobically under agitation at 35°C. During the exponential growth phase (*i.e.*, after 18–20 h of incubation), the *C. albicans* cells were washed twice with phosphate-buffered saline (PBS; pH 7.2). Suspensions were prepared in YNB medium without (control group) or with carbohydrate supplementation by 1% starch (starch group), 1% starch +1% sucrose (starch+sucrose group), 1% glucose (glucose group), or 1% sucrose (sucrose group). The suspensions were standardized spectrophotometrically (Spectronic 20; Bausch & Lomb, Rochester, NY, US) to ~10^7 cells/mL (OD=0.25 at 520 nm).

Biofilm Development

To mimic the oral cavity, the specimens were coated with human salivary pellicle prior to biofilm development. Saliva was collected from a healthy volunteer, who provided written formal consent according to a protocol approved by the Ethics Committee in Research of Piracicaba Dental School (105/2011). Saliva was collected in an ice-chilled polypropylene tube during masticatory stimulation with a flexible film and clarified by centrifugation (10,000×*g* for 5 min at 4°C). The supernatant was filter-sterilized and used immediately.

Under aseptic conditions, each specimen was placed in a presterilized 24-well culture plate. Two milliliters of saliva were added to each well. The plate was incubated aerobically under agitation for 1 h at 35°C to form the salivary pellicle. Saliva-coated specimens were transferred to a 24-well culture plate, and the prepared standard suspensions were added to each well. The sets were incubated aerobically under agitation at 35°C for 1.5 h (adhesion phase). The specimens were carefully washed with PBS and transferred to new presterilized 24-well culture plates containing culture medium without or with carbohydrate supplementation. At the end of each 24-h period, the specimens were washed with PBS, and fresh medium was added. The plates were incubated aerobically under agitation for 72 h at 35°C.

Cell Count

Biofilm-containing specimens were washed twice with PBS, and, afterwards, were immersed in PBS, and sonicated (7 W, for 30 s) to disrupt the biofilm structure. The sonicated solutions were serially diluted in PBS, and samples were plated in triplicate onto SDA. The plates were incubated aerobically for

24 h at 35°C. Yeast cells were counted (in cells/mL) on a stereomicroscope (Coleman Comp. Imp., Santo André, São Paulo, Brazil).

Metabolic Activity

The metabolic activity was determined with an adapted XTT assay [24]. Briefly, biofilm-containing specimens were placed in a 24-well culture plate with XTT solution (PBS supplemented with 200 mM glucose, 1 mg/mL XTT, and 0.4 mM menadione). The plates were protected from light and incubated under agitation for 3 h at 35°C. Colorimetric changes in the supernatant were analyzed spectrophotometrically at 492 nm.

Secretion of Proteinase and Phospholipase

Enzyme secretion assays were performed on sonicated suspensions of biofilm cells only after 72 h of biofilm development because of the large amounts of biofilm that were needed for these analyses. The proteinase activity was determined as described previously [25]. Briefly, the supernatant was mixed with 1% azocasein at 1:9 (v/v) for 1 h at 35°C. The reaction was stopped by adding 500 μL of 10% trichloroacetic acid. The mixture was centrifuged at 10,000×*g* for 5 min. Next, 500 μL of the supernatant were mixed with an equal volume of 0.5 M NaOH and incubated for 15 min. The proteinase activity was evaluated spectrophotometrically at 440 nm. The specific proteinase activity was defined as the amount of enzyme that elicited an increase of 0.001 units of absorbance per minute of digestion by biofilm dry weight [25], [26].

The phospholipase activity was determined as described by Taniguchi et al. [27]. Briefly, the biofilm supernatants were mixed and incubated with an equal volume of phosphatidylcholine substrate for 1 h at 35°C. The phospholipase activity was evaluated spectrophotometrically at 630 nm. The specific phospholipase activity was established as the absorbance shift per minute of reaction by biofilm dry weight [26], [27].

Cell Surface Energy

The cell surface energy was evaluated after 72 h of biofilm development because of the large amount of biofilm that was needed for this analysis. Biofilm-containing specimens were washed twice, immersed in PBS, and sonicated to disrupt the biofilm structure. The sonicated solution was filtered through a 0.45-μm cellulose acetate membrane, which had been previously soaked with 10 mL of distilled water, to obtain a homogeneous cell layer.

The filters with cell layers were placed on a glass slide and maintained in a Petri dish containing 1% (w/v) agar and 10% (v/v) glycerol for 3 h at 35°C [28].

The cell surface energy was determined by an acid-base technique [23] with goniometer (Ramé-Hart Instrument Co.). Briefly, droplets (15 μL) of distilled water, formamide, and 1-bromonaphthalene were dispensed on the filters with cell layers. The contact angle was automatically measured. The surface free energy was measured by the DropImage software package (Ramé-Hart Instrument Co.), which allowed the hydrophobicity of fungal cells to be estimated.

Polysaccharide Extraction

For polysaccharide extraction, 400 μL of the previously obtained sonicated biofilm suspension were centrifuged at 10,000×g for 5 min at 4°C. The supernatant, which contain the soluble extracellular polysaccharides (SEPs), was transferred to a tube named SEPs. An aliquot of 1 M NaOH was added to the pellet to extract the insoluble extracellular polysaccharides (IEPs) [29]. The tube was agitated for 15 min, centrifuged, and the supernatant was transferred to another tube named IEPs. Finally, an aliquot of 1 M NaOH was added to the residual pellet to extract the intracellular polysaccharide (IPs) [30]. The tube was heated for 15 min at 100°C, centrifuged, and the supernatant was transferred to another tube named IPs.

Three volumes of cold ethanol were added to the tubes containing SEPs, IEPs, and IPs. The tubes were maintained for 30 min at −20°C. Next, the tubes were centrifuged, and the pellets were washed twice with cold 75% ethanol. The precipitated polysaccharides were resuspended in 1 M NaOH. The total carbohydrate was estimated by the phenol sulfuric method [31]. The results were normalized by dry weight of biofilm.

Confocal Structural Analysis

The biofilm structure was evaluated at the adhesion phase and after 24, 48, or 72 h of biofilm development by confocal laser-scanning microscopy (CLSM) (Leica Microsystems CMS, Mannheim, Germany). Biofilm-containing specimens were washed twice with PBS and stained by SYTO-9 and propidium iodide with the Live/Dead *BacLight* viability kit (Invitrogen-Molecular Probes, Eugene, OR, USA). The biofilms were incubated for 20 min at 35°C, with care taken to protect the samples from light. A series of optical sections was taken by CLSM through the full depth of the biofilm at 1-μm intervals in the *z*-axis. At least five representative random optical fields were examined for each specimen. The resulting series of images was analyzed by the COMSTAT software program [17] to quantify structural parameters, such as the biovolume ($\mu m^3/\mu m^2$), average thickness (μm), and roughness coefficient (μm).

Statistical Analysis

The results were statistically analyzed by the SAS/LAB software package (SAS Software, version 9.0; SAS Institute Inc., Cary, NC, USA). Assumptions of the equality of variances and the normal distribution of errors were checked. Data were transformed as suggested by the software, as follows: cell counts, IEPs, and IPs (transformed by log 10); metabolic activity, phospholipase and proteinase activities, and SEP (transformed by exponentiation, y^2), average thickness (transformed by exponentiation, y^3), and roughness coefficient (transformed by exponentiation, y^5).

The proteinase and phospholipase activities, cell surface energy, and polysaccharide production were analyzed by one-way ANOVA followed by Tukey's HSD test, considering the carbohydrates or their combination as the study factor. The cell count, metabolic activity, biovolume, average thickness, and roughness coefficient were analyzed by two-way ANOVA followed by Tukey's HSD test. In this case, the time-point of biofilm development (adhesion, 24, 48, or 72 h) and the carbohydrates or their combination were considered as study factors.

RESULTS

No statistically significant differences were observed between the adhesion phase and 24 h, or between 48 and 72 h for any of the parameters investigated during biofilm development ($p>0.05$). However, during the early stages of colonization (adhesion phase and 24 h of development), the glucose group showed the highest cell counts and metabolic activity among the groups ($p<0.05$). No differences were observed among the other groups at these time points ($p>0.05$, data not shown). Similarly, for the mature biofilms (48 and 72 h of development), glucose, sucrose, or starch+sucrose exposure led to higher cell counts and metabolic activity compared to starch or no carbohydrate exposure ($p<0.001$). Mature biofilms developed in the presence of starch showed cell counts and metabolic activity that were similar to those of the control group during these experimental periods ($p>0.05$, Fig. 1A, B).

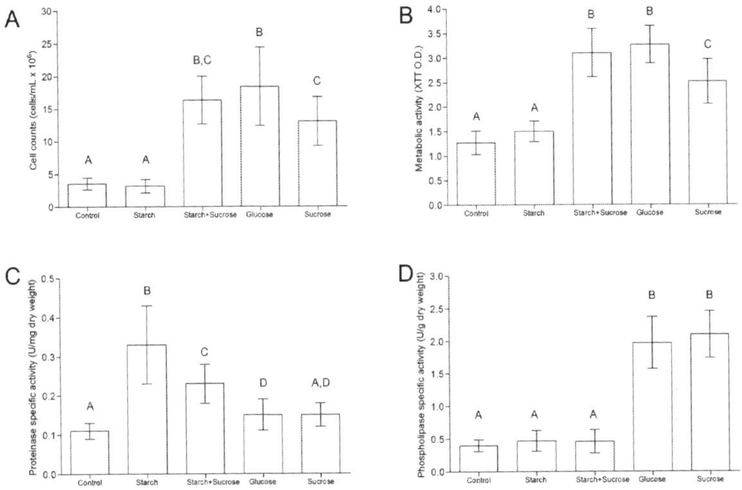

Figure 1. *C. albicans* **biofilms developed under exposure to dietary carbohydrates.** (A) Cell count and (B) metabolic activity of mature biofilms of *C. albicans* (48 and 72 h). (C) Proteinase-specific activity and (D) phospholipase-specific activity of *C. albicans* biofilms (72 h).

Biofilms developed on starch secreted the highest amount of proteinase ($p<0.001$), followed by biofilms developed on starch+sucrose (Fig. 1C). The phospholipase activity was higher in the glucose and sucrose groups compared to the control, starch, and starch+sucrose groups ($p<0.05$, Fig. 1D).

The starch+sucrose group showed the highest production of SEPs followed by the starch group ($p<0.05$, Fig. 2A). No significant difference in SEPs production was found between the glucose and sucrose groups, which both presented greater SEPs amounts than the control group ($p<0.001$). The IEPs production was similar among biofilms exposed to starch+sucrose, sucrose, and glucose ($p>0.05$). The starch+sucrose and glucose groups produced the highest amounts of IPs ($p<0.05$, Fig. 2A).

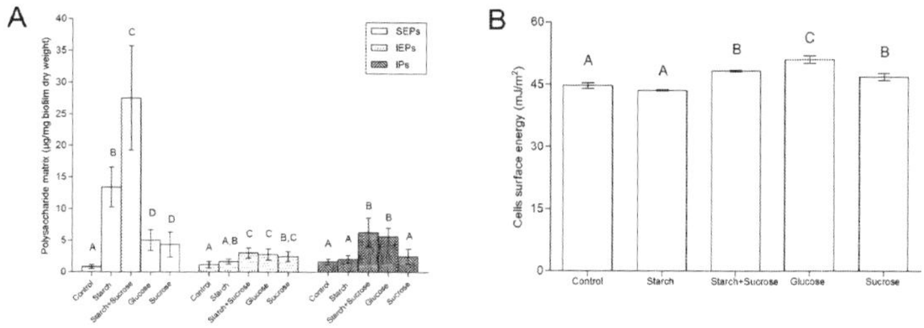

Figure 2. *C. albicans* **biofilms (72 h) developed under exposure to dietary carbohydrates.** (A) Polysaccharide matrix and (B) cell surface energy.

The cell surface energy varied depending on the carbohydrate source. The glucose group displayed cells with the highest hydrophilicity ($p<0.05$). The starch and control groups resulted in cells with more hydrophobic behaviors ($p<0.05$, Fig. 2B).

Regarding the structural parameters obtained by CLSM images, during the early stages of colonization, no differences in the biovolume, average thickness, and roughness coefficient were observed between the groups ($p>0.05$, data not shown). Also, biofilms developed for 48 or 72 h did not differ from each other, but they differed from the early-stage biofilms. Mature biofilms exposed to glucose presented the highest biovolume ($p<0.05$, Fig. 3A). The control group showed the lowest biovolume and average thickness values ($p<0.001$, Fig. 3A, B). The glucose and sucrose groups showed the lowest roughness coefficient values ($p<0.05$, Fig. 3C).

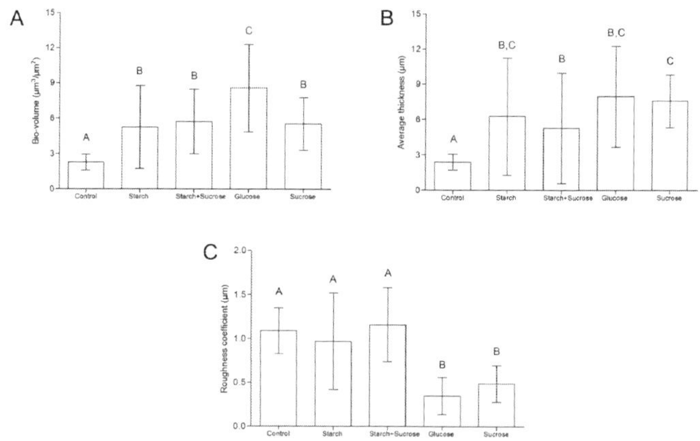

Figure 3. Structural parameters of *C. albicans* **biofilms (48 and 72 h) developed under exposure to dietary carbohydrates.** (A) Biovolume, (B) average thickness, and (C) roughness coefficient.

Figure 4 shows representative CLSM images of *C. albicans* biofilms at 72 h. These images reveal the presence of black spaces and hyphae forms on biofilms developed on starch (Fig. 4B) or starch+sucrose (Fig. 4C). In addition, higher amounts of dead cells can be observed on the control (Fig. 4A) and starch (Fig. 4B) groups compared to the other groups.

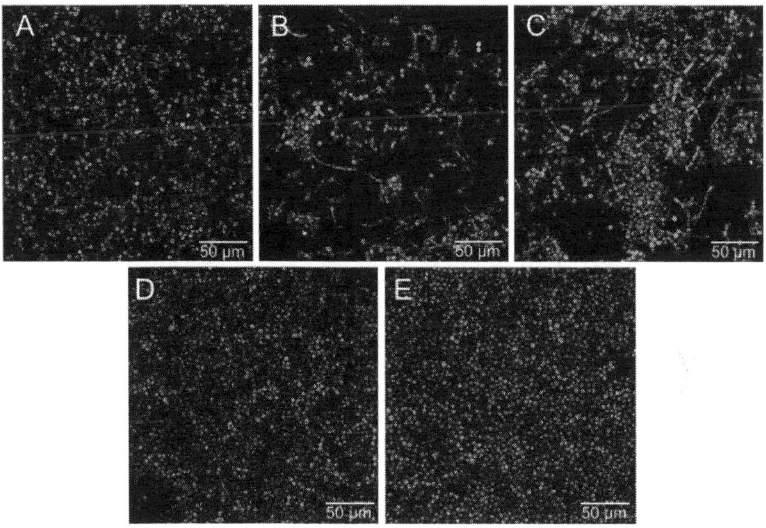

Figure 4. Representative confocal images of *C. albicans* **biofilms (72 h) developed under exposure to dietary carbohydrates.** (A) Control, (B) starch, (C) starch+sucrose, (D) glucose, and (E) sucrose. Live cells appear green and dead cells appear red. A considerable number of hyphae forms can be observed in images B and C.

DISCUSSION

In this study, we investigated whether dietary carbohydrates or their combinations could modulate the development of *C. albicans* biofilms by affecting the cell count, metabolic activity, proteinase and phospholipase secretion, hydrophobicity, polysaccharide production, and structural features.

During the early stages of biofilm development, the cell count and metabolic activity were significantly increased in the presence of glucose. This result may be explained by considering that when yeast cells are colonizing a surface and are exposed to different carbon sources, they preferentially utilize monosaccharides, such as glucose, before assimilating others carbon sources [32]. For the mature biofilms, the glucose, starch+sucrose, and sucrose groups showed the highest cell counts and metabolic activity. After achieving an advanced stage of organization and metabolism, it may be that the yeast cells are able to hydrolyze more complex carbon sources, such as sucrose, which could be a critical factor during biofilm formation [33]. This hierarchical mechanism of carbon assimilation is regulated by a complex glucose-signaling network, which involves the coordinated action of AMP kinase, protein kinase A, and sugar repressor signaling pathways [34].

Besides glucose and sucrose, starch is commonly found in a carbohydrate-rich diet [19]. Although starch is classified as a polysaccharide and can be a good nutrient source, our results suggested that it was not easily degraded by yeast cells. Actually, starches are rapidly hydrolyzed into maltose, malto-dextrins, and other oligosaccharides by salivary α-amylase in the oral environment [20]. Although our biofilms were developed on a salivary pellicle, the salivary pellicle did not appear to be sufficient to hydrolyze the starch. Consequently, the exposure to starch resulted in lower cell counts and metabolic activity during all experimental periods, which were similar to levels of the control group.

In this study, it is possible to suggest that the control and starch groups represented nutrient-limited conditions. These environmental conditions resulted in lower structural bases and higher amounts of dead cells, as clearly seen in the CLSM images. Whereas the biofilms developed in the presence of glucose, sucrose, and starch+sucrose formed robust 3D structures, the presence of nutrient-limited conditions resulted in negligible biofilm formation. This behavior may be attributed to a possible "starved" condition, in which the metabolic capacity of *C. albicans* may have been greatly depressed and the growth and proliferation inhibited [35]. This possibility was confirmed by our cell count, metabolic activity, biovolume, and average thickness results.

Biofilm dispersion could also be induced under nutrient-limited conditions [36]. In this regard, the high roughness coefficient shown by the "starved" groups by COMSTAT analysis can be considered as an indicator of heterogeneity [17], reflecting their irregularity and dispersion on the substratum surface. Overall, it is supposed that these metabolic and structural changes are beneficial for survival under nutrient-limited conditions. Future investigations should consider the exact molecular mechanisms that allow fungal cells to survive in the starvation environment.

The adhesion of *Candida* to the denture surface and the coaggregation between cells are critical virulence factors for biofilm formation [9], [10]. Dietary sugars may affect adherence by altering the components of the fungal cell wall [5], [8], [11], [12]. *C. albicans* possesses many cell-wall proteins [37] that vary considerably both in quantity and quality under different environmental conditions [1], [2], [35]. Because these cell-wall proteins are closely related to the cell-surface properties, profound changes are expected to be observed in the hydrophobicity of fungal cells when they are exposed to different carbohydrate sources [38].

In our data, biofilms exposed to starch or no carbohydrate had more hydrophobic behavior. In contrast, glucose exposure, followed by sucrose and starch+sucrose exposure, resulted in cells that were more hydrophilic. These data support the idea that the lowest biofilm formation occurred under the "starved" conditions. A previous study demonstrated that the salivary pellicle coating the denture surface had many hydrophilic proteins [39]. In addition, the adherence and biofilm development of hydrophilic species (*e.g.*, *C. albicans*) are known to be increased on more hydrophilic surfaces [40], [41]. Thus, changes in the cell-wall proteins under nutrient-limited conditions may account for the higher hydrophobicity and, consequently, reduced the biofilm formation.

C. albicans produces numerous short-chain carboxylic acids during carbohydrate hydrolysis [13], which are related to the reduction in the environmental pH [42]. This condition may activate other virulence factors, such as the yeast-hyphae transition and the secretion of proteinases and phospholipases [10], [14]. According to the CLSM images, biofilms developed in glucose, sucrose, or without carbohydrate supplementation were predominantly composed of only blastospore cells, whereas the starch and starch+sucrose groups also displayed a filamentous growth mode. Interestingly, the biofilms with hyphae also showed the highest proteinase activity. A previous study demonstrated that some proteinase isoenzymes were preferentially expressed during hyphae formation [16], which may help to explain our results.

On the other hand, the phospholipase activity was greater when biofilms were exposed to glucose or sucrose. It is possible that this secretion was modulated by environmental conditions, such as the pH of the medium [15], rather than by morphological aspects. Phospholipases are more active in pH ranging from 2.5 to 3.5 [43]. The pH condition observed in the glucose and sucrose groups after biofilm maturation (pH ~3.2, data not shown) was lower than that of the other groups.

Our biochemical analyses demonstrated that the presence of starch+sucrose resulted in the highest amounts of soluble and insoluble polysaccharides, consistent with previous reports using bacterial biofilm models [18]. Additionally, the energetic reserve represented by the IPs was also increased for this group. A previous study showed that the proportion of β-glucans and mannoproteins, the main fungal matrix components, could vary depending on the nature of the fermentative carbon source [44]. This result seems to be very important, given that the ECM creates a 3D environment that is important for both the biofilm integrity and its resistance to antifungal agents [9]. It may be that the combination of starch+sucrose can trigger *C. albicans* responses at the transcriptional level [20], modulating the expression of genes associated with ECM formation. Nevertheless, the exact mechanism by which dietary carbohydrates and their combinations enhance ECM production by *C. albicans* biofilms requires future investigation.

CONCLUSION

Within the limitations of this study, it can be concluded that dietary carbohydrates can modulate biofilm development on the denture surface by affecting virulence factors and structural features.

REFERENCES

1. Ene IV, Adya AK, Wehmeier S, Brand AC, MacCallum DM, et al. (2012) Host carbon sources modulate cell wall architecture, drug resistance and virulence in a fungal pathogen. Cell Microbiol 14: 1319–1335.
2. Ene IV, Heilmann CJ, Sorgo AG, Walker LA, de Koster CG, et al. (2012) Carbon source-induced reprogramming of the cell wall proteome and secretome modulates the adherence and drug resistance of the fungal pathogen Candida albicans. Proteomics 12: 3164–3179.
3. Seneviratne CJ, Zhang T, Fang HH, Jin LJ, Samaranayake LP (2009) Distribution coefficients of dietary sugars in artificial Candida biofilms. Mycopathologia 167: 325–331.
4. Samaranayake LP (1986) Nutritional factors and oral candidosis. J Oral Pathol 15: 61–65.
5. Jin Y, Samaranayake LP, Samaranayake Y, Yip HK (2004) Biofilm formation of Candida albicans is variably affected by saliva and dietary sugars. Arch Oral Biol 49: 789–798.
6. Gendreau L, Loewy ZG (2011) Epidemiology and etiology of denture stomatitis. J Prosthodont 20: 251–260.
7. Lyon JP, de Resende MA (2006) Correlation between adhesion, enzyme production, and susceptibility to fluconazole in Candida albicans obtained from denture wearers. Oral Surg Oral Med Oral Pathol Oral Radiol Endod 102: 632–638.
8. Samaranayake LP, MacFarlane TW (1982) The effect of dietary carbohydrates on the in-vitro adhesion of Candida albicans to epithelial cells. J Med Microbiol 15: 511–517.
9. Seneviratne CJ, Jin L, Samaranayake LP (2008) Biofilm lifestyle of Candida: a mini review. Oral Dis 14: 582–590.
10. Jayatilake JA (2011) A review of the ultrastructural features of superficial candidiasis. Mycopathologia 171: 235–250.
11. McCourtie J, Douglas LJ (1984) Relationship between cell surface composition, adherence, and virulence of Candida albicans. Infect Immun 45: 6–12.
12. Nikawa H, Egusa H, Makihira S, Yamashiro H, Fukushima H, et al. (2001) Alteration of the coadherence of Candida albicans with oral bacteria by dietary sugars. Oral Microbiol Immunol 16: 279–283.

13. Samaranayake LP, MacFarlane TW (1985) On the role of dietary carbohydrates in the pathogenesis of oral candidosis. FEMS Microbiology Letters 27: 1–5.
14. Konno N, Ishii M, Nagai A, Watanabe T, Ogasawara A, et al. (2006) Mechanism of Candida albicans transformation in response to changes of pH. Biol Pharm Bull 29: 923–926.
15. Ghannoum MA (2000) Potential role of phospholipases in virulence and fungal pathogenesis. Clin Microbiol Rev 13: 122–143.
16. Naglik J, Albrecht A, Bader O, Hube B (2004) Candida albicans proteinases and host/pathogen interactions. Cell Microbiol 6: 915–926.
17. Heydorn A, Nielsen AT, Hentzer M, Sternberg C, Givskov M, et al. (2000) Quantification of biofilm structures by the novel computer program COMSTAT. Microbiology 146 (Pt 10): 2395–2407.
18. Aires CP, Del Bel Cury AA, Tenuta LM, Klein MI, Koo H, et al. (2008) Effect of starch and sucrose on dental biofilm formation and on root dentine demineralization. Caries Res 42: 380–386.
19. Lingstrom P, van Houte J, Kashket S (2000) Food starches and dental caries. Crit Rev Oral Biol Med 11: 366–380.
20. Klein MI, Duarte S, Xiao J, Mitra S, Foster TH, et al. (2009) Structural and molecular basis of the role of starch and sucrose in Streptococcus mutans biofilm development. Appl Environ Microbiol 75: 837–841.
21. Moura JS, da Silva WJ, Pereira T, Del Bel Cury AA, Rodrigues Garcia RC (2006) Influence of acrylic resin polymerization methods and saliva on the adherence of four Candida species. J Prosthet Dent 96: 205–211.
22. Quirynen M, Marechal M, Busscher HJ, Weerkamp AH, Darius PL, et al. (1990) The influence of surface free energy and surface roughness on early plaque formation. An in vivo study in man. J Clin Periodontol 17: 138–144.
23. Combe EC, Owen BA, Hodges JS (2004) A protocol for determining the surface free energy of dental materials. Dent Mater 20: 262–268.
24. da Silva WJ, Seneviratne J, Parahitiyawa N, Rosa EA, Samaranayake LP, et al. (2008) Improvement of XTT assay performance for studies involving Candida albicans biofilms. Braz Dent J 19: 364–369.
25. Pande M, Dubey VK, Yadav SC, Jagannadham MV (2006) A novel serine protease cryptolepain from Cryptolepis buchanani: purification and biochemical characterization. J Agric Food Chem 54: 10141–10150.

26. Goncalves LM, Del Bel Cury AA, Sartoratto A, Garcia Rehder VL, Silva WJ (2012) Effects of undecylenic acid released from denture liner on Candida biofilms. J Dent Res 91: 985–989.
27. Taniguchi L, de Fatima Faria B, Rosa RT, de Paula ECA, Gursky LC, et al. (2009) Proposal of a low-cost protocol for colorimetric semi-quantification of secretory phospholipase by Candida albicans grown in planktonic and biofilm phases. J Microbiol Methods 78: 171–174.
28. Busscher HJ, Weerkamp AH, van der Mei HC, van Pelt AW, de Jong HP, et al. (1984) Measurement of the surface free energy of bacterial cell surfaces and its relevance for adhesion. Appl Environ Microbiol 48: 980–983.
29. Cury JA, Rebello MA, Del Bel Cury AA (1997) In situ relationship between sucrose exposure and the composition of dental plaque. Caries Res 31: 356–360.
30. Tenuta LM, Ricomini Filho AP, Del Bel Cury AA, Cury JA (2006) Effect of sucrose on the selection of mutans streptococci and lactobacilli in dental biofilm formedin situ. Caries Res 40: 546–549.
31. Dubois M, Gilles K, Hamilton JK, Rebers PA, Smith F (1951) A colorimetric method for the determination of sugars. Nature 168: 167.
32. Johnston M (1999) Feasting, fasting and fermenting. Glucose sensing in yeast and other cells. Trends Genet 15: 29–33.
33. Askew C, Sellam A, Epp E, Hogues H, Mullick A, et al. (2009) Transcriptional regulation of carbohydrate metabolism in the human pathogen Candida albicans. PLoS Pathog 5: e1000612.
34. Kim JH, Johnston M (2006) Two glucose-sensing pathways converge on Rgt1 to regulate expression of glucose transporter genes in Saccharomyces cerevisiae. J Biol Chem 281: 26144–26149.
35. Ning Y, Hu X, Ling J, Du Y, Liu J, et al... (2012) Candida albicans survival and biofilm formation under starvation conditions. Int Endod J.
36. Uppuluri P, Chaturvedi AK, Srinivasan A, Banerjee M, Ramasubramaniam AK, et al. (2010) Dispersion as an important step in the Candida albicans biofilm developmental cycle. PLoS Pathog 6: e1000828.
37. Nett JE, Sanchez H, Cain MT, Ross KM, Andes DR (2011) Interface of Candida albicans biofilm matrix-associated drug resistance and cell wall integrity regulation. Eukaryot Cell 10: 1660–1669.

38. Sosinska GJ, de Groot PW, Teixeira de Mattos MJ, Dekker HL, de Koster CG, et al. (2008) Hypoxic conditions and iron restriction affect the cell-wall proteome of Candida albicans grown under vagina-simulative conditions. Microbiology 154: 510–520.
39. Gocke R, Gerath F, von Schwanewede H (2002) Quantitative determination of salivary components in the pellicle on PMMA denture base material. Clin Oral Investig 6: 227–235.
40. Kang SH, Lee HJ, Hong SH, Kim KH, Kwon TY (2012) Influence of surface characteristics on the adhesion of Candida albicans to various denture lining materials. Acta Odontol Scand.
41. Minagi S, Miyake Y, Inagaki K, Tsuru H, Suginaka H (1985) Hydrophobic interaction in Candida albicans and Candida tropicalis adherence to various denture base resin materials. Infect Immun 47: 11–14.
42. Samaranayake LP, Weetman DA, Geddes DAM, MacFarlane TW (1983) Carboxylic acids and pH of denture plaque in patients with denture stomatitis. Journal of Oral Pathology 12: 84–89.
43. Samaranayake LP, Raeside JM, MacFarlane TW (1984) Factors affecting the phospholipase activity of Candida species in vitro. Sabouraudia 22: 201–207.
44. Aguilar-Uscanga B, Francois JM (2003) A study of the yeast cell wall composition and structure in response to growth conditions and mode of cultivation. Lett Appl Microbiol 37: 268–274.

10

Carbohydrate Availability Regulates Virulence Gene Expression in Streptococcus suis

M. Laura Ferrando[1,2], Peter van Baarlen[1], Germano Orrù[3], Rosaria Piga[1], Roger S. Bongers[4], Michiel Wels[4], Astrid De Greeff[5], Hilde E. Smith[5] and Jerry M. Wells[1]

[1] Host-Microbe Interactomics, Animal Sciences, Wageningen University, Wageningen, The Netherlands,
[2] Department of Medical Microbiology, Academic Medical Center, Amsterdam, The Netherlands,
[3] Oral Biotechnology Laboratory, University of Cagliari, Cagliari, Italy,
[4] NIZO Food Research B.V., Ede, The Netherlands,
[5] Central Veterinary Institute, Animal Sciences Group, Wageningen University, Lelystad, The Netherlands,
University of Oklahoma Health Sciences Center, United States of America,

ABSTRACT

Streptococcus suis is a major bacterial pathogen of young pigs causing worldwide economic problems for the pig industry. *S. suis* is also an emerging pathogen of humans. Colonization of porcine oropharynx by *S. suis* is

---

Citation: Carbohydrate Availability Regulates Virulence Gene Expression in Streptococcus suis. Ferrando ML, van Baarlen P, Orru` G, Piga R, Bongers RS, et al. (2014) PLoS ONE 9(3): e89334. doi:10.1371/journal.pone.0089334

considered to be a high risk factor for invasive disease. In the oropharyngeal cavity, where glucose is rapidly absorbed but dietary α-glucans persist, there is a profound effect of carbohydrate availability on the expression of virulence genes. Nineteen predicted or confirmed *S. suis* virulence genes that promote adhesion to and invasion of epithelial cells were expressed at higher levels when *S. suis* was supplied with the α-glucan starch/pullulan compared to glucose as the single carbon source. Additionally the production of suilysin, a toxin that damages epithelial cells, was increased more than ten-fold when glucose levels were low and *S. suis* was growing on pullulan. Based on biochemical, bioinformatics and *in vitro* and *in vivo* gene expression studies, we developed a biological model that postulates the effect of carbon catabolite repression on expression of virulence genes in the mucosa, organs and blood. This research increases our understanding of *S. suis* virulence mechanisms and has important implications for the design of future control strategies including the development of anti-infective strategies by modulating animal feed composition.

INTRODUCTION

Streptococcus suis is a major bacterial pathogen of young pigs and a worldwide economic problem for the pig industry. Furthermore, *S. suis* is emerging as a zoonotic pathogen associated with meningitis and septicaemia in humans [1], [2]. In pigs, invasive disease is thought to be caused by translocation of *S. suis* across the mucosal epithelium in the upper respiratory tract [3], [4]. The ecological conditions that promote adhesion to, and invasion of the host mucosa by *S. suis* are unknown and probably depend on the environmental conditions and the bacterial genotype including presence of virulence genes. *In vitro* studies on adhesion and invasion by *S. suis* have often been performed in medium containing glucose as a carbon source which does not accurately reflect the situation *in vivo*. In the oropharyngeal cavity including the saliva, glucose may be present but concentrations usually diminish rapidly (within 30 min) after ingestion [5], [6] as glucose is readily absorbed by the host and metabolized by commensal bacteria. In contrast, starch α-glucans, large polymers of D-glucose that are present in large amounts in animal feeds [7] can persist in high concentrations in the oropharynx of humans and pigs [8]–[11]. A second type of carbohydrate that may promote proliferation of pathogenic bacteria is animal glycogen released from damaged or lysed cells. Suilysin, a hemolytic toxin encoded by the *S. suis sly* gene, may release, through its cytotoxic effect on epithelial cells, cellular glycogen [12], [13] that may serve as an important substrate for pathogen growth during the early stages of infection.

Currently, little is known about the expression and regulation of *S. suis* virulence factors such as suilysin. Streptococcal pathogens contain genes required for efficient utilization of α-glucans including amylases and/or pullulanases which cleave α-1,4 and α-1,6 glycosidic bonds in starch or glycogen [14]–[18]. Previously, we have shown that a cell wall anchored amylopullulanase of *S. suis* serotype 2 (*apuA*-SSU1849) was necessary to support bacterial proliferation on the α-glucan starch/pullulan (an α-1,6; α-1,4 linked glucose polymer) as a single carbon source and to promote adhesion of *S. suis* to porcine tracheal epithelial cells [19].

As in all Gram-positive bacteria, in *S. suis* the expression of carbohydrate metabolic enzymes are under the control of the global transcriptional regulator, catabolite control protein A (CcpA) that mediates carbon catabolite control (CCC) in presence of a preferred sugar, usually glucose. CcpA can repress or activate transcription by binding to *cis*-acting catabolite response element (*cre*) sites in the gene promoter regions [20], [21].

Given the relative abundance of dietary complex carbohydrates such as α-glucans and low level of glucose in the porcine upper oropharyngeal niche occupied by *S. suis*, we compared the genome-wide effects of growth on glucose or the α-glucan starch/pullulan on *S. suis* metabolism and virulence gene expression. We first constructed a metabolic map for *S. suis* and used it to analyze the key metabolic pathways altered by growth on glucose or pullulan. In addition to genes that play roles in (carbohydrate) metabolic pathways, 19 virulence genes were differentially expressed, seven of which were strongly induced (ratios>10) by growth in pullulan compared to glucose and nine were shown to contain a consensus *cre* site in their promoter sequences. The regulation of *apuA*, the virulence factor most strongly induced in pullulan compared to glucose, was investigated in detail using qPCR analysis of gene transcripts in bacteria grown in different carbon sources, promoter mapping and binding studies with CcpA and a newly identified transcriptional regulator, ApuR. The biological consequences of carbohydrate metabolism and virulence gene expression were also assessed in an *in vitro* porcine tracheal cell model using qPCR, hemolytic assays and adhesion and invasion assays. Based on these findings we propose a model for the transcriptional regulation of production of virulence factors during different stages of infection dependent on a CcpA-mediated, carbon catabolite control-dependent mechanism. To verify the predictions of this model, the *in vivo* expression of *apuA* and *sly* was measured for *S. suis* serotype 2 recovered from the blood and organs of pigs infected under controlled conditions.

MATERIALS AND METHODS

Bacterial Strains, Plasmids and Culture Conditions

The virulent *S. suis* serotype 2 strain S10 [22] and S735-pCOM1-V10 [23] were used in this study. The genome of *S. suis* S10 is more than 99% identical to the genome of *S. suis* 2 strain P1/7, a sequenced reference strain of which the genome had been annotated previously (NCBI Genome and NCBI BioProject http://www.ncbi.nlm.nih.gov/genome/?term=Streptococcus%20 suis) [24]. *S. suis* was grown in Todd-Hewitt broth (THB) (Difco) or on Columbia agar plates with 6% sheep blood (Oxoid) at 37°C under 5% CO_2 for 18 hr. A complex medium (CM) (Text S1) was prepared as previously described [19], [25]. The carbohydrates were added separately and sterilized by autoclaving at 100°C for 10 min (pullulan) or filtration with 0.45 μM pore size filter (glucose, lactose and maltotriose). We previously demonstrated that *S. suis* only grows to high density in CM when exogenous carbohydrates are added [19]. Growth in complex medium was determined by measurement of turbidity at OD_{600} using a SpectraMax M5 reader (Molecular Devices LLC).

RNA Extraction from *in vitro* grown *S. suis* and Quantitative PCR (qPCR)

For RNA extraction, *S. suis* S10 was grown to exponential (e) and early stationary (s) phase as indicated in Figure S1. Ten ml of culture was collected and centrifugated for each time point. The pellet was immediately frozen in liquid nitrogen until further handling. The frozen pellet was dissolved in 600 μl RA1 reagent (Macherey-Nagel) plus β-mercaptoethanol and lysed using a FastPrep-24 (MP. Biomedicals, Solon, OH) for 6.0 m/sec at 20 sec. Total RNA was purified using NucleoSpin RNA II (Macherey-Nagel). The quality and the concentration of RNA were assessed with an Experion System (Bio-Rad) and by analysis of the A_{260}/A_{280} ratio (NanoDrop 8000 UV-Vis Spectrophotometer). For qPCR, cDNA was synthesized using SuperScript VILO (Invitrogen). Primers were designed using Oligo Program version 6 (MedProbe, Oslo, Norway) (Table S1). Quantitative PCR was performed using a LightCycler 4.0 V and the LightCycler FastStart DNA Master SYBR

Green I Kit (Roche). Constitutive gene expression in complex media was determined as a ratio of target gene *vs* reference gene *proS* (Text S1) [26]. The level of *proS* expression was constant at all the time points analyzed (data not shown). Two replicates of all samples and primer pairs were included and the experiment was performed in triplicate. Non-template controls were included for each gene in each run.

Microarray Transcriptome Analysis

An *S. suis* oligoarray (8×15 K) containing *in situ* synthesized 60-mers was produced by Agilent Technologies (Santa Clara, USA), based on the genome sequence of *S. suis* P1/7 [24]. A total of 7651 unique 60-mers having a theoretical melting temperature of approximately 81°C and representing 1960 ORFs were selected as described [27]. Genes were represented by 4 (91%), 3 (4%), 2 (2%), or 1 probe (3%). Twenty-five putative genes were not represented on the array because no unique probe satisfying the selection criteria could be selected. RNA (1 µg) from *S. suis* samples was labeled using the Cyanine 5 (Cy5) labeling reaction (Amersham Biosciences, Buckinghamshire, UK). Co-hybridization with labeled cDNA probes was performed on these oligonucleotide arrays at 42°C for 16 h in Slidehyb#1 (Ambion, Austin, USA). The data were normalized using Lowess normalization [28] as available in MicroPrep [29] and corrected for inter-slide differences on the basis of total signal intensity per slide using Postprep [29]. Significance of differential gene expression was based on FDR values lower than 0.05. All microarray data are MIAME compliant and available in the NCBI GEO database (http://www.ncbi.nlm.nih.gov/geo/) under accession number GSE40658. Details of the custom-made 60-mer oligonucleotide array design (Agilent Biotechnologies, Amstelveen, The Netherlands) are deposited in the ArrayExpress database (http://www.ebi.ac.uk/arrayexpress/) under accession number A-MEXP-1671.

Bioinformatic tools: Microarray Analysis and *cre* Motifs Search

Differential gene expression of *S. suis* bacteria grown in CM supplemented with pullulan (Pul) or glucose (Glc) and harvested at early exponential (e) or early stationary (s) phase was cross-compared in different combinations (Pul_e vs Glc_s, Pul_s vs Glc_s, Pul_e vs Pul_s and Glc_e vs Glc_s). Overlapping and unique differentials were visualized using Venn diagrams at http://bioinfogp.cnb.csic.es/tools/venny/index.html.

For all genes and proteins identified in the *S. suis* P1/7 genome, Gene Ontology (GO, http://www.geneontology.org) and KEGG pathway annotations were obtained using the BLAST2GO software (www.blast2go.org) [30] including annotations based on terms obtained from EBI using the InterPROScan feature [31] that is part of BLAST2GO.

BLAST2GO was used to annotate all known *S. suis* genes and proteins according to standard Gene Ontology (GO) (www.geneontology.org) nomenclature. BLAST2GO uses the integrated Gossip package [30] for statistical assessment of differences in GO term abundance between two sets of sequences. The GO enrichment analysis feature of BLAST2GO was then used to identify functional GO terms that were statistically over- or underrepresented in the set of genes differentially expressed in pullulan vs glucose in exponential and early stationary phases of growth (Fisher›s Exact Test $p<0.05$).

The MEME (http://meme.sdsc.edu/meme/meme.html) software suite (version 4.1.0) was used for the identification of motifs OM1 and OM2 [32]. As input, we used the promoter sequences of the *MdxE* and *MalE* genes, with well-characterized *cre* sites, from *Bacillus* and *Listeria* (TGWAARCGYTWNCW [W=A or T; R=A or G; Y=C or T; N=any base]). A range of motif widths (15 nt in length) and zero or one motif per sequence were specified in our queries and FIMO, part of MEME, was used to search for this motif in the genome of *S. suis* P1/7. The algorithm in MAST, also part of the MEME suite, calculates position scores for the motif at each possible position within a sequence [33]; motif hits with a position-specific goodness-of-fit *P* value below 10^{-4} were considered to identify putative CcpA binding sites. *In silico* searches and comparisons of predicted *cre* sites within the *S. suis* P1/7 genome sequence and reconstruction of *cre* locations in the predicted operons were conducted using the corresponding databases provided by the MicrobesOnline database web server (http://microbesonline.org) and RegPrecise database [34].

5′-Rapid Amplification of cDNA Ends (5′-RACE)

The 5′-rapid amplification of cDNA ends (RACE) system (Invitrogen) was used to determine the transcription start site of the *apuA* gene. Briefly, the first strand cDNA was reverse transcribed from RNA from 1 μg of total RNA *S. suis* grown in CM plus pullulan using the specific primer ASP1 (Table S1). A homopolymeric tail was added to the 3′-end of the cDNA using terminal deoxynucleotidyl transferase (TdT) and the deoxynucleotide dCTP. The tailed cDNA was amplified in nested-PCR with Abridged Anchor Primer

(AAP) and a second *apuA* specific primer ASP2 primer upstream ASP1. The resulting 5′-RACE product of ~380 bp was sequenced and analysed by Vector NTI software (Invitrogen).

Recombinant Regulators Production and Infrared EMSA

The genes encoding the regulators ApuR (amino acids 2 to 312) and CcpA (amino acids 2 to 333) were amplified by PCR from *S. suis* 2 S10 genomic DNA using GoTaq (Promega) with primers ApuR_F/R and CcpA_F/R respectively (Table S1). The recombinant ApuR (rApuR) and CcpA (rCcpA) proteins were cloned in pTrcHis TOPO2 TA and purified by HPLC affinity chromatography (HisTrap affinity column, Amersham Pharmacia Biotech) as previously described (Text S1) [19]. Fractions containing purified fusion proteins of the expected size (approx. 38 kDa for ApuR and 40 kDa for CcpA) were collected and dialyzed against buffer (500 mM NaCl, 50 mM Tris-HCl, pH 7.4) and stored at −80°C with 10% of glycerol. Protein concentrations were measured using a BCA Protein Assay kit (Thermo Scientific). Further details can be found in the Text S1.

Infrared electrophoretic mobility shift assays (EMSA) were performed utilizing three pairs of fluorescent InfraRed-dye 800 (IRdye 800) labelled primer on the 5′ ends (Biolegio, The Netherlands). The IRdye-primers used to PCR-amplify three DNA fragments of ca. 120 bp (Pr1-Pr3) contained overlapping regions to cover the full length of the *S. suis* P_{apuA} promoter sequence (Table S1). The IRdye-PCR fragments were purified with QIAquick PCR Purification Kit (Qiagen) and used for the binding reaction at a concentration of around 50 nM. DNA binding reactions were performed in 20 μl of binding buffer containing 10 mM Tris-HCl pH 8.5, 50 mM NaCl, 10 mM EDTA, 0.5% Tween-20, 10 mM DTT, and 1 mg of poly(dI-dC)-labelled IRdye-PCR fragments at room temperature for 30 min. For the specific and non-specific competition assays, D(+)-glucose 6-phosphate (30 mM) (Sigma) was added to the binding buffer in the EMSA as CcpA cofactor [35]. Purified rApuR and rCcpA proteins (from 0.5 to 4.0 μM) were incubated, in separate experiments, with the fragment Pr2 and a non-specific competitor fragment that was obtained by PCR amplification of the gene SSU0879 (from 25 to 150 nM). Two 95 bp oligonucleotides complementary to Pr2 fragment lacking the two predicted binding sites (ΔOM1 and ΔOM2/*cre*) were synthesized (Eurogentec, The Netherlands), PCR-amplified with labelled Pr2F/R primers and incubated with the proteins. Free and bound DNAs were separated on 5% Tris-Borate-EDTA (TBE) native gels for 30 min at 10 cm/v. The gels were visualized using a LI-COR Odyssey Imager and scanned at 800 nm wavelengths.

Titration of Hemolytic Activity

The hemolytic activity was assayed as previously described [36]. Two different independent assays were carried out in triplicate. Briefly, *S. suis* bacteria were grown in CM+Glc and CM+Pul at three different growth stages: lag (OD_{600} 0.1–0.2), exponential (OD_{600} 0.2–0.5) and stationary (OD_{600} 0.5–0.7). The supernatant was collected from 1 ml for each culture by centrifugation at 12000 *g* for 1 min. Serial twofold dilutions (150 µl) of test samples were prepared in polystyrene deep-well titer plates (Beckman) with 10 mM Tris-buffered saline (PBS, pH 7.4). Subsequently, 150 µl of a 2% washed horse erythrocyte suspension in 10 mM Tris-buffered saline containing 0.5% BSA was added to each well. After the wells were sealed, the plates were incubated on a Coulter mixer for 2 h at 37°C. Unlysed erythrocytes were sedimented by centrifugation (1500 *g* for 10 min), 150 µl portions of the supernatant were transferred to a polystyrene flat-bottom microtiter plate and measured at 540 nm with a microELISA (enzyme-linked immunosorbent assay) reader (SpectraMax M5, Molecular Devices LLC). A 100% lysis reference sample was obtained by lysing cells with 1% Triton-X and the background lysis was subtracted before calculation of hemolytic activity.

Adherence and Invasion Assays using NPTr Cell Line

Culturing of Newborn pig tracheal cells (NPTr) [37] and adhesion assays were performed as previously described [19]. For the invasion assays, *S. suis* was added to the cell culture at an multiplicity of infection (m.o.i.) of ~50 and incubated for 2 h at 37°C with 5% CO_2 to allow cellular invasion by the bacteria. The monolayers were then washed three times with PBS; 1 ml of cell culture medium containing 100 µl/ml gentamycin and 5 µg/ml penicillin G was added to each well, and the plates were incubated for 2 h at 37°C, 5% CO_2 to kill extracellular and surface-adherent bacteria. The monolayers were washed three times with PBS, and cells were disrupted by the addition of 800 µl of ice-cold Milli-Q water and repeated up-and-down pipetting to release intracellular bacteria. To enumerate the viable bacteria, serial dilutions of each cell lysate were plated in triplicate on Columbia sheep blood agar plates and incubated at 37°C for 24 h. The rate of invasion was expressed as percentage of initial inoculum that was recovered per well. Two independent assays were performed in triplicate.

Experimental Infection of Pigs

The animal experiment described in this paper was approved by the ethical committee of the Central Veterinary Institute of Wageningen UR, Lelystad, The Netherlands under proposal number 2010113c in accordance with Dutch legislation (The Experiments on Animals Act, 1997) and the general principles governing the use of animals in experiments of the European Communities (Directive 86/609/EEC). To reduce unnecessary suffering of animals, humane end points were used: piglets were euthanized as soon as *S. suis*-specific symptoms occurred. Caesarean-derived, colostrum-deprived (CDCD) piglets were infected i.v. with 1×10^6 CFU of S735-pCOM1-orf2 [23]. The strain used to experimentally infect pigs was different to that used in our *in vitro* qPCR assays but was selected because it was known to be effectively recovered from the blood and organs in experimental pig infections. Genes encoding *apuA* and *sly* are both present in the genome of this strain (not shown). Animals were euthanized when specific *S. suis* symptoms (meningitis, arthritis, sepsis) were observed, or at the end of the experiment (6 days post infection). After animals were euthanized; organs and specific sites of *S. suis* infection were examined macroscopically and bacteriologically. Tissue samples were collected during post-mortem observation and snap-frozen in liquid nitrogen.

Quantitative PCR using Bacterial-Enriched RNA isolated from *in vivo* Tissues and Blood

Blood collected in PAXgene tubes was treated as recommended by the manufacturer (PreAnalytix/Qiagen). Samples taken from heart, brain and joints consisted of a mixed pellet of porcine cells combined with bacteria. All pellets from *in vivo* samples were thawed on ice, resuspended in 600 μl of Trizol (Invitrogen, Carlsbad, CA, USA) and subjected to 40 seconds of 6.5 m s^{-2} in the Fastprep-24 (MP Biomedicals, Solon, OH, USA) to disrupt bacteria. The mixture was extracted with 120 μl of chloroform, mixed for 15 seconds, incubated for 3 min at RT and centrifuged for 15 min at 20.000 *x g*. Supernatant was removed and extracted with 1 volume of chloroform. RNA in the supernatant was precipitated with 1 volume of isopropanol. After incubation for 30 min at RT or 16 h at −20°C, RNA was collected by centrifugation, and washed with 70% ethanol. The pellet was resuspended in water. Subsequently, RNA was purified as described above. To remove eukaryotic RNA all samples were treated using the MicrobEnrich kit (Ambion, Austin, Tx, USA). Bacterial-enriched RNA was amplified using the Ovation PicoSl WTA system v2 (Nugen, San Carlos, CA, USA). cDNA was purified using MinElute spin columns (Qiagen, Hilden, Germany) and diluted 25 times for qPCR analysis. Primers were designed using PrimerExpress software (Applied Biosystems, Foster City, CA, USA). Each reaction contained 12.5 pmol forward primer, 12.5 pmol reverse primer and POWR SYBR Green PCR Master Mix (Applied Biosystems). qPCR was performed using an ABI7500 (Applied Biosystems). As the bacterial RNA was amplified we used a more stringent GeNorm method [38] for normalization of the real-time qPCR data. GeNorm utilizes multiple internal control genes for normalization, in this case *gyrA*, *proS*, and *mutS* that were most stably expressed among 7 tested genes (data not shown). In each run a standard curve was incorporated consisting of seven 10-fold dilutions of a vector containing the cloned target of the PCR. In this way, both target genes and the reference genes could be related to a standard line. For each reaction, negative water controls were included. Analysis was performed using the ABI7500 Software (Applied Biosystems).

Statistical Analysis

The results obtained in the *in vitro* studies were analyzed using GraphPad Prism version 5.0 software (San Diego California, USA). All qPCR experiments and hemolytic activity assay were reproduced at least two times in triplicates and, where indicated, representative experiments are shown. Two-way ANOVA tests were carried out using Bonferroni›s post hoc test. The adhesion and invasion assays were performed at least two times using triplicate samples. All numerical data presented here are expressed as means ± standard error of the mean (SEM). Statistical significance was determined using a two-tailed unpaired Student›s t test. Differences were considered significant at $p<0.05$. Statistical significance was indicated as follows: * $p<0.05$; ** $p<0.01$; *** $p<0.001$.

RESULTS

The switch from Glucose to α-Glucan starch Fermentation has Pleiotropic Effects on Gene Expression

To investigate the global effects of a shift in carbohydrate metabolism on *S. suis* gene expression we compared genome-wide transcriptomic data from exponential (e) and stationary phase (s) cultures of *S. suis* in complex media

(CM) supplemented with 1% of starch/pullulan (α-1,4; α-1,6 glucan) (Pul) or glucose (Glc) as carbon sources.

The numbers of genes differentially expressed during growth in pullulan versus glucose were 1028 (52% of annotated genes) during exponential growth and 1015 (51% of annotated genes) during early stationary growth (Figure 1A). In total 738 (37% of annotated genes) genes were differentially regulated in pullulan compared to glucose, irrespective of the growth phase. In starch/pullulan, 209 genes were differentially regulated between the exponential and early stationary phases of growth; in glucose, 432 genes were differentially regulated for the same comparison (Figure 1A).

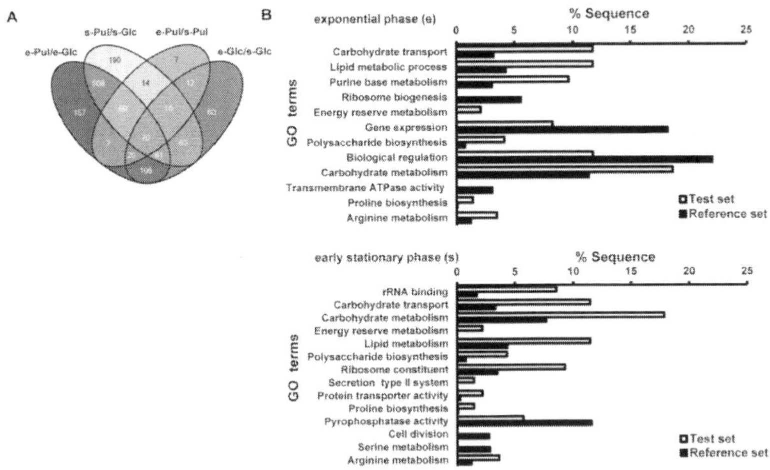

Figure 1. Comparison of genome-wide transcriptomic data for exponential (e) and early stationary phase (s) cultures of *S. suis* **grown in pullulan (Pul) or glucose (Glc). A**. Venn diagram of *S. suis* genes differentially regulated during growth in pullulan (Pul) vs glucose (Glc) in the exponential (e) or stationary (s) phase. In each sector the numbers of shared or unique differentially expressed genes is indicated. **B**. GO term distribution of *S. suis* genes differentially regulated in pullulan vs glucose in early exponential and early stationary phase. GO Enrichment analysis was performed using BLAST2GO (P=0.05, two-tailed Fisher›s Exact test).

To link gene expression data to changes in bacterial metabolic and physiological pathways, we obtained gene ontology (GO) functional gene annotations of all differentially expressed *S. suis* P1/7 genes using BLAST2GO (see Methods) (Figure 1B). Similar analyses were also performed for a subset of differentials, namely all upregulated genes within the GO category "carbohydrate metabolism. The highest number of genes differentially expressed in both growth phases participated in "carbohydrate metabolism" (18%) and "carbohydrate transport" (11%). Genes in the GO category "energy reserve metabolism" (2% for both growth phases) and lipid metabolism that generates the precursors of lipoteichoic acids and membrane phospholipids (11% for both growth phases) were also enriched when bacteria were grown in starch/pullulan vs glucose (Figure 1B). Other enriched GO categories included amino acid metabolic pathways for arginine and proline.

These analyses showed that culturing *S. suis* in presence of these different carbon sources leads to changes of 35–50% of the *S. suis* transcriptome, affecting not only carbohydrate metabolism but also basal metabolic and stress survival pathways. To summarize these broad changes in global transcriptomes, we generated a visual representation of the main carbohydrate pathway genes for six growth conditions comparisons according to their expression ratios (see Methods) (Figure 2A).

A

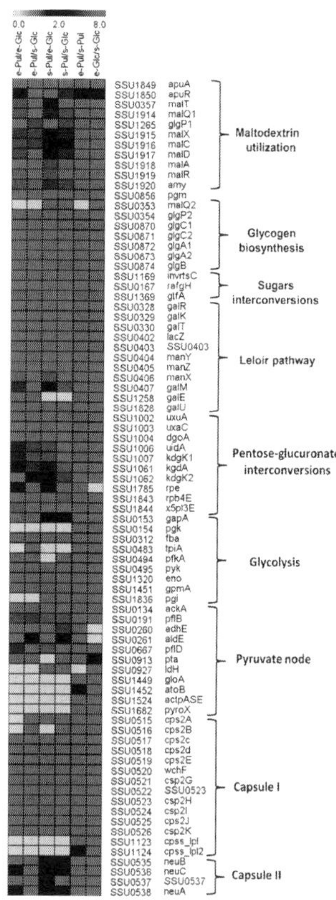

B

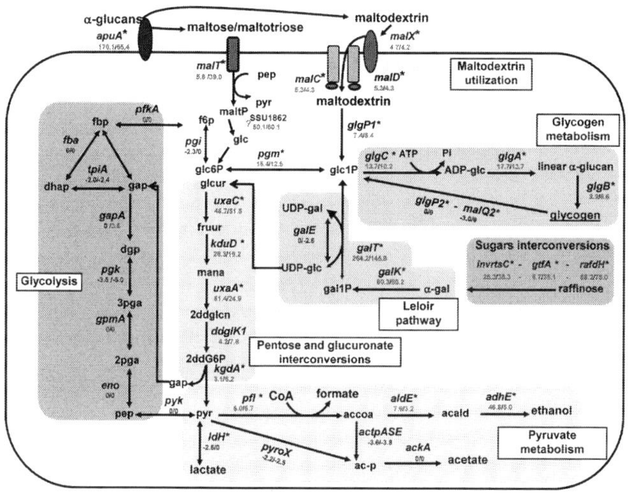

Figure 2. Transcriptome analysis of *S. suis* **metabolism** in presence of pullulan. **A**. Heatmap showing the effect of starch/pullulan on the transcription of genes involved in carbohydrate metabolism and capsule production. Expression (ratios) of genes participating in different pathways (indicated at the right of the heatmap) are shown for 6 different comparisons (indicated at the top of the heatmap). At the top of the Figure 2A, a color scale depicts the ratio of expression during growth in pullulan vs. glucose. Red indicates induction (upregulation) of the respective genes and blue indicates repression (downregulation) of the respective genes for each comparison. For each gene, the *S. suis* P1/7 locus tag and the gene name is depicted on the right. **B**. Schematic representation of *S. suis* metabolic pathways differentially regulated in pullulan vs glucose. α-glucans (i.e. starch/pullulan) are degraded by extracellular amylopullulanase (*apuA*) and the end degradation products, maltose/maltotriose and maltodextrins, are transported by PTS for maltose/maltotriose (*malT*) and maltodextrin ABC transport inside the bacteria (*malX, malC* and *malD*). Maltodextrins and maltose are most lilkely converted to glucose-1-phosphate (Glc1P) or α-glucose by 4-α-glucanotransferase and maltodextrin phosphorylase (*malQ*1 and *glgP*1 respectively). Glc1P can be metabolized in different pathways: phosphoglucomutase (*pgm*) isomerize glc1P to glucose-6-phosphate (glc6P) which may enter glycolysis (violet box) where it is consequently oxidated to pyruvate (pyr). Homolactic fermentation reduces pyruvate into lactate, whereas heterofermentative growth leads to other products, such as formate, acetate and ethanol (pyruvate metabolism, yellow box). The excess of glc1P that cannot enter in glycolysis may be used for synthesis of glycogen as an energy reserve (light blue box). The genome of *S. suis* is predicted to encode the enzymes sucrose phosphorylase

gtfA, α-fructofuranosidase (interconvertase) *invrtsC*, and raffinose galactohydrolase, *rafgH* for the interconversion of raffinose- like sugars. These enzymes participate in the starch and galactose Leloir pathway. Part of Leloir pathway (e.g. galactose-1-phosphate uridylyltransferase, *galT*, and galactokinase, *galK*) was induced more strongly in starch/pullulan. *GalT* interconverts galactose-1-phosphate (gal1-P) and UDP-Glucose (UDP-glc) to UDP-galactose (UDP-gal) and glc1P. Alternatively, UDP-glc may be converted into glucuronic acid (glcur) by UDP-D-glucuronate (UDP-glcur) to enter in an alternative (to glycolysis) pathway for pyruvate (pyr) production. Pathway predictions were reconstructed based on genome information, literature and database surveys (KEGG, MetaCyc). The following gene annotation was downloaded from NCBI: *galM*, aldose 1-epimerase; *galK*, galactokinase; *galE*, UDP-glucose 4-epimerase; *galT*, galactose 1-phosphate uridylyltransferase; *pgm*, Phosphoglucomutase/ phosphomannomutase; *pfkA*, 6-phosphofructokinase; *fba*, fructose bisphosphate aldolase; *tpiA*, triosephosphate isomerase; *gapA*, glyceraldehyde-3-phosphate dehydrogenase; *pgk*, phosphoglycerate kinase; *gpmA*, phosphoglyceromutase; *eno*, phosphopyruvate hydratase; *pyk*, pyruvate kinase; *ldh*, L-lactate dehydrogenase; *pyroX*, pyruvate oxidase; *ackA*, acetate kinase; *pfl*, pyruvate formate-lyase; *adlE* acetaldehyde-CoA dehydrogenase; *adhE* alcohol dehydrogenase; *glgB*, *glgA* glycogen synthase; *glgC* glucose-1-phosphate adenylyltransferase; *glgP* glycogen phosphorylase.

Comparative transcriptome analysis of *S. suis* grown in starch/ pullulan vs glucose revealed that growth in pullulan induced expression of the maltodextrins utilization gene cluster that includes *apuA* and the corresponding phosphotransferase systems (PTS) and ATP-binding cassette (ABC) transporters. The expression of the glycolysis pathway genes did not change significantly during growth in pullulan compared to glucose, suggesting that activity of this pathway was not altered during growth in both carbon sources. Rather, it appeared that excess intracellular glucose was converted in glycogen as energy reserves (glycogen biosynthesis pathways) or had entered into hexose-pentose carbohydrate pathways as an alternative to glycolysis (sugar interconversion and Leloir pathways) (Figure 2A). The biochemical links between the differentially expressed metabolic pathways are shown in Figure 2B; for each pathway, an extensive overview of the main genes involved and their functions can be found in the SI Text .

Links between Carbohydrate Metabolism and Virulence Gene Regulation

We analyzed the transcriptome data to see whether the differential fermentation of starch/pullulan and glucose affected expression of known or predicted *S. suis* virulence genes. Seven genes possibly involved in the invasion of mucosal tissues or avoidance of host defenses by streptococcal pathogens were highly upregulated (expression ratio>10) in starch/pullulan compared to glucose (Table 1). As CcpA in Gram-positive bacteria controls carbohydrate metabolism, one of the most profoundly altered categories in our microarray experiment, we investigated the potential role of CcpA in regulating virulence gene expression. We first used the MEME [39] and MAST [33] algorithms to mine the genome sequence of *S. suis* P1/7 for catabolite-responsive element (*cre*) sites, short DNA regions that can be bound by CcpA, using the consensus *B. subtilis cre* sequence [40]–[42].

Table 1. Confirmed and proposed *S. suis* **virulence factors differentially expressed in pullulan (Pul) compared to glucose (Glc).**

Annotation *S. suis* P 1/7	Protein	Function	Virulence	Pul/Glc[1]	Biblio
Galactosyl/rhamnosyl transferase - SSU0520	CpsE/F	CPS biosynthesis	Attenuated-pig		[71]
Tyrosine-protein kinase Wze - SSU0517	CpC	CPS biosynthesis	Attenuated-pig		[71]
N-acetylneuraminic acid synthase - SSU0535	NeuB	Sialic acid synthesis	Attenuated-pig	U	[72]
Peptidoglycan GlcNAc deacetylase - SSU1448	PgdA	Peptidoglycan	Attenuated-pig	D	[73]
D-alanine-poly ligase - SSU0554	DltA	LTA D-alanylation	Attenuated-pig		[74]
Fibronectin-fibrinogen binding - SSU1311	FbpS	Adhesion ECM	Attenuated-pig		[75]
Enolase - SSU1320	Eno	Adhesion ECM	no Mutant		[76]
Glyceraldehyde-3-P-dehydrog - SSU0153	GAPDH	Adhesion ECM	no Mutant	U	[77]
Di-peptidyl peptidase IV - SSU0187	DppIV	Adhesion ECM	Attenuated-mouse	U	[78]
Amynoacyl histidine peptidase - SSU1215	PepD	Subtilisin- protease	No Mutant	U>10	[79]

6-phosphogluco-nate-dehydrogen - SSU1541	6-PGD	Adhesion epithelium	No Mutant		[80]
Amylopullulanase - SSU1849	ApuA	Adhesion epithelium	Not tested	U>10	[19]
Glutamine synthetase - SSU0157	GlnA	Adhesion epithelium	Attenuated-mouse	D	[81]
Streptococcal adhesin P - SSU0253	SadP	Adhesion epithelium	no Mutant	U	[82]
Arginine deaminase - SSU0580	ArcB	Resistance to acidity	Not tested	U>10	[83]
Anchored DNA nuclease - SSU1760	SsnA	DNA degradation	Not tested	U	[84]
Cell envelope proteinase - SSU0757	SspA	Subtilisin- protease	Attenuated-mouse	U	[85]
Metallo-serine protease - SSU1773	IgAP	IgA1 protease	Attenuated-pig	U>10	[86]
Suilysin - SSU1231	Sly	Haemolysin	Unaffected-pig	U>10	[36]
Hyaluronate lyase - SSU1050	Hyl	hyaluronidase	Not tested	U>10	[87]
putative oligohyaluronate lyase - SSU1048	HepI/III	hyaluronidase	Not tested	U>10	[88]
Sortase A - SSU0925	SrtA	Protein sorting	Attenuated-pig		[89]
Serum opacity-like factor - SSU1474	OFS	Serum opacification	Attenuated-pig	D	[90]
S-ribosyl homocysteinase - SSU0376	LuxS	Quorum sensing	Attenuated zebrafish	D	[91]
Muramidase released protein - SSU0706	MRP	Unknown	Unaffected-pig	U	[22]
Extracellular protein factor - SSU0171	Ef	Unknown	Unaffected-pig	U	[92]

Pul/Glc1 upregulated (U) or downregulated (D) expression when *S. suis* was grown in pullulan (Pul) compared to glucose (Glc).

38 potential *cre* sites located upstream of the start codon of predicted proteins (P) or in proximity of the gene transcription start site (G) were predicted to control expression of 172 genes through interaction with CcpA. Of the 172 genes in the predicted CcpA regulon, 145 (84%) were differentially regulated during growth in starch/pullulan compared to glucose. We show 38 *cre* sites and their downstream genes or operons and if these were differentially expressed in pullulan compared to glucose (our study) or in an *S. suis* serotype 2 Δ*ccpA* mutant compared to wild-type [43].

As expected, *cre* sites were commonly associated with the predicted promoters of carbohydrate PTS and ABC transporters (17%) and enzymes for carbohydrate metabolism (25%). Additionally, *cre* sites were identified in the promoters of regulators (10%) and 9 (8%) out of the 19 virulence genes (47%) that were differentially expressed in pullulan compared to glucose, including *sly* and *apuA*. As *apuA* is essential for growth on pullulan and was shown to play a role in colonization of mucosal epithelia *in vitro* [19] and suilysin is a major virulence factor, we sought to understand the regulation of the encoding genes in more detail and explore the role of CcpA in transcriptional regulation during the switch from glucose to starch fermentation.

Expression of *apuA* and *sly* is Repressed by CCR

We assessed the relative expression of *apuA* and *sly* by qPCR when *S. suis* was grown in complex media supplemented with glucose, lactose, starch/pullulan or maltotriose, all of which supported efficient growth (Figure 3 panels A). Compared to growth in glucose or lactose, *apuA* and *sly* transcription was induced by growth in pullulan (up to 5.2 and 1.3 fold after 4 hours respectively; $P<0.001$) and maltotriose (up to 2.2 and 0.23 after two 2.5 hours respectively; $P<0.01$) (Figure 3 panels B and C).

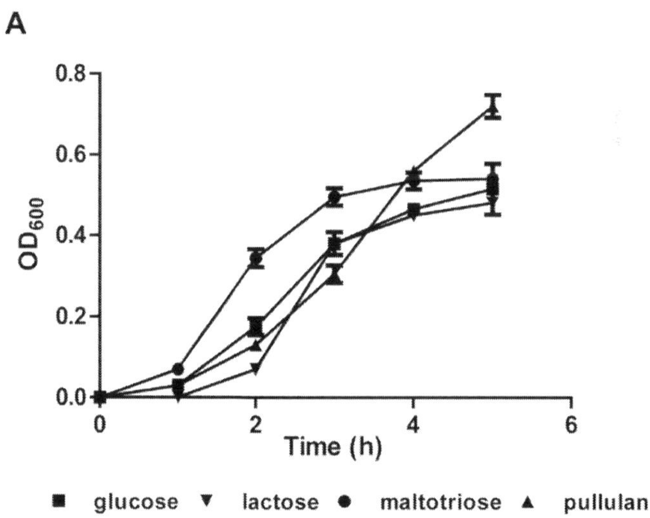

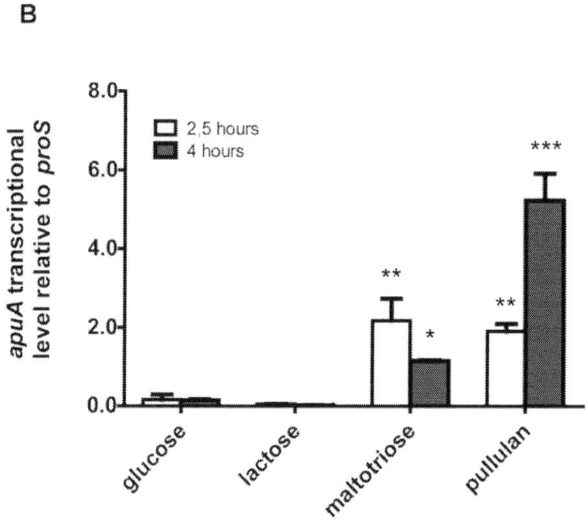

Figure 3. Transcriptional regulation of *apuA* **and** *sly* **grown in the presence of different carbon sources.** A. *S. suis* S10 growth curve at 37°C in CM containing 1% (w/v) of different sugars as indicated. The graph shows the means and standard deviations from two independent experiments. B. and C. Relative expression of *apuA* and *sly* in *S. suis* grown in CM containing 1% (w/v) different sugars was determined by qPCR. The transcript levels of *apuA* were measured after 2.5 hours and 4 hours of growth relative to the reference gene *proS*, which is constitutively expressed at similar levels during growth in different sugars (data not shown). The height of the bars represent mean values for the relative expression data ± SEM from 2 independent experiments (n=3). Statistical significance was calculated using a two-way ANOVA test followed by Bonferroni's post hoc test (* $p<0.05$; ** $p<0.01$; *** $p<0.001$.).

As *apuA* and *sly* contain a conserved *cre* in the promoter region, we predicted that their transcription might be repressed during growth on glucose but not lactose. However, the relative level of *apuA* transcription in lactose was comparable to that of transcription in glucose suggesting possible regulation by a second transcriptional regulator. To test this hypothesis we added starch/pullulan or maltotriose to *S. suis* growing exponentially in either lactose (Figure 4A) or glucose (Figure 4B) and quantified *apuA* transcription at different time points (Figure 4 panels C and D). In medium containing lactose, both maltotriose and pullulan strongly induced expression of *apuA* after 30 min (4.5 fold and 2.2 fold respectively; P<0.001; Figure 4C). In medium containing glucose the addition of maltotriose or starch/pullulan had a small effect on the expression of *apuA*; only a slight increase (1.6

fold; P<0.05) of *apuA* expression was observed 30 min after addition of maltotriose (Figure 4D). These results show that when *S. suis* is growing in glucose for 30 min, *apuA* transcription is hardly induced by pullulan or maltotriose but when grown in lactose for 30 min, *apuA* expression is strongly induced by pullulan and maltotriose.

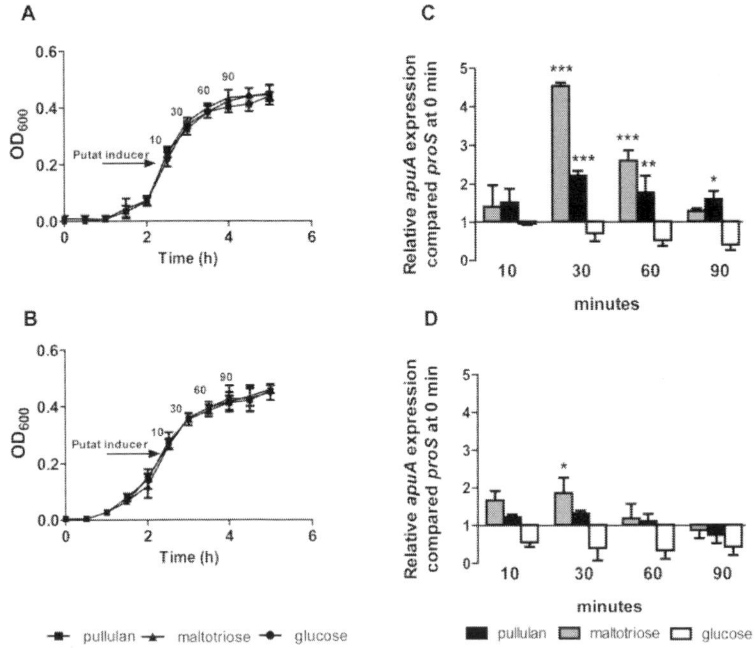

Figure 4. Induction of *apuA* **expression by putative inducers in the presence of glucose or lactose. A**. Growth curves of *S. suis* S10 in CM containing 1% w/v lactose or **B**. 1% w/v glucose before and after addition of 0.25% w/v putative inducers (arrow) i.e. maltotriose, pullulan or glucose. The graphs show the means and standard deviations from two independent experiments. **C**. Relative expression of *apu*A genes following addition of putative inducers to *S. suis* growing in CM plus lactose. **D**. Relative expression of *apu*A genes following addition of putative inducers. The relative expression of *apuA*, was measured by qPCR 10, 30, 60 and 90 minutes after addition of the putative inducers, The height of the bars shows the mean (n=3) fold change in expression ± SEM from two independent experiments. Statistical significance was calculated using a two-way ANOVA test followed by Bonferroni's post hoc test (* p<0.05; ** p<0.01; *** p<0.001.).

CcpA and ApuR bind to conserved operator motifs in the *apuA* promoter

We hypothesized that the induction of *apuA* expression by maltotriose was due to ApuR, a putative transcriptional regulator upstream of *apuA* that possesses its own promoter and a predicted rho-independent downstream terminator (Figure 5A). Homology searches indicated that ApuR was a LacI/GalR type regulator containing an N-terminus DNA-binding and a C-terminus ligand-binding domain that can be bound by a specific sugar. Comparison of ApuR with protein sequences in the UniProt database revealed similarities to several transcriptional regulators of operons which comprising maltodextrin utilization gene clusters. In the top ranking were listed for highest similarities: BL23 YvdE (*Lactobacillus casei*; 53% identity) [44], MdxR (syn. YyvdE; *B. subtilis* 168; 49% identity) [45], [46], the activator EGD-e Lmo2128 (*Listeria monocytogenes*; 47% identity) [47] (Figure 5 E–F) and the activator MdxR (*Enterococcus faecium* E1162; 49% identity) [48].

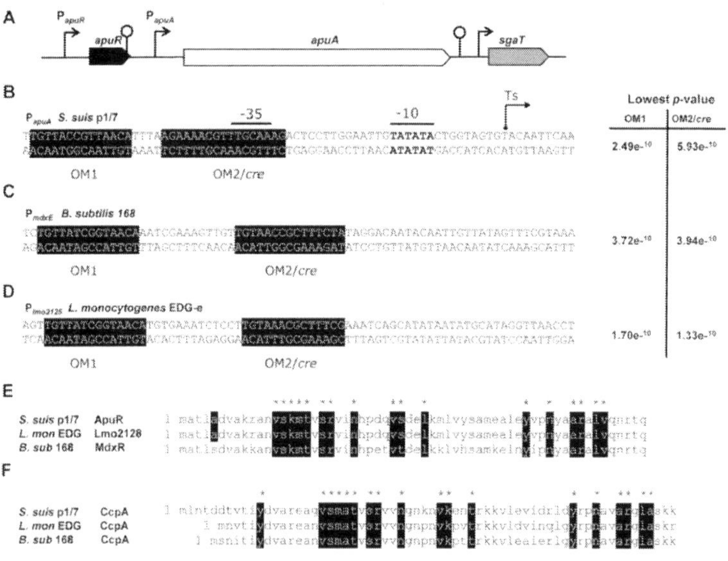

Figure 5. Identification of conserved operator binding motifs for ApuR (OM1) and CcpA (OM2) in *S. suis* **P1/7,** *B. subtilis* **168 and** *L. monocytogenes* **EDG-e. A**. The 6 kb amylopullulanase gene *apuA* is located downstream of *apuR* which encodes a putative transcriptional regulator of the LacI/GalR family. Located downstream of *apuA* are a cluster of genes predicted to be involved in uptake and fermentation of ascorbate (*sgaT*, *sgaB*). For each gene, the direction of transcription is indicated by an arrow, the size of which is proportional to the length

of the corresponding open reading frame. Putative promoters are represented by arrows and transcription terminators by loops. The operator motifs OM1 and OM2/cre (shaded sequences in B, C and D) were determined using the MEME software suite and their relative probability p-values are indicated. **B**. The apuA promoter based on the experimentally determined transcription start site (Ts arrow) **C**. The *B. subtilis mdxE* (BSU34610) promoter **D**. The *L. monocytogenes lmo2125* promoter **E**. and **F**. Sequence alignment of the DNA binding domains of the ApuR and CcpA proteins of *S. suis, L. monocytogenes* and *B. subtilis*. Conserved amino acid sequences are indicated in black.

As *apuA* is considered a relevant virulence factor and its regulation by different carbohydrates could be of biologic relevance, we decided to investigate the binding of CcpA and ApuR to operator sites in the *apuA* promoter. To elucidate the *apuA* promoter regulatory modules we determined the transcription start site, conducted homology searches for binding motifs of known regulators *in silico*, and performed *in vitro* promoter binding assays with purified transcription factors (TFs). The transcriptional start site of the *apuA* transcript was determined experimentally by 5′-RACE to be 31 nt upstream of the start codon. The consensus −35 element (*5′-TTGCAA-3′*) for RpoD (sigma 70) and the −10 element (*5′-TATATA-3′*) required for interaction with the RNA polymerase and transcription initiation were found near the expected positions upstream of the transcription start site (Figure 5B). Visual inspection of the *apuA* promoter region revealed the presence of two potential operator motifs (OMs) that were also listed in the RegPrecise database [34]. One *mdxR*-like operator motif, designated here as OM1, was found 13 bp upstream of the −35 element; a second operator motif predicted *cre* site designated OM2/*cre*, located at position −30 nt, overlapping the conserved −35 element; this overlap could interfere with binding of the RNA polymerase (Figure 5 C–F).

To demonstrate binding of ApuR and CcpA to the predicted operator motifs, we expressed and purified these proteins in *E. coli* with a C-terminal His-tag and tested DNA binding in electrophoretic mobility shift assays (EMSA) using three overlapping fragments of the P_{apuA} promoter region (Pr 1–3; approx. 120 bp in length). A single rApuR-DNA complex was observed with promoter fragments Pr2 and Pr3 suggesting the binding motif for ApuR lies within the 64 bp overlapping region (panels A and B). The binding of ApuR to Pr2 was shown to be concentration dependent in the range from 0.5 to 4.0 μM of rApuR. The specificity of rApuR binding to Pr2 was shown in a competition EMSA using unlabeled competitor DNA (i.e. fragment Pr2) or an unlabeled non-specific competitor lacking the two predicted OM1 and OM2/*cre* operator binding motifs. The non-specific competitor DNA fragment had no effect on Pr2 complex formation whereas the specific competitor (Pr2) substantially reduced rApuR binding (panels D and E). To investigate whether ApuR binds to OM1 or OM2/*cre*, promoter fragments lacking these motifs were synthesized and tested in the EMSA (i.e. ΔOM1 and ΔOM2/*cre*). In three independent experiments we observed a lower amount of the rApuR-DNA complex with fragments lacking ΔOM1 suggesting that ApuR binds most strongly to this motif. In some experiments, complex formation was slightly reduced when ΔOM2/*cre* was deleted, possibly due to low affinity binding of ApuR to ΔOM2/*cre* at higher protein concentrations.

Similarly, binding of CcpA to the P_{apuA} OM2/*cre* that overlaps with the predicted −35 promoter element was demonstrated by EMSA using a fluorescent IRdye-Pr2 fragment containing the putative *cre* site (panels G–J). An increasing amount of an rCcpA-DNA complex was observed with an increasing concentration of purified rCcpA (1.5 to 5 μM). The complex could be outcompeted by addition of unlabeled Pr2 but not with non-specific competitor DNA indicating that CcpA binds specifically to P_{apuA} (panels H and I). Recombinant CcpA also appears to bind to the ΔOM2/*cre* fragment although less DNA/transcription factor complex is observed suggesting that CcpA may also bind OM1 or other sequences in Pr2 with lower affinity.

Taken together, these results show that *apuA* expression may be regulated via repression through CcpA-mediated carbon catabolite control and via transcriptional activation through a dedicated regulator encoded by the *apuR* gene.

Relief from CCR increases adhesion and invasion of *S. suis* to porcine epithelial cells

As our microarray data showed increased transcription of *apuA* and seven other genes predicted to play a role in *S. suis* adhesion and invasion in starch/pullulan compared to glucose, we hypothesized that culturing *S. suis* in pullulan as single carbon source, thus in absence of glucose might increase its adhesion and invasion capacity. Exponentially growing *S. suis* grown in CM supplemented with 1% of glucose or pullulan were incubated with NPTr cells for 2 h at m.o.i. of ~50 bacteria/cell. To maintain similar conditions during the period of co-culture, the cell culture medium was replaced with glucose-free DMEM supplemented with either 1% glucose or pullulan during the 2 h incubation with *S. suis*. In agreement with previous adhesion studies using the NPTr cell line [19], we found that adherence of *S. suis* bacteria grown in CM plus glucose was 19.4±1.0% of original inoculum (averaged over 3 independent replicates). The adherence of *S. suis* was significantly increased (24.9±1.7%; P<0.05; 3 replicates) after growth in 1% pullulan (Figure 6A). In accordance with previous studies using a human Hep-2 cell line [49], *S. suis* 2 S10 showed low invasion capacity (0.05% of original inoculum) in glucose. Invasiveness of *S. suis* was nearly 9-fold higher when grown in pullulan (0.40%±0.01) (P<0.01) (Figure 6B), corresponding to approximately 8.0×10^2 cfu/ml when grown in glucose, to 7.2×10^3 cfu/ml in pullulan. In conclusion, *S. suis* grown in starch/pullulan showed a small but significant increase in adherence to, and a strongly increased invasiveness of NPTr cells compared to *S. suis* grown in glucose.

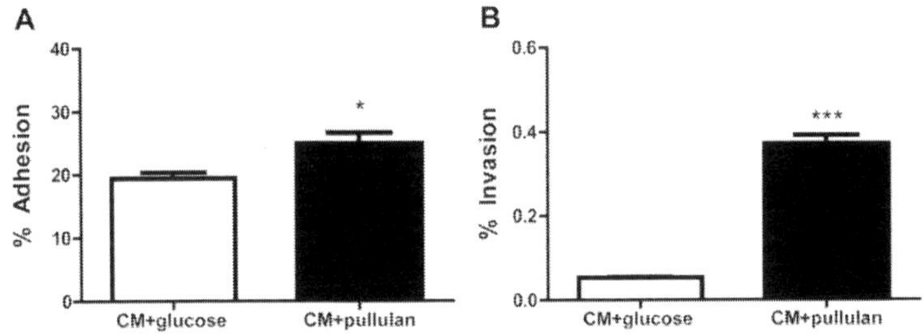

Figure 6. Comparison of adherence A. and invasion B. of *S. suis* **after growth in CM+1% w/v pullulan (black bars) vs. CM+1% w/v glucose (white bars).**

NPTr confluent monolayers were co-cultivated for 2 h with *S. suis* S10 bacteria grown in CM plus pullulan or glucose. Adherence and invasion are shown as mean % values of the initial inoculum from two independent experiments in triplicate. Error bars indicate the SD.

Carbon Catabolite Control of Suilysin Expression

The expression of *sly* was strongly induced in pullulan compared to glucose, both in exponential and stationary phase (+18.1/+17.9 ratio in pullulan vs glucose). The presence of a *cre* site in the *sly* promoter region suggested that the induction of *sly* during growth in pullulan was due to relief from CcpA-mediated carbon catabolite repression rather than a specific induction by starch/pullulan. As suilysin has been proposed to compromise the integrity of the host epithelium and facilitate entry into the body we measured erythrocyte hemolytic activity (HA) of culture supernatants of *S. suis* grown in glucose or pullulan. HA was significantly higher in supernatants of S. *suis* grown in pullulan compared to glucose over a wide range (ca. 0.5) of OD600 values (P<0.001; Figure 7). In early stationary phase, HA for pullulan and glucose cultures was approx. 91% and 18%, respectively (Figure 7).

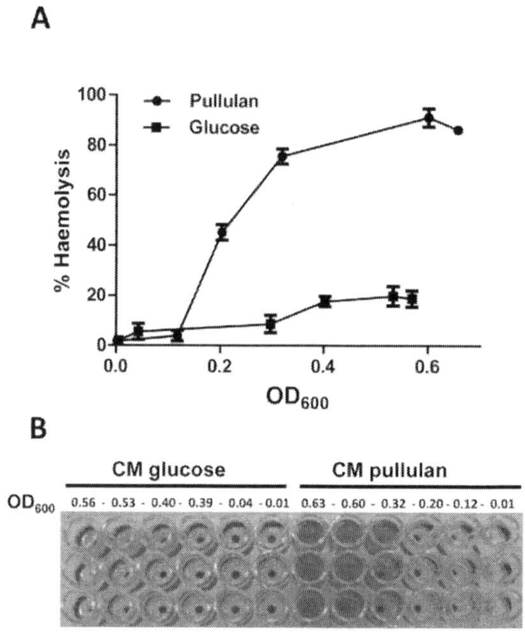

Figure 7. Hemolysis assay of *S. suis* **growing in two different carbon sources.** **A**. The hemolysis production was measured by analyzing the supernatant of *S. suis* grown in CM plus 1% w/v of glucose or pullulan in the lag, exponential and stationary phase (OD_{600} values 0 to 0.56). **B**. Deep-well titer plate showing hemolytic activity of supernatants collected from *S. suis* grown in CM supplemented with glucose or pullulan.

In vivo Expression of the Virulence Factors apuA and sly

Based on the results from this study so far, we predicted that expression of *apuA* and *sly* would be substantially higher in the mucosal colonization stage of the infection when glucose is scarce than in the bloodstream where glucose levels are sufficient to support growth and induce CCR. *S. suis* invasive disease often leads to infection of the joints, heart and brain. To test our hypothesis *S. suis* was recovered from the blood, synovial joints, heart and brains of infected piglets immediately after euthanasia and RNA was extracted from the blood and organs. As hypothesized we found transcription of *sly* and *apuA* to be significantly higher in the synovial joints, heart and brain than in the blood (Figure 8). These results show that our regulatory model (Figure 9) was informative for some aspects of porcine infection with *S. suis*.

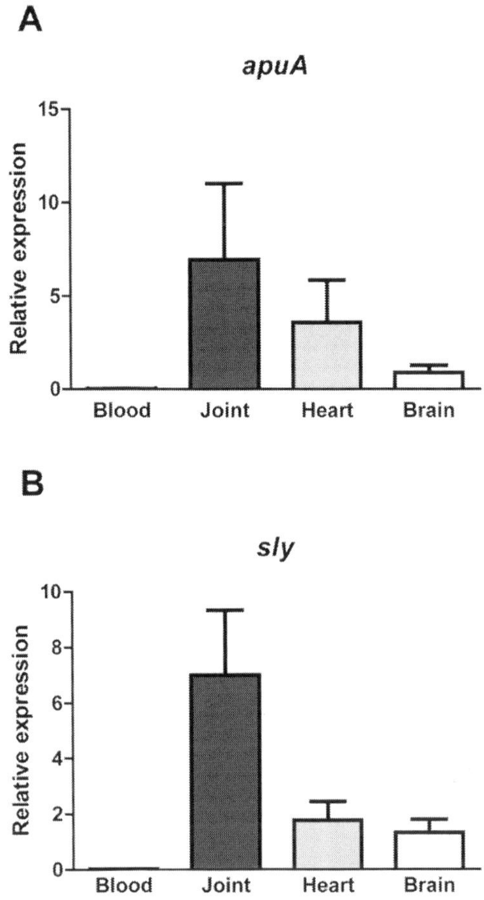

Figure 8. Expression of *apuA* **and** *sly in S. suis recovered from blood and tissues of experimentally infected piglets.*

The relative expression of *apuA* and *sly* in *S. suis* blood, joints, heart and brain recovered from intravenously infected piglets calculated using the GeNorm method [38] using three housekeeping genes for data normalization. **A**. The relative expression of apuA ($\times 10^7$) are shown for *S. suis* recovered from blood and different organs. **B**. Relative expression of sly ($\times 10^5$) in blood and different organs.

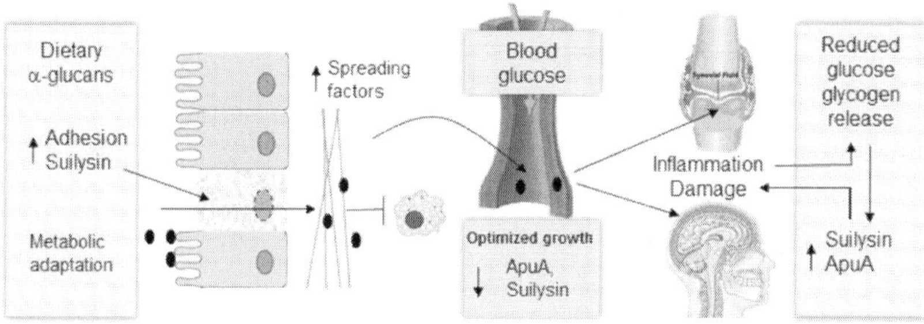

Figure 9. Links between carbohydrate metabolism and virulence in *Streptococcus suis*.

At the mucosal surfaces a high ratio of a-glucans to glucose upregulates expression of several sugar transport systems and metabolic pathways associated with starch metabolism. Additionally, several virulence factors involved in adherence to host cells, degradation of connective tissue (spreading factors), and avoidance of phagocytic killing, including ApuA and suilysin are upregulated when glucose is diminished. Suilysin may facilitate dispersion of bacteria in mucosal tissues due to loss of barrier integrity. Once *S. suis* reaches the bloodstream metabolism is adapted for optimal growth on glucose and the expression of virulence factors is reduced by CcpA mediated-repression. In infected organs glucose levels are lower than in the blood and are further reduced by inflammation and utilization by *S. suis* leading to upregulation of ApuA, suilysin and other virulence factors. In the organs and tissues, glycogen released from damaged cells is degraded by ApuA to generate maltodextrins which sustain growth of *S. suis*.

DISCUSSION

Asymptomatic carriage of *S. suis* is common in adult pigs but young piglets that become colonized by *S. suis* may develop invasive disease. Host susceptibility to *S. suis* infection may occur due to inadequate host immune responses but may also be due to environmental factors such as the availability of glucose or other carbohydrates. Here we investigated whether the availability of different types of carbohydrates, notably starch and glucose, could influence the switch from an asymptomatic to a pathogenic association of *S. suis* with the host. To search for the mechanism controlling such a switch, we initially performed an *in vitro* transcriptomics study on *S. suis* grown in CM medium supplemented with pullulan or glucose; the latter mediates carbon catabolite control in streptococci via catabolite control protein A (CcpA). Pullulan, a starch, was chosen since we had previously found that growth of *S. suis* in presence of this carbohydrate induces expression of *apuA*, a gene known to be involved in carbohydrate metabolism [19].

A switch from growth in glucose to starch/pullulan resulted in a change of ca. 50% of its transcriptome, involving multiple GO function categories (Figure 1). As anticipated many genes annotated with GO term "carbohydrate metabolism" were differentially expressed, including the α-glucan-starch-degrading amylopullulanase *apuA* [19] and downstream pathway genes required for maltodextrin metabolism (Figure 2).

In other streptococcal pathogens CcpA and CCC have been shown to play a role in virulence gene expression as well as colonization [50], [51] and virulence in an animal model [51]. To predict which *S. suis* genes might be regulated by CCC we searched for conserved *cre* sites in the genome of *S. suis* P1/7 and identified 172 genes, some of which were organized in operons. Of these 172 genes, 145 (84%) were differentially regulated during growth in pullulan compared to glucose. The difference in these two values may be due to incorrect *cre* site predictions or to the fact that expression of certain genes or operons are controlled by additional transcription factors, for example suilysin may be also co-regulated by the two-component system CovS/CovR (SSU1190-SSU1191) [52]. The 145 genes that appeared to be controlled by CCC were also compared to a published microarray data set generated by comparison of an *S. suis* serotype 2 Δ*ccpA* mutant with the corresponding wild-type strain [43]. In total, 99 genes comprising 38 operons were differentially regulated in both datasets (68% overlap).

The CCC-regulon included genes encoding transcriptional regulators, carbohydrate metabolic enzymes and, importantly, putative or known virulence factors (8%). The most highly up-regulated virulence gene in pullulan compared to glucose was *apuA* encoding a cell-surface amylopullulanase required for growth on starch or glycogen that also contributes to adherence to porcine epithelial cells *in vitro* and thus may be relevant to mucosal colonization [19]. Expression of *apuA* was highly induced during growth in starch/pullulan or maltotriose, the most abundant degradation product of pullulan, and repressed during growth in glucose (Figure 3 and and4).4). An analysis of the promoter region of *apuA* revealed the presence of two transcription binding motifs (Figure 5) and EMSA studies showed that *apuA* expression was co-regulated by CcpA and a second regulator, ApuR. The similar location of the conserved

OM in the promoters regulated by ApuR and the known transcriptional activator of maltodextrin utilization cluster Lmo2128 in *L. monocytogenes* [47] and MdxR in *E. faecium* [48], with high sequence similarity to ApuR, suggested that ApuR might also be a transcriptional activator. Addition of maltotriose to *S. suis* exponentially growing in medium containing lactose as a carbon source, significantly induced expression of *apuA* and, to a lesser extent, *apuR* suggesting that ApuR, like Lmo2128, might be allosterically regulated by maltotriose (Figure 4). In contrast, CcpA binding repressed *apuA* transcription, most likely through binding to an OM2/*cre* site overlapping with the −35 promoter element, since addition of maltotriose to *S. suis* bacteria precultured in glucose as sole carbon source did not lead to significant induction of *apuA*. This dual regulatory model postulates that *apuA* is maximally expressed when glucose level is low, thus allowing relief from CcpA-mediated catabolite repression, and when substrates that can be degraded by *apuA* are present.

In addition to *apuA*, eighteen other predicted or known virulence genes were differentially regulated in pullulan vs glucose, seven of which were highly upregulated in pullulan compared to glucose (Table 1). One of these genes was *sly* encoding suilysin, a pore-forming toxin which was secreted in 5-fold higher amounts in pullulan compared to glucose (Figure 7); its increased expression was confirmed in qPCR assays (Figure 3). Suilysin plays an important role in damaging host epithelial [12], [13], [53], endothelial [54], [55] and immune cells [56], suggesting that suilysin could play roles *in vivo* in damaging and penetration of different cell and tissue types, promoting tissue invasion and inhibition or killing of leukocytes [3], [57]. Of note, virulent *S. suis* strains that do not produce Sly may still be invasive in pigs, possibly best exemplified by Allen *et al* (2001) [58] who reported that an isogenic *S. suis sly* mutant, lacking of the hemolytic characteristics, was nearly as invasive as the wild-type parental strain in a pig infection studies after bacteria intravenous injection. The authors of the study by Allen *et al* (2001) [58] proposed that Sly is relevant for translocation across epithelia and during the infection stages prior to dispersion via the blood, and proposed that production of Sly may correlate with increased severity of clinical symptoms and the capacity to reach higher colonization of organs. This notion is supported by the study of King *et al* (2001) [59] who reported that *sly* was present in a significantly higher number of isolates from pigs with meningitis, septicemia, and arthritis than isolates from pigs with pneumonia. Conversely, *sly* is also significantly overrepresented in asymptomatic carriage strains of pigs [59], showing that presence of *sly* does not exclusively correlate with invasive isolates. In addition to suilysin, it is likely that other virulence factors are relevant for the invasion of connective tissue, such as the hyaluronidase and heparinase II/III-like proteins. Note that the function of these putative virulence proteins still has to be demonstrated. In contrast to our results, *sly* expression was not increased in a recent study which compared gene expression of a wild-type strain with a Δ*ccpA* mutant grown to stationary phase [43]. The reason for these different findings may be related to the use of THB growth medium [43], a rich medium containing glucose, instead of the supplemented CM medium that we used, or different growth stages or different physiological states of the bacteria in the different media. Lastly, it is also possible that expression of *sly* is controlled by a CcpA-independent carbon catabolite repression mechanism although a *cre* site is present in the *sly* promoter region.

We propose that at mucosal surfaces the high ratios of starch to glucose promotes production of extracellular ApuA to support growth of *S. suis* in the oropharyngeal cavity. Additionally relief from CCR leads to higher expression of other virulence genes involved in mucosal infection (Figure 9). Indeed, in an *in vitro* experiment employing tracheal epithelial cells and *S. suis* bacteria grown with starch/pullulan or glucose as sole carbon source, a nearly 10-fold increase of epithelial cell invasion was observed when bacteria were grown with pullulan. We observed that in presence of starch, *S. suis* induced expression of the metabolic pathway enzymes required for transport and metabolism of maltodextrins and the glycogen biosynthesis enzymes which generate bacterial glycogen energy reserves. Several studies have linked bacterial glycogen metabolism to environmental survival, symbiotic performance, and colonization and virulence [60]–[67]. Glycogen reserves may be important for survival and fitness of *S. suis* when carbon sources are scarce. During colonization, relief from CCR would increase production of suilysin and may facilitate dispersion of *S. suis* into the deeper tissues due to loss of barrier integrity [4], [53], [58]. We hypothesized that once *S. suis* would reach the bloodstream, higher glucose concentrations (from 4.4 to 6.6 mM) would repress expression of *apuA* and *sly* by CCR (Figure 9). Indeed, gene expression analysis of bacteria isolated from the blood or organs of acutely infected pigs showed significantly lower expression of *apuA* and *sly* in blood than in the organs (Figure 8) supporting occurrence of CCR. In contrast, *S. suis* isolated from infected joints, heart or brain tissue expressed significantly higher levels of *sly* and *apuA* than bacteria isolated from blood. These results are probably due to the lower levels of glucose in the tissues (0.25–0.55 mM in normal synovial fluid and joint cavities) than in the blood (from 4.4 to 6.6 mM) [68], [69] and thus, relief from CCR when bacteria were located in the organs. In inflamed tissues, glucose levels may have been reduced further due to the consumption of glucose by neutrophils and macrophages. It is tempting to speculate that induction of suilysin expression in the tissues will release host glycogen from damaged cells that would be degraded by ApuA and metabolized to further support growth of *S. suis* (Figure 9). Additionally *S. suis* may be able to metabolize host glucans such as hyaluronan which is present in high concentrations (3–4 mg/ml) in synovial fluid [70].

Taken together, our results clearly demonstrate that the availability of glucose and other carbohydrates such as starch serves as an environmental cue to regulate the expression of *apuA* and other virulence genes. We propose a schematic model of how carbohydrate content of different tissues could modulate *S. suis* metabolism at the different stages of infection (Figure 9). Our novel findings demonstrate a regulatory mechanism dependent on relief from CcpA repression that links carbohydrate metabolism and virulence at least *in vitro* and that might play roles *in vivo*, at different stages of *S. suis* infection. Awareness of the role of carbohydrate content of the *S. suis* environment may lead to new strategies for combating this important disease, for instance via modulation of carbohydrate content and composition of animal feeds, or the inhibition of *S. suis* enzymes required for metabolism of starch.

ACKNOWLEDGMENTS

The authors are grateful to Michiel Kleerebezem for valuable discussions and insights on heterofermentative metabolism.

REFERENCES

1. Lun ZR, Wang QP, Chen XG, Li AX, Zhu XQ (2007) *Streptococcus suis*: an emerging zoonotic pathogen. Lancet Infect Dis 7: 201–209.
2. Wertheim HF, Nghia HD, Taylor W, Schultsz C (2009) *Streptococcus suis*: an emerging human pathogen. Clin Infect Dis 48: 617–625.
3. Gottschalk M, Segura M (2000) The pathogenesis of the meningitis caused by *Streptococcus suis*: the unresolved questions. Vet Microbiol 76: 259–272.
4. Fittipaldi N, Segura M, Grenier D, Gottschalk M (2012) Virulence factors involved in the pathogenesis of the infection caused by the swine pathogen and zoonotic agent *Streptococcus suis* . Future Microbiol 7: 259–279.
5. Meurman JH, Rytomaa I, Kari K, Laakso T, Murtomaa H (1987) Salivary pH and glucose after consuming various beverages, including sugar-containing drinks. Caries Res 21: 353–359.
6. Gough H, Luke GA, Beeley JA, Geddes DA (1996) Human salivary glucose analysis by high-performance ion-exchange chromatography and pulsed amperometric detection. Arch Oral Biol 41: 141–145.
7. Bakker GC, Dekker RA, Jongbloed R, Jongbloed AW (1998) Non-starch polysaccharides in pig feeding. Vet Q 20 Suppl 3: S59–64.
8. Mormann JE, Muhlemann HR (1981) Oral starch degradation and its influence on acid production in human dental plaque. Caries Res 15: 166–175.
9. Mormann JE, Amadò R, Neukom H (1982) Comparative Studies on the in vitro alpha-Amylolysis of Different Wheat Starch Products. Starch - Stärke 34: 121–124.
10. Taravel FR, Datema R, Woloszczuk W, Marshall JJ, Whelan WJ (1983) Purification and characterization of a pig intestinal alpha-limit dextrinase. Eur J Biochem 130: 147–153.
11. Shelburne SA 3rd, Granville C, Tokuyama M, Sitkiewicz I, Patel P, et al. (2005) Growth characteristics of and virulence factor production by group A *Streptococcus* during cultivation in human saliva. Infect Immun 73: 4723–4731.
12. Norton PM, Rolph C, Ward PN, Bentley RW, Leigh JA (1999) Epithelial invasion and cell lysis by virulent strains of *Streptococcus suis* is enhanced by the presence of suilysin. FEMS Immunol Med Microbiol 26: 25–35.

13. Lalonde M, Segura M, Lacouture S, Gottschalk M (2000) Interactions between *Streptococcus suis* serotype 2 and different epithelial cell lines. Microbiology 146 Pt 8: 1913–1921.
14. Walker GJ, Builder JE (1971) Metabolism of the reserve polysaccharide of *Streptococcus mitis*. Properties of branching enzyme, and its effect on the activity of glycogen synthetase. Eur J Biochem 20: 14–21.
15. Eisenberg RJ, Elchisak M, Lai C (1974) Glycogen accumulation by pleomorphic cells of Streptococcus sanguis. Biochem Biophys Res Commun 57: 959–966.
16. Bongaerts RJ, Heinz HP, Hadding U, Zysk G (2000) Antigenicity, expression, and molecular characterization of surface-located pullulanase of *Streptococcus pneumoniae* . Infect Immun 68: 7141–7143.
17. Shelburne SA 3rd, Keith DB, Davenport MT, Horstmann N, Brennan RG, et al. (2008) Molecular characterization of group A *Streptococcus* maltodextrin catabolism and its role in pharyngitis. Mol Microbiol 69: 436–452.
18. Abbott DW, Higgins MA, Hyrnuik S, Pluvinage B, Lammerts van Bueren A, et al. (2010) The molecular basis of glycogen breakdown and transport in *Streptococcus pneumoniae* . Mol Microbiol 77: 183–199.
19. Ferrando ML, Fuentes S, de Greeff A, Smith H, Wells JM (2010) ApuA, a multifunctional alpha-glucan-degrading enzyme of *Streptococcus suis*, mediates adhesion to porcine epithelium and mucus. Microbiology 156: 2818–2828.
20. Moreno MS, Schneider BL, Maile RR, Weyler W, Saier MH Jr (2001) Catabolite repression mediated by the CcpA protein in *Bacillus subtilis*: novel modes of regulation revealed by whole-genome analyses. Mol Microbiol 39: 1366–1381.
21. Deutscher J, Francke C, Postma PW (2006) How phosphotransferase system-related protein phosphorylation regulates carbohydrate metabolism in bacteria. Microbiol Mol Biol Rev 70: 939–1031.
22. Vecht U, Arends JP, van der Molen EJ, van Leengoed LA (1989) Differences in virulence between two strains of *Streptococcus suis* type II after experimentally induced infection of newborn germ-free pigs. Am J Vet Res 50: 1037–1043.
23. Smith HE, Buijs H, Wisselink HJ, Stockhofe-Zurwieden N, Smits MA

(2001) Selection of virulence-associated determinants of *Streptococcus suis* serotype 2 by in vivo complementation. Infect Immun 69: 1961–1966.

24. de Greeff A, Wisselink HJ, de Bree FM, Schultsz C, Baums CG, et al. (2011) Genetic diversity of *Streptococcus suis* isolates as determined by comparative genome hybridization. BMC Microbiol 11: 161.
25. Santi I, Pezzicoli A, Bosello M, Berti F, Mariani M, et al. (2008) Functional characterization of a newly identified group B *Streptococcus pullulanase* eliciting antibodies able to prevent alpha-glucans degradation. PLoS One 3: e3787.
26. Pfaffl MW (2001) A new mathematical model for relative quantification in real-time RT-PCR. Nucleic Acids Res 29: e45.
27. Saulnier DM, Santos F, Roos S, Mistretta TA, Spinler JK, et al. (2011) Exploring metabolic pathway reconstruction and genome-wide expression profiling in *Lactobacillus reuteri* to define functional probiotic features. PLoS One 6: e18783.
28. Yang YH, Dudoit S, Luu P, Lin DM, Peng V, et al. (2002) Normalization for cDNA microarray data: a robust composite method addressing single and multiple slide systematic variation. Nucleic Acids Res 30: e15.
29. van Hijum SA, Garcia de la Nava J, Trelles O, Kok J, Kuipers OP (2003) MicroPreP: a cDNA microarray data pre-processing framework. Appl Bioinformatics 2: 241–244.
30. Conesa A, Gotz S, Garcia-Gomez JM, Terol J, Talon M, et al. (2005) Blast2GO: a universal tool for annotation, visualization and analysis in functional genomics research. Bioinformatics 21: 3674–3676.
31. Zdobnov EM, Apweiler R (2001) InterProScan–an integration platform for the signature-recognition methods in InterPro. Bioinformatics 17: 847–848.
32. Bailey TL, Boden M, Buske FA, Frith M, Grant CE, et al. (2009) MEME SUITE: tools for motif discovery and searching. Nucleic Acids Res 37: W202–208.
33. Bailey TL, Gribskov M (1998) Combining evidence using p-values: application to sequence homology searches. Bioinformatics 14: 48–54.
34. Novichkov PS, Rodionov DA, Stavrovskaya ED, Novichkova ES, Kazakov AE, et al. (2010) RegPredict: an integrated system for regulon inference in prokaryotes by comparative genomics approach. Nucleic

Acids Res 38: W299–307.

35. Chaptal V, Gueguen-Chaignon V, Poncet S, Lecampion C, Meyer P, et al. (2006) Structural analysis of *B. subtilis* CcpA effector binding site. Proteins 64: 814–816.
36. Jacobs AA, Loeffen PL, van den Berg AJ, Storm PK (1994) Identification, purification, and characterization of a thiol-activated hemolysin (suilysin) of *Streptococcus suis* . Infect Immun 62: 1742–1748.
37. Ferrari M, Losio MN, Bernori E, Lingeri R (1993) Established thyroid cell line of newborn pig (NPTh). New Microbiol 16: 381–384.
38. Vandesompele J, De Preter K, Pattyn F, Poppe B, Van Roy N, et al. (2002) Accurate normalization of real-time quantitative RT-PCR data by geometric averaging of multiple internal control genes. Genome Biol 3: RESEARCH0034.
39. Bailey TL, Elkan C (1994) Fitting a mixture model by expectation maximization to discover motifs in biopolymers. Proc Int Conf Intell Syst Mol Biol 2: 28–36.
40. Miwa Y, Nakata A, Ogiwara A, Yamamoto M, Fujita Y (2000) Evaluation and characterization of catabolite-responsive elements (cre) of *Bacillus subtilis* . Nucleic Acids Res 28: 1206–1210.
41. Lorca GL, Chung YJ, Barabote RD, Weyler W, Schilling CH, et al. (2005) Catabolite repression and activation in *Bacillus subtilis*: dependency on CcpA, HPr, and HprK. J Bacteriol 187: 7826–7839.
42. Fujita Y (2009) Carbon catabolite control of the metabolic network in *Bacillus subtilis* . Biosci Biotechnol Biochem 73: 245–259.
43. Willenborg J, Fulde M, de Greeff A, Rohde M, Smith HE, et al. (2011) Role of glucose and CcpA in capsule expression and virulence of *Streptococcus suis* . Microbiology 157: 1823–1833.
44. Monedero V, Yebra MJ, Poncet S, Deutscher J (2008) Maltose transport in *Lactobacillus casei* and its regulation by inducer exclusion. Research in microbiology 159: 94–102.
45. Schonert S, Seitz S, Krafft H, Feuerbaum EA, Andernach I, et al. (2006) Maltose and maltodextrin utilization by *Bacillus subtilis* . Journal of bacteriology 188: 3911–3922.
46. Shim JH, Park JT, Hong JS, Kim KW, Kim MJ, et al. (2009) Role of maltogenic amylase and pullulanase in maltodextrin and glycogen metabolism of *Bacillus subtilis* 168. Journal of bacteriology 191:

4835–4844.

47. Gopal S, Berg D, Hagen N, Schriefer EM, Stoll R, et al. (2010) Maltose and maltodextrin utilization by *Listeria monocytogenes* depend on an inducible ABC transporter which is repressed by glucose. PloS one 5: e10349.
48. Zhang X, Rogers M, Bierschenk D, Bonten MJ, Willems RJ, et al. (2013) A LacI-family regulator activates maltodextrin metabolism of *Enterococcus faecium* . PLoS One 8: e72285.
49. Benga L, Goethe R, Rohde M, Valentin-Weigand P (2004) Non-encapsulated strains reveal novel insights in invasion and survival of *Streptococcus suis* in epithelial cells. Cell Microbiol 6: 867–881.
50. Iyer R, Baliga NS, Camilli A (2005) Catabolite control protein A (CcpA) contributes to virulence and regulation of sugar metabolism in *Streptococcus pneumoniae* . J Bacteriol 187: 8340–8349.
51. Shelburne SA 3rd, Keith D, Horstmann N, Sumby P, Davenport MT, et al. (2008) A direct link between carbohydrate utilization and virulence in the major human pathogen group A *Streptococcus* . Proc Natl Acad Sci U S A 105: 1698–1703.
52. Pan X, Ge J, Li M, Wu B, Wang C, et al. (2009) The orphan response regulator CovR: a globally negative modulator of virulence in *Streptococcus suis* serotype 2. J Bacteriol 191: 2601–2612.
53. Seitz M, Baums CG, Neis C, Benga L, Fulde M, et al. (2013) Subcytolytic effects of suilysin on interaction of *Streptococcus suis* with epithelial cells. Vet Microbiol 167: 584–591.
54. Vanier G, Segura M, Gottschalk M (2007) Characterization of the invasion of porcine endothelial cells by *Streptococcus suis* serotype 2. Can J Vet Res 71: 81–89.
55. Charland N, Nizet V, Rubens CE, Kim KS, Lacouture S, et al. (2000) *Streptococcus suis* serotype 2 interactions with human brain microvascular endothelial cells. Infect Immun 68: 637–643.
56. Lecours MP, Gottschalk M, Houde M, Lemire P, Fittipaldi N, et al. (2011) Critical role for *Streptococcus suis* cell wall modifications and suilysin in resistance to complement-dependent killing by dendritic cells. J Infect Dis 204: 919–929.
57. Gottschalk M, Xu J, Calzas C, Segura M (2011) *Streptococcus suis*: a new emerging or an old neglected zoonotic pathogen? Future Microbiol 5: 371–391.

58. Allen AG, Bolitho S, Lindsay H, Khan S, Bryant C, et al. (2001) Generation and characterization of a defined mutant of *Streptococcus suis* lacking suilysin. Infect Immun 69: 2732–2735.
59. King SJ, Heath PJ, Luque I, Tarradas C, Dowson CG, et al. (2001) Distribution and genetic diversity of suilysin in *Streptococcus suis* isolated from different diseases of pigs and characterization of the genetic basis of suilysin absence. Infect Immun 69: 7572–7582.
60. Bonafonte MA, Solano C, Sesma B, Alvarez M, Montuenga L, et al. (2000) The relationship between glycogen synthesis, biofilm formation and virulence in *Salmonella enteritidis* . FEMS Microbiol Lett 191: 31–36.
61. Bourassa L, Camilli A (2009) Glycogen contributes to the environmental persistence and transmission of *Vibrio cholerae* . Mol Microbiol 72: 124–138.
62. Chang DE, Smalley DJ, Tucker DL, Leatham MP, Norris WE, et al. (2004) Carbon nutrition of *Escherichia coli* in the mouse intestine. Proc Natl Acad Sci U S A 101: 7427–7432.
63. Fabich AJ, Jones SA, Chowdhury FZ, Cernosek A, Anderson A, et al. (2008) Comparison of carbon nutrition for pathogenic and commensal *Escherichia coli* strains in the mouse intestine. Infect Immun 76: 1143–1152.
64. Henrissat B, Deleury E, Coutinho PM (2002) Glycogen metabolism loss: a common marker of parasitic behaviour in bacteria? Trends Genet 18: 437–440.
65. Jones SA, Jorgensen M, Chowdhury FZ, Rodgers R, Hartline J, et al. (2008) Glycogen and maltose utilization by *Escherichia coli* O157:H7 in the mouse intestine. Infect Immun 76: 2531–2540.
66. Sambou T, Dinadayala P, Stadthagen G, Barilone N, Bordat Y, et al. (2008) Capsular glucan and intracellular glycogen of Mycobacterium tuberculosis: biosynthesis and impact on the persistence in mice. Mol Microbiol 70: 762–774.
67. Wilson WA, Roach PJ, Montero M, Baroja-Fernandez E, Munoz FJ, et al. (2010) Regulation of glycogen metabolism in yeast and bacteria. FEMS Microbiol Rev 34: 952–985.
68. Pitsillides AA, Skerry TM, Edwards JC (1999) Joint immobilization reduces synovial fluid hyaluronan concentration and is accompanied by changes in the synovial intimal cell populations. Rheumatology

(Oxford) 38: 1108–1112.

69. Praest BM, Greiling H, Kock R (1997) Assay of synovial fluid parameters: hyaluronan concentration as a potential marker for joint diseases. Clin Chim Acta 266: 117–128.
70. Hui AY, McCarty WJ, Masuda K, Firestein GS, Sah RL (2012) A systems biology approach to synovial joint lubrication in health, injury, and disease. Wiley Interdiscip Rev Syst Biol Med 4: 15–37.
71. Smith HE, Damman M, van der Velde J, Wagenaar F, Wisselink HJ, et al. (1999) Identification and characterization of the cps locus of *Streptococcus suis* serotype 2: the capsule protects against phagocytosis and is an important virulence factor. Infect Immun 67: 1750–1756.
72. Smith HE, de Vries R, van't Slot R, Smits MA (2000) The cps locus of *Streptococcus suis* serotype 2: genetic determinant for the synthesis of sialic acid. Microb Pathog 29: 127–134.
73. Fittipaldi N, Sekizaki T, Takamatsu D, de la Cruz Dominguez-Punaro M, Harel J, et al. (2008) Significant contribution of the pgdA gene to the virulence of *Streptococcus suis* . Mol Microbiol 70: 1120–1135.
74. Fittipaldi N, Sekizaki T, Takamatsu D, Harel J, Dominguez-Punaro Mde L, et al. (2008) D-alanylation of lipoteichoic acid contributes to the virulence of *Streptococcus suis* . Infect Immun 76: 3587–3594.
75. de Greeff A, Buys H, Verhaar R, Dijkstra J, van Alphen L, et al. (2002) Contribution of fibronectin-binding protein to pathogenesis of *Streptococcus suis* serotype 2. Infect Immun 70: 1319–1325.
76. Esgleas M, Li Y, Hancock MA, Harel J, Dubreuil JD, et al. (2008) Isolation and characterization of alpha-enolase, a novel fibronectin-binding protein from *Streptococcus suis* . Microbiology 154: 2668–2679.
77. Brassard J, Gottschalk M, Quessy S (2004) Cloning and purification of the *Streptococcus suis* serotype 2 glyceraldehyde-3-phosphate dehydrogenase and its involvement as an adhesin. Vet Microbiol 102: 87–94.
78. Ge J, Feng Y, Ji H, Zhang H, Zheng F, et al. (2009) Inactivation of dipeptidyl peptidase IV attenuates the virulence of *Streptococcus suis* serotype 2 that causes streptococcal toxic shock syndrome. Curr Microbiol 59: 248–255.

79. Jobin MC, Grenier D (2003) Identification and characterization of four proteases produced by *Streptococcus suis* . FEMS Microbiol Lett 220: 113–119.

80. Tan C, Fu S, Liu M, Jin M, Liu J, et al. (2008) Cloning, expression and characterization of a cell wall surface protein, 6-phosphogluconate-dehydrogenase, of *Streptococcus suis* serotype 2. Vet Microbiol 130: 363–370.

81. Si Y, Yuan F, Chang H, Liu X, Li H, et al. (2009) Contribution of glutamine synthetase to the virulence of *Streptococcus suis* serotype 2. Vet Microbiol 139: 80–88.

82. Kouki A, Haataja S, Loimaranta V, Pulliainen AT, Nilsson UJ, et al. (2011) Identification of a novel streptococcal adhesin P (SadP) protein recognizing galactosyl-alpha1-4-galactose-containing glycoconjugates: convergent evolution of bacterial pathogens to binding of the same host receptor. J Biol Chem 286: 38854–38864.

83. Winterhoff N, Goethe R, Gruening P, Rohde M, Kalisz H, et al. (2002) Identification and characterization of two temperature-induced surface-associated proteins of *Streptococcus suis* with high homologies to members of the Arginine Deiminase system of *Streptococcus pyogenes* . J Bacteriol 184: 6768–6776.

84. Fontaine MC, Perez-Casal J, Willson PJ (2004) Investigation of a novel DNase of *Streptococcus suis* serotype 2. Infect Immun 72: 774–781.

85. Bonifait L, Vaillancourt K, Gottschalk M, Frenette M, Grenier D (2010) Purification and characterization of the subtilisin-like protease of *Streptococcus suis* that contributes to its virulence. Vet Microbiol 148: 333–340.

86. Zhang A, Mu X, Chen B, Han L, Chen H, et al. (2010) IgA1 protease contributes to the virulence of *Streptococcus suis* . Vet Microbiol 148: 436–439.

87. Allen AG, Lindsay H, Seilly D, Bolitho S, Peters SE, et al. (2004) Identification and characterisation of hyaluronate lyase from *Streptococcus suis* . Microb Pathog 36: 327–335.

88. Tonnaer EL, Hafmans TG, Van Kuppevelt TH, Sanders EA, Verweij PE, et al. (2006) Involvement of glycosaminoglycans in the attachment of pneumococci to nasopharyngeal epithelial cells. Microbes Infect 8: 316–322.

89. Vanier G, Sekizaki T, Dominguez-Punaro MC, Esgleas M, Osaki M, et al. (2007) Disruption of srtA gene in *Streptococcus suis* results in decreased interactions with endothelial cells and extracellular matrix proteins. Vet Microbiol
90. Baums CG, Kaim U, Fulde M, Ramachandran G, Goethe R, et al. (2006) Identification of a novel virulence determinant with serum opacification activity in *Streptococcus suis* . Infect Immun 74: 6154–6162.
91. Wen ZT, Burne RA (2002) Functional genomics approach to identifying genes required for biofilm development by *Streptococcus mutans* . Appl Environ Microbiol 68: 1196–1203.
92. Smith HE, Vecht U, Wisselink HJ, Stockhofe-Zurwieden N, Biermann Y, et al. (1996) Mutants of *Streptococcus suis* types 1 and 2 impaired in expression of muramidase-released protein and extracellular protein induce disease in newborn germfree pigs. Infect Immun 64: 4409–4412.

INDEX

C

D

E